T0328287

Biotic Stress Resistance in Millets

Biotic Stress Resistance in Millets

Edited by

I.K. Das and P.G. Padmaja

ICAR-Indian Institute of Millets Research,
Hyderabad, India

AMSTERDAM • BOSTON • HEIDELBERG • LONDON
NEW YORK • OXFORD • PARIS • SAN DIEGO
SAN FRANCISCO • SINGAPORE • SYDNEY • TOKYO

Academic Press is an imprint of Elsevier

Academic Press is an imprint of Elsevier
125 London Wall, London EC2Y 5AS, United Kingdom
525 B Street, Suite 1800, San Diego, CA 92101-4495, United States
50 Hampshire Street, 5th Floor, Cambridge, MA 02139, United States
The Boulevard, Langford Lane, Kidlington, Oxford OX5 1GB, United Kingdom

Notices
Knowledge and best practice in this field are constantly changing. As new research and experience broaden our
understanding, changes in research methods, professional practices, or medical treatment may become necessary.

Practitioners and researchers must always rely on their own experience and knowledge in evaluating and using any
information, methods, compounds, or experiments described herein. In using such information or methods they should be
mindful of their own safety and the safety of others, including parties for whom they have a professional responsibility.

To the fullest extent of the law, neither the Publisher nor the authors, contributors, or editors, assume any liability for any
injury and/or damage to persons or property as a matter of products liability, negligence or otherwise, or from any use or
operation of any methods, products, instructions, or ideas contained in the material herein.

British Library Cataloguing-in-Publication Data
A catalogue record for this book is available from the British Library

Library of Congress Cataloging-in-Publication Data
A catalog record for this book is available from the Library of Congress

ISBN: 978-0-12-804549-7

For Information on all Academic Press publications
visit our website at https://www.elsevier.com

 Working together
to grow libraries in
developing countries

www.elsevier.com • www.bookaid.org

Publisher: Nikki Levy
Acquisition Editor: Nancy Maragioglio
Editorial Project Manager: Billie Jean Fernandez
Production Project Manager: Nicky Carter
Designer: Matthew Limbert

Typeset by MPS Limited, Chennai, India

Contents

Contributors ix
Preface xi

PART I INTRODUCTION TO MILLETS

CHAPTER 1 Millets, Their Importance, and Production Constraints 3
I.K. Das and S. Rakshit

  1.1 Introduction 3
      1.1.1 Origin and Distribution...4
      1.1.2 Area, Production, and Productivity 5
  1.2 Importance of Millets...8
      1.2.1 Dryland Agriculture..8
      1.2.2 Food and Nutritional Security 10
      1.2.3 Bioenergy Production 12
      1.2.4 Climate Resilient Agriculture...12
  1.3 Major Production Constraints 13
      1.3.1 Biotic Constraints 14
      1.3.2 Abiotic Constraints 15
      1.3.3 Socioeconomic Factors 16
  1.4 Conclusions 16
  Acknowledgments 17
  References 17

PART II DISEASES AND INSECT PEST RESISTANCE

CHAPTER 2 Disease Resistance in Sorghum 23
I.K. Das and P. Rajendrakumar

  2.1 Introduction 23
  2.2 Disease, Biology, and Epidemiology..24
      2.2.1 Fungal Diseases 24
      2.2.2 Bacterial Diseases..36
      2.2.3 Viral Diseases 37
  2.3 Host-Plant Resistance...39
      2.3.1 Screening for Resistance 39

2.3.2 Sources of Resistance 44
2.3.3 Mechanisms of Resistance 48
2.3.4 Genetics of Resistance 49
2.3.5 Utilization of Host Resistance 51
2.4 Conclusions 56
2.5 Future Research Need 56
Acknowledgments 57
References 57

CHAPTER 3   Disease Resistance in Pearl Millet and Small Millets 69
A. Nagaraja and I.K. Das

3.1 Introduction 69
3.2 Pearl Millet 70
3.2.1 Disease, Biology and Epidemiology 71
3.2.2 Host-Plant Resistance 76
3.3 Small Millets 81
3.3.1 Disease, Biology and Epidemiology 82
3.3.2 Host-Plant Resistance 90
3.4 Conclusions 94
3.5 Future Priorities..95
References 95

CHAPTER 4   Insect Pest Resistance in Sorghum 105
P.G. Padmaja

4.1 Introduction 105
4.2 Pest Biology 105
4.2.1 Seedling Pests 105
4.2.2 Leaf Feeders... 106
4.2.3 Borers 107
4.2.4 Sucking Pests 108
4.2.5 Panicle Pests 109
4.2.6 Root Pests 110
4.3 Host-Plant Resistance..111
4.3.1 Host Finding and Orientation 111
4.3.2 Screening Techniques 113
4.3.3 Sources of Pest Resistance 115
4.3.4 Genetics and Inheritance of Resistance 117
4.3.5 Mechanism of Resistance 117
4.3.6 Development and Use of Pest-Resistant Cultivar......................... 130

4.4 Conclusions 132
4.5 Future Priorities..133
References...133

CHAPTER 5 Insect Pest Resistance in Pearl Millet and Small Millets........... 147
G.S. Prasad and K.S. Babu

5.1 Introduction 147
5.2 Pest Biology 147
    5.2.1 Seedling Pests 147
    5.2.2 Foliage Pests 149
    5.2.3 Sucking Pests 149
    5.2.4 Stemborer..150
    5.2.5 Earhead Pests 152
    5.2.6 Soil Dwelling Insects..154
5.3 Host-Plant Resistance...155
    5.3.1 Sources of Resistance 155
    5.3.2 Genetics of Resistance..155
    5.3.3 Mechanisms of Resistance 159
    5.3.4 Utilization of Resistance...161
5.4 Conclusions 161
5.5 Future Priorities..162
References...162

PART III *STRIGA* AND WEEDS

CHAPTER 6 *Striga*: A Persistent Problem on Millets 173
B.A. Kountche, S. Al-Babili and B.I.G. Haussmann

6.1 Introduction 173
6.2 Importance and Biology..174
    6.2.1 Importance 174
    6.2.2 Biology...176
6.3 Host-Plant Resistance and Heredity 178
    6.3.1 Host Finding and Orientation: The Key Role of Strigolactones 178
    6.3.2 Sources of Resistance 181
    6.3.3 Mechanisms of Resistance 182
    6.3.4 Nature and Genetic Basis of Resistance 183
    6.3.5 Development and Use of *Striga*-Resistant Millet Cultivars 184
    6.3.6 Integrated *Striga* Management 190

6.4 Conclusions 191
6.5 Future Perspectives and Priorities 192
Acknowledgments 192
References..192

CHAPTER 7   Weed Problem in Millets and Its Management **205**
J.S. Mishra

7.1 Introduction 205
7.2 Weeds of Millets and Their Importance................................205
    7.2.1 Weeds of Millets...205
    7.2.2 Losses Due To Weeds 206
    7.2.3 Critical Period of Crop−Weed Competition 209
    7.2.4 Climate Change and Weed Competition..........................209
7.3 Management Strategies..210
    7.3.1 Mechanical Methods...210
    7.3.2 Cultural Methods 211
    7.3.3 Chemical Methods 211
    7.3.4 Intercropping...214
    7.3.5 Sequence Cropping/Double Cropping..............................214
    7.3.6 Management of *Striga* spp. 215
7.4 Herbicide Resistance..215
7.5 Conclusions 216
7.6 Future Thrusts 216
References..217

Index 221

Contributors

S. Al-Babili
King Abdullah University of Science and Technology (KAUST), Thuwal,
Kingdom of Saudi Arabia

K.S. Babu
ICAR-Indian Institute of Millets Research, Hyderabad, India

I.K. Das
ICAR-Indian Institute of Millets Research, Hyderabad, India

B.I.G. Haussmann
University of Hohenheim, Stuttgart, Germany

B.A. Kountche
King Abdullah University of Science and Technology (KAUST), Thuwal,
Kingdom of Saudi Arabia

J.S. Mishra
ICAR Research Complex for Eastern Region, Patna, India

A. Nagaraja
UAS-Gandhi Krishi Vignana Kendra, Bengaluru, India

P.G. Padmaja
ICAR-Indian Institute of Millets Research, Hyderabad, India

G.S. Prasad
ICAR-Indian Institute of Millets Research, Hyderabad, India

P. Rajendrakumar
ICAR-Indian Institute of Millets Research, Hyderabad, India

S. Rakshit
ICAR-Indian Institute of Millets Research, Hyderabad, India

Preface

Millets are important crops in the developing countries in the semiarid tropics. Incidentally, these are the regions where most of the world's poor people live. These crops provide high-energy, nutritious, and healthy food, recommended for children, convalescents as well as elders. Millets can be consistently grown under extreme agricultural conditions (low precipitation, high temperature, and poor soils), where other cereals fail to produce an acceptable harvest. Under the changing climatic scenario millets are being assigned as the crops for future attention.

Biotic stresses particularly diseases, insect pests, weeds, and birds are major constraints in the way of realizing the potential yield of millets. In the future these problems are likely to increase because of the perceptible changes in the global climate. Management of biotic stresses in millets is attempted more through resistant cultivars and less or negligibly through the use of chemicals. Host-plant resistance is the most economical and ecosafe method for management of biotic stress and the only affordable method for the poor farmers. A lot of information has been generated on this aspect over time by numerous millet researchers. However, the information is scattered through the literature and there is hardly any publication that has discussed the diseases, insect pests, weeds, *Striga*, and birds resistance of all the millets including sorghum, pearl millet, and small millets in a single book.

In this book, a sincere attempt has been made to present updated information on the subject with emphasis on literature published in the 21st century. All aspects of host-plant resistance including screening techniques, sources, mechanisms and genetics of resistance, and utilization of host resistance by conventional and molecular breeding methods have been discussed in depth. In addition to biology, epidemiology, and resistance, other aspects of millets, such as the origin, distribution, production, and their importance, have been covered for the beginners.

The book will be highly useful for researchers and research-planners who are involved in plant protection and production research on millets and related cereals. This book, which is authored by experts in the respective field of studies, will primarily act as a guide for the researchers to formulate suitable research strategies on plant protection. Industry sectors like the "millets seed industry" will benefit from information on biotic stress-related issues that are constraints in seed production. It is also expected to attract global readership among the students in advanced courses.

Many scientists contributed by their constructive suggestions and by spending valuable time in reviewing the chapters. We wish to thank all of them and the subject matter experts who accepted to be part of this book. Our sincere thanks are due to Billie Jean Fernandez and Nancy Maragioglio at Elsevier for their continuous support and technical assistance.

I.K. Das and P.G. Padmaja

INTRODUCTION TO MILLETS

MILLETS, THEIR IMPORTANCE, AND PRODUCTION CONSTRAINTS

I.K. Das and S. Rakshit

ICAR-Indian Institute of Millets Research, Hyderabad, India

1.1 INTRODUCTION

Millets include a group of highly variable small-grained, annual grasses of taxonomically different genera. They are grown mostly in developing countries (McDonough et al., 2000) located in Africa (Niger, Nigeria, Sudan, Mali, Burkina Faso) and Asia (India, China, Pakistan, Myanmar, Nepal) for food, feed, and fodder. These crops have an advantage of being able to grow in the harshest environments (scarce rainfall, drought, low soil fertility), where there is limited scope for growing other crops. Millets provide major sources of energy and protein for the people in sub-Saharan Africa. There are about nine millets that are predominantly cultivated. In order of worldwide production, the most widely cultivated millets are sorghum (*Sorghum bicolor* (L.) Moench), pearl millet (*Pennisetum glaucum* (L.) R.Br.), foxtail millet (*Setaria italica* (L.) Beauv), proso millet (*Panicum miliaceum* L.), and finger millet (*Eleusine coracana* (L.) Gaertn) (Crawford and Lee, 2003). Other millets are little millet (*Panicum sumentranse* Roth.ex Roem. and Schultz), kodo millet (*Paspalum scrobiculatum* L.), Indian barnyard millet (*Echinochloa frumentacea* Roxb., Link), Japanese barnyard millet (*Echinochloa utilis* Ohwi et Yabuno), tef (*Eragrostis tef*), and fonio (*Digitaria* spp.). In Africa, the most widely grown millets are sorghum, pearl millet, finger millet, tef, and fonio. The area covered by these millets, excluding sorghum, are 76% for pearl millet, 19% for finger millet, 9% for tef, and 4% for fonio (Obilana, 2003). Sorghum and pearl millet are the two most important global members in the millets group, and these two have the largest area among the millets and occupy the fifth and sixth positions in total global crops, respectively, after rice, wheat, maize, and barley. Cultivated sorghum has the following four agronomic types: grain sorghum, sweet sorghum, sudangrass, and broomcorn (Berenji and Dahlberg, 2004). Millets other than sorghum and pearl millet are called minor or small millets. Among the small millets, the foxtail millet, proso millet, little millet, kodo millet, and barnyard millet are mostly concentrated in Eurasian or Asian countries, including India, China, countries in the Middle East, and South East Asia (Seetharam et al., 1986). This chapter describes the origin and domestication of these crops in different parts of the world, their global importance, and the production constraints in present-day agriculture.

Biotic Stress Resistance in Millets. DOI: http://dx.doi.org/10.1016/B978-0-12-804549-7.00001-9

1.1.1 ORIGIN AND DISTRIBUTION

Different members of the millets group originated and were later domesticated in different parts of the world. Sorghum originated and was domesticated in northeast Africa. However, there are different reports regarding the exact geographic location of origin within northeast Africa. Relying on archeological, palaeobotanical, anthropological, and botanical evidences, Harlan and de Wet (1972) believed that the center of origin of sorghum extended from near Lake Chad in Africa, where a diversity and abundance of wild and weedy species are represented as well as the primitive race of *Sorghum bicolor.* According to Mann et al. (1983) the area north of the equator and east of 10° E latitude in northeastern Africa, was considered as the center of origin and domestication of sorghum, approximately 5000 years ago. Reports based on carbonized seeds of sorghum with radiocarbon dates of 8000 years indicated that it was Nabta Playa near the Egyptian–Sudanese border, where the oldest evidence of sorghum exists (Wendorf et al., 1992; Dahlberg and Wasylikowa, 1996). The secondary center of origin of sorghum is the Indian subcontinent. It might have reached India not earlier than 1500 BC and China by AD 900. However, the introduction of cultivated sorghums in the Americas and Australia took place only about 100 years ago (Peacock and Wilson, 1984). Broomcorn, a special type of sorghum, was believed to have originated in the Mediterranean region, specifically Italy (Dahlberg et al., 2011). Presently, sorghum is distributed all over the world in about 105 countries. Among these countries, the top 10 sorghum producers are the United States, India, Mexico, Nigeria, Sudan, Ethiopia, Australia, Brazil, China, and Burkina Faso (Rakshit et al., 2014).

Pearl millet has been widely grown in Africa and the Indian subcontinent since prehistoric times. There is a lack of clarity about the evolutionary history of this crop. Several authors agree that the domestication of pearl millet took place in Africa along the Sahelian zone, from Mauritania to Sudan (Harlan, 1975; Porteres, 1976; Marchais and Tostain, 1993). The Sahel zone of West Africa is considered to be the center of diversity and the suggested area of domestication for this crop. Recent archaeobotanical research has confirmed the presence of domesticated pearl millet on the Sahel zone of northern Mali between 2500 and 2000 BC (Katie et al., 2010). Molecular studies on wild and cultivated accessions originating from the entire pearl millet distribution area in Africa and Asia suggested that West Africa was the most likely center of origin of cultivated pearl millet (Oumar et al., 2008). Its cultivation subsequently spread and moved overseas to India. The earliest archeological records on pearl millet in India have been dated to around 2000 BC (Fuller, 2003). Presently, the crop is cultivated mostly in Africa (~14 million ha) and Asia (~12 million ha). In Asia, India has the largest area (~10 million ha) under this crop.

Finger millet is native to the Ethiopian highlands (D'Andrea et al., 1999). The crop has been found in the archeological record of early African agriculture and was introduced to India at least 3000 years ago (Vishnu-Mittre, 1968). Finger millet has five races (*coracana*, *vulgaris*, *elongeta*, *plana*, and *compacta*), of which *coracana* is particularly well adapted to agriculture in the eastern highlands of Africa and the Ghats of India. The most common race of finger millet found in Africa and Asia is *vulgaris*, which is grown from Uganda to Ethiopia (in Africa) and from India to Indonesia (in Asia). Finger millet is adaptable to higher elevations and is grown in the Himalayas up to an altitude of 2300 m. In Africa, it is mostly cultivated in Eastern, Southern, and Central Africa: Uganda, Western Kenya, Sudan, and Eritrea; Zimbabwe, Zambia, Malawi, and Madagascar; and Rwanda and Burundi.

Foxtail millet is the most important millet in East Asia. It could have originated from and been domesticated in Eurasia. It has the longest history of cultivation among the millets, having being grown in China since 5000 BC. The earliest evidence of its cultivation comes from the Peiligang culture of China (Zohary and Hopf, 2000).

Proso or common millet appeared as a crop in both Transcaucasia and China about 7000 years ago. This crop might have been domesticated independently in each of the above areas. Later (>3000 years ago), it was probably introduced into Europe. It was introduced to India long ago and is extensively cultivated in India, the Middle East, Romania, Russia, Turkey, and Ukraine. Kodo millet, little millet, and barnyard millet are indigenous to India. They are grown primarily in India, but also found in Indonesia, the Philippines, Thailand and Vietnam, and in West Africa. The fonio is cultivated in West Africa, mostly in Mali, Burkina Faso, Guinea Conakry, and Nigeria. Tef is grown in the highlands of Ethiopia.

1.1.2 AREA, PRODUCTION, AND PRODUCTIVITY

Global area, production, and productivity data for millets are available separately under two categories, that is, "sorghum" and "millets" in FAOSTAT. Here, the millets include pearl millet and small millets. Therefore, this section will be discussed in two parts, that is, "sorghum" and "pearl millet plus small millets."

Sorghum is one of the most important crops in semiarid tropics covering a harvested area of around 42.3 million ha across the globe. Major areas under sorghum cultivation are in Africa, which holds around 63.1% of the total harvested area in the world. Asia shares around 18.5% and the Americas have 16.2% of the total harvested area in the world (Fig. 1.1) (FAOSTAT, 2013). With respect to acreage the top 10 countries include seven in Africa (Sudan, Nigeria, Niger, Ethiopia, Burkina Faso, Mali, and Tanzania), one in Asia (India) and two in the Americas (United States and Mexico) (Table 1.1). The contribution of these countries to the world acreage are as follows: Sudan (17%); India (15%); Nigeria (13%); Niger (7%); United States (6%); Ethiopia, Burkina Faso, and Mexico (4% each); and Mali and Tanzania (2% each). India has the largest acreage (6.2 million ha) in Asia followed by China (0.6 million ha) and Yemen (0.5 million ha). The global production of sorghum is estimated to be around 61.5 million tonnes. The top 10 sorghum producers are the United States, Nigeria, Mexico, India, Sudan, Ethiopia, Argentina, Australia, Brazil, and China. These countries together represent nearly 77% of the global sorghum production and 70% of the harvested area. Africa and Asia together constitute nearly 82% of the area and produce around 56% of sorghum grains. Among the top 10 sorghum producers, Argentina ($>4000 \text{ kg ha}^{-1}$) recorded the highest productivity followed by Australia, Mexico, United States, and China ($3000-4000 \text{ kg ha}^{-1}$), Brazil and Ethiopia ($2000-3000 \text{ kg ha}^{-1}$), Nigeria (around 1200 kg ha^{-1}), and India and Sudan ($<1000 \text{ kg ha}^{-1}$). However, the average sorghum productivity of the world is around 1452 kg ha^{-1} (FAOSTAT, 2013). Recently, an analysis of the trends in area and yield gains and associated changes in yield stability in the top 10 sorghum-producing countries from 1970 to 2009 by Rakshit et al. (2014) revealed that the Asian countries and the USA recorded the largest drop in harvested area. Grain yield levels increased substantially in all the countries except Sudan. Relative to yield level of 1970, productivity increased annually at $0.96\% \text{ year}^{-1}$ across the top 10 countries.

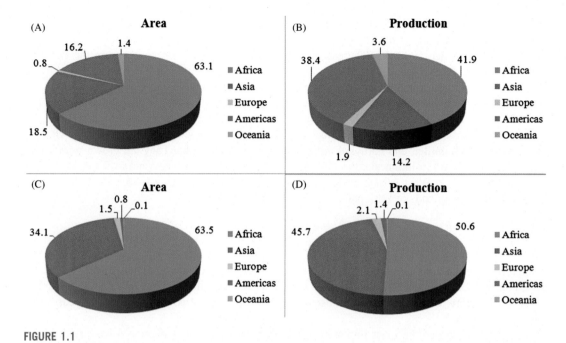

FIGURE 1.1

Percentage share of area and production of sorghum (A and B) and pearl millet plus small millets (C and D) in different continents.

From http://faostat3.fao.org/faostat-gateway/go/to/download/Q/QC/E (accessed 18 August 2014).

Pearl millet and small millets are produced in an area of 32.9 million ha globally with a production of 29.9 million tonnes. About 63.5% of this area is in Africa, while Asian countries occupy 34.1% of the total area. European countries cover 1.5%, North America 0.8%, and Oceania 0.1% of the millet area (Fig. 1.1) (FAOSTAT, 2013). With respect to acreage, the top 10 countries include seven in Africa (Niger, Nigeria, Sudan, Mali, Burkina Faso, Chad, and Senegal) and three in Asia (India, China, and Pakistan) (Table 1.1). The contribution of these countries to the world acreage are as follows: India (28%); Niger (22%); Nigeria (12%); Sudan (8%); Mali, Burkina Faso (4% each); Chad, China, Senegal (2% each); and Pakistan (1%). In Africa, the production of pearl millet and small millets is distributed differently among a large number of African countries; the largest producer being Nigeria (17% of world production) followed by Niger (10%), Mali, Burkina Faso, Sudan (4% each), Ethiopia (3%), Senegal and Chad (2% each). India ranks first in the world, in terms of both harvested area (9.2 million ha) and production (10.9 million tonnes). The highest productivity is recorded in China (2250 kg ha^{-1}), followed by Ethiopia (1870 kg ha^{-1}), Nigeria (1316 kg ha^{-1}), and India (1186 kg ha^{-1}). Pearl millet is cultivated mostly in Africa (~14 million ha) and Asia (~12 million ha) (Khairwal et al., 2007). India has an area of about 9.0 million ha under pearl millet and produces more than half the world's pearl millet. The productivity of pearl millet in India is 991 kg ha^{-1}. Globally, India is the leading producer of small millets with about 20% of the area under these crops. The annual planting area under small millets in India is around 2.5 million ha.

Table 1.1 Area, Production, and Productivity of Millets in Different Countries

Country	Sorghum			Pearl Millet and Small Millets		
	Area (000, ha)[a]	Production (000, tonne)	Yield (kg ha^{-1})	Area (000, ha)[a]	Production (000, tonne)	Yield (kg ha^{-1})
Angola	192	46	242	195	39	198
Argentina	890	3636	4085	6	11	1829
Australia	595	2230	3747	35	40	1143
Benin	110	115	1046	27	23	836
Bolivia	175	408	2338	nr	nr	nr
Botswana	50	23	460	6	2	328
Brazil	773	2073	2682	nr	nr	nr
Burkina Faso	1800	1940	1078	1300	1109	853
Cameroon	800	1150	1438	70	97	1386
Chad	850	745	876	800	582	728
China	552	2019	3657	720	1621	2250
Congo	8	8	905	62	48	774
Cote d'Ivoire	68	48	706	65	50	769
Egypt	141	749	5305	nr	nr	nr
El Salvador	109	144	1321	nr	nr	nr
Eritrea	250	80	320	55	20	364
Ethiopia	1847	4338	2348	432	807	1870
France	51	279	5435	12	40	3333
Gambia	33	25	758	125	119	952
Ghana	230	277	1204	185	200	1081
Guinea	37	50	1351	318	230	723
Haiti	127	108	853	nr	nr	nr
India	6180	5280	854	9200	10910	1186
Kazakhstan	180	400	2222	52	54	1039
Kenya	210	139	660	100	64	641
Malawi	89	86	965	49	39	805
Mali	938	820	874	1437	1152	802
Mauritania	200	94	470	10	1	68
Mexico	1689	6308	3735	2	2	943
Mozambique	625	188	301	100	48	480
Myanmar	228	215	941	210	185	881
Namibia	16	7	432	230	33	143
Nepal	nr	nr	nr	274	306	1114
Niger	3100	1287	415	7100	2995	422
Nigeria	5500	6700	1218	3800	5000	1316
North Korea	22	38	1727	68	107	1574
Pakistan	200	125	625	465	325	699
Russia	120	172	1436	355	419	1178

(Continued)

Table 1.1 Area, Production, and Productivity of Millets in Different Countries *Continued*

Country	Sorghum			Pearl Millet and Small Millets		
	Area (000, ha)[a]	Production (000, tonne)	Yield (kg ha^{-1})	Area (000, ha)[a]	Production (000, tonne)	Yield (kg ha^{-1})
Rwanda	109	157	1443	5	9	1667
Saudi Arabia	85	265	3118	4	11	2933
Senegal	140	98	696	714	572	801
Somalia	270	231	854	nr	nr	nr
South Africa	60	150	2500	14	7	479
Sudan	7136	4524	634	2782	1090	392
Tanzania	900	850	944	260	297	1142
Togo	250	285	1141	95	64	678
Uganda	350	299	854	180	228	1267
Ukraine	68	284	4176	78	102	1308
United States	2643	9882	3739	258	418	1620
Venezuela	240	500	2083	nr	nr	nr
Yemen	504	439	873	123	83	677
Zimbabwe	230	69	300	230	55	239

[a]*Countries with either sorghum area or pearl millet plus small millets area ≥50,000 ha are listed; nr, report not available.*
From http://faostat3.fao.org/faostat-gateway/go/to/download/Q/QC/E (accessed 18 August 2014).

Over the last four decades, there have been substantial changes in harvested area, production, and productivity of pearl millet and small millets globally. A comparative analysis of the harvest area, production, and productivity in different continents during the period from 1970 to 2013 revealed that the harvested area and production have decreased in Asia (61% and 41%) and Europe (82% and 68%) whereas they have increased in Africa (55% and 47%) and the Americas (13% and 35%) (Fig. 1.2). During the last four decades the harvested area and production under pearl millet and small millets have decreased globally by 27% and 10%, respectively. However, there was an increase in the productivity in all the continents during this period. The percentage increase in productivity was the highest in Europe (726−1282 kg ha^{-1}), followed by Asia (802−1221 kg ha^{-1}), and the Americas (1125−1620 kg ha^{-1}).

1.2 IMPORTANCE OF MILLETS

1.2.1 DRYLAND AGRICULTURE

Dryland areas cover over 40% of the global terrestrial area and are home to more than two billion people (Millennium Ecosystem Assessment, 2005). These areas are mostly inhabited by the world's poorest people for whom millets constitute the main staple food. Millets are mostly grown under agricultural conditions where other cereals fail to consistently produce an acceptable harvest.

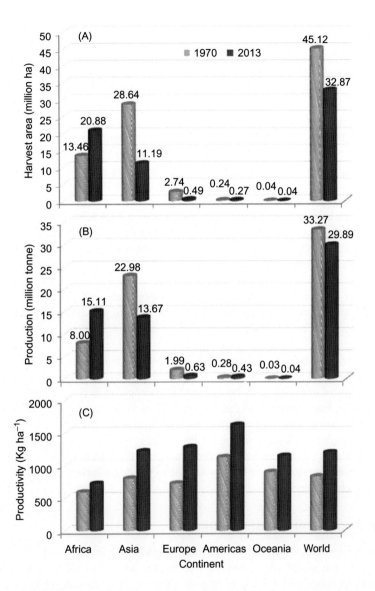

FIGURE 1.2

Changes in area (A), production (B), and productivity (C) of pearl millet and small millets in different continents over last four decades.

From http://faostat3.fao.org/faostat-gateway/go/to/download/Q/QC/E (accessed 18 August 2014).

The geographic areas having average annual rainfall of 200–600 mm in Africa and South Asia are the major places of cultivation of millets (Bidinger et al., 1982). These zones are generally characterized by a short rainy season (2–4 months), high mean temperatures and evapotranspiration rates, and arid shallow and sandy soils (Kowal and Kassam, 1978). Millets are favored due to their short growing season and acceptable productivity under dry, high temperature and poor soil conditions. In the semiarid tropics, where low precipitation and poor soils limit the cultivation of other major food crops, millets can grow and give reasonable yield.

Sorghum is an excellent rainfed crop for both grain as well as fodder. There is hardly any cereal that can beat sorghum in its productivity under rainfed conditions. The entire sorghum growing areas in eastern and southern Africa receive low rainfalls (43−180 mm month^{-1}) and experience a moderate temperature (16−26°C) during the cropping season. About 35% of these are drought prone areas, where the combined effects of warm temperature (>20°C) and low rainfall (<120 mm month^{-1}) create a water-scarce condition (Wortmann et al., 2009). In India, the post-rainy sorghum (mid-September−mid-February) is grown in an area of more than 5 million ha with an annual rainfall of 400−600 mm. These areas are characterized by shallow−medium depth of soil (45−75 cm), limited rainfall (200−250 mm), cooler temperature (12−20°C), and shorter days during the cropping season. Even though the postrainy sorghum heavily relies on the stored moisture from the preceding rainy season and precipitation received during the growing season (mostly received early in the growing season), it produces a reasonable harvest. There is hardly any crop, other than sorghum, that can survive and produce an acceptable harvest under these conditions.

Pearl millet is well adapted to the growing areas characterized by drought (rainfall 300−600 mm year^{-1}), low soil fertility, and high temperature (>35°C). It performs well in saline as well as acidic soils. Due to its tolerance to difficult growing conditions, it can be grown in areas where other cereals, such as maize or wheat, would not survive. A large area in western Rajasthan, Haryana, and Gujarat characterized by sandy soils, little rainfall (<400 mm year^{-1}), and an arid-zone climate, constitutes the major pearl millet growing region in India.

Small millets have a wide adaptability and can be grown in extremes of soil and climatic conditions. Finger millet is considered to be an excellent crop for dryland conditions, since it has a remarkable capacity for recovery and can be grown with minimum water (400 mm year^{-1}). Proso millet is a short duration crop, well adapted to diverse soil and climatic conditions, and its water requirement is probably the lowest among all the cereals. It is an excellent crop for dryland no-till farming. Kodo millet is a hardy and drought-tolerant crop that can survive on marginal soils, where other crops may not survive. It has a large potential to provide nourishing food to subsistence farmers in Africa and elsewhere.

1.2.2 FOOD AND NUTRITIONAL SECURITY

Despite the improvement in agriculture and food production over the years, the number of people fighting hunger in the world has exceeded one billion. Ever increasing demands for food have formed a tough challenge in the meeting of food security. On a global perspective, the crops such as rice, wheat, and maize are major cereals in terms of food security, followed by barley, sorghum, pearl millet, oat, and rye. The other cereals are known as minor millets. The share of sorghum and millets in the total food grain production in the world are 1.08% and 2.21%, respectively (FAOSTAT, 2013). However, the importance of millets in food and nutritional security cannot be judged by their poor share in total food grain production. Their importance lies in the fact that these crops can utilize the agricultural lands that otherwise would remain fallow. Millets constitute an important staple in a large number of countries in the semiarid tropics, where the cultivation of other major food crops is limited due to low rainfall and poor soil fertility. Therefore, sorghum and millets are not only important in terms of world food production, but indispensable as food crops in the rainfed agro-ecosystems. Millets are sources of high-energy and nutritious food, especially suggested for children as well as adults. In general, millets are rich in fibers, minerals, and B-vitamins, and hence are called nutricereals. Among the millets, foxtail millet, pearl millet, and sorghum are rich in protein (10.4−12.3 g per 100 g); barnyard, kodo, foxtail, and proso millets are rich in fiber (7.6−9.8 g per 100 g); finger millet is rich in calcium (344 mg per 100 g); proso and pearl millets are rich in iron (8.0−9.3 mg per 100 g); and proso, barnyard, and pearl millets are rich in zinc (3.0−3.7 mg per 100 g) (Gopalan et al., 1989). They are also rich in phytochemicals that are beneficial to health. Several types of food are prepared from millets in different countries, the most predominant being porridge or pancake-like flat bread.

Sorghum is utilized for various purposes such as food, feed, and fodder and more recently, as a source of biofuels. In Africa, sorghum grain is processed into a variety of nutritious traditional foods such as semileavened bread, dumplings, couscous, and different types of porridges. The farmers in central and southern parts of India use sorghum flour for making *jowar roti* (called "*bhakri*"). Many ready-to-cook and ready-to-eat products of sorghum have been developed and are available in the Indian market. Value-added products of sorghum such as rawa, flakes, vermicelli, pasta, and biscuits are quite popular in India (Patil et al., 2013). Apart from human food, sorghum has many industrial uses. Grains of sorghum are used as feed for poultry, birds, and animals. Grains are also used for alcohol production in the distilleries and starch-based products in the starch industry. In developed countries like the United States and Australia, sorghum is mainly used as animal feed. At present, the poultry feed sector in India is using approximately 1.30 million tonnes, the animal feed sector about 0.45 million tonnes, and alcohol distillers about 0.09 million tonnes of grains annually. The postrainy sorghum grown in the semiarid regions of India is highly valued for its fodder. Food and nutritional security of animals in this region is largely dependent on the availability of sorghum fodder since sorghum is the sole source of fodder as other crops rarely produce a reasonable fodder-yield under the harsh growing conditions. Sorghum and pearl millet are the main staples in a large region of northern Nigeria, Niger, Mali, and Burkina Faso. Pearl millet is an important food across the Sahel. Its grains are ground into flour for the preparation of large balls, which are sometimes liquefied into a watery paste using fermented milk and consumed as a beverage. This beverage called "fora" in Hausa is a popular drink in northern Nigeria and southern Niger. Pearl millet stems are used for a wide range of purposes such as the construction of walls and thatches and the making of brooms, mats, and baskets.

Small millets produce small-sized grains, which are highly nutritious. The grains have good storability that ensures their availability for a long period so that poor farmers can use them during a year of crop failure. Finger millet is especially valuable as it contains an important amino acid, methionine, which is lacking in the diets of people who live on starchy staples like cassava, plantain, and maize meal, etc. Its grains are ground and cooked for making cakes, puddings, or porridge. Grains can be used for the preparation of high value foods due to its excellent malting property. In Nepal and many parts of Africa, the grains are used for making a fermented and flavored drink. It is an important millet because of the excellent storage properties of its grains and its nutritive value, which is higher than that of rice and similar to that of wheat. In addition, finger millet is a good source of micronutrients like calcium, iron, phosphorus, zinc, and potassium. Proso millet grains lack gluten and are sold as a health food. It can be included in the diets of people who are suffering from gluten intolerance. Kodo millet grain is nutritious, an excellent source of fiber, and also contains high amount of polyphenols, an antioxidant compound (Hedge and Chandra, 2005). The other minor millets like tef and fonio are mostly used for food (porridges and flat breads) and they possess a low malting potential. In the developed world, millets are less important as food. For example, the only significant millet crop in the United States is proso millet, which is mostly grown for bird's seed (McDonough et al., 2000). Besides supplying food to human beings, millets are good sources of fodder for animals in the semiarid regions. Domestic animals in the dry areas are highly dependent on millets for fodder, especially sorghum and pearl millet, since no other crops can be grown due to the insufficient water in soil.

1.2.3 BIOENERGY PRODUCTION

The world is increasingly focusing on bioenergy with the intention of having a renewable alternative to fossil fuels. Several annual and perennial grasses (eg, Sorghum, Guinea grass, Napier grass, Switch grass) and oilseed plants (*Jatropha*) are being researched intensively for their economic viability in terms of biofuel production. Among millets, a specially adapted type of sorghum called "sweet sorghum" is highly promising in this regard since it can accumulate high levels of sugar in the juice-rich stalk. This unique attribute of sweet sorghum is considered favorable for its eventual emergence as a bioenergy crop. Some sweet sorghum lines can produce a juice yield of about 78% of its total biomass as well as yielding a considerable amount of grain (Reddy et al., 2008). Sweet sorghum juice extracted from the stalks is high in sugars (15−23%) that are readily fermentable. It is reported that sweet sorghum has the potential of producing 530−700 gal ethanol ac^{-1} (University of Kentucky, 2013). This sorghum has some advantages, such as lower cultivation costs than sugarcane, superior quality ethanol with less sulfur, high octane rating, and automobile-friendly. Bagasse obtained after juice extraction is rich in micronutrients and can be used for power generation. Sorghum bagasse produced from one hectare of land can generate a power of about 2.5 MW (Prabhakar et al., 2015). Sorghum can produce about 3−4 tonnes jaggery and 4 tonnes syrup per hectare. Sweet sorghum can be used as both first and second generation (lignocellulosic) biofuels due to its high content of both soluble as well as structural sugars (obtained from cellulose and hemicellulose). Cellulose and hemicellulose content in the high biomass sorghum lines range from 27% to 52% and 17% to 23%, respectively. Recently, brown midrib lines have attracted focus for second generation biofuel production as they contain less lignin (Oliver et al., 2005). The rainy season sorghum grain, which is often damaged by mold infection, has got good potential as a raw material for the production of good quality potable alcohol with a competitive cost over molasses. This may provide new opportunities for agribusiness and employment generation in the sorghum growing regions in the semiarid tropics. Active research is now underway in several countries to make sweet sorghum into an industry-friendly commercial crop for ethanol production.

1.2.4 CLIMATE RESILIENT AGRICULTURE

Impacts of climate change are being felt increasingly throughout the world. Agriculture is one of the most impacted areas of climate change because of its high dependence on the weather. Enhancing the resilience of agriculture to cope with climate variability and climate change is imperative to the livelihood and security of millions of farmers. Warmer temperatures, lesser and erratic rainfall along with fewer numbers of rainy days are some of the indicators of climatic changes. As a result of these, several regions in the semiarid tropics are expected to experience scarce rainfall, long dry spells, and warmer temperatures. A change in rainfall pattern makes it difficult for the people who are dependent on rainfed farming to decide on the sowing and harvesting timings. There is also a chance of occurrence of new pests and diseases due to changes in temperature and humidity regimes. Under such conditions, climate resilient crops that yield near normal productivity under changing climate are required and millets have more promising prospects than any other crops under this situation. Pearl millet, sorghum, and small millets are basically drought-tolerant crops which can grow under poor and marginal soil, giving a substantial yield. In addition to drought-tolerance, sorghum can also withstand periods of water logging (Taylor, 2015).

Pearl millet is one of the most suited crops for climate resilient agriculture. The combination of its short duration and heat tolerance makes it adaptable to harsh stress conditions. It can perform better than maize and sorghum in regions with high temperature and low soil fertility. It can be grown in areas where rainfall is not sufficient (200−600 mm) for maize and sorghum. Pearl millet is often referred to as the "Camel" of the crops due to its exceptional ability to tolerate drought. It can produce heat shock proteins when the seedlings are exposed to heat stress (35−45°C). Compared to many other cereals pearl millet has a higher threshold temperature, the temperature beyond which physiological or growth processes stop. This enables it to produce viable pollens at a temperature, at which the crops like rice, groundnut, and sorghum stop producing them.

Sorghum and pearl millet have a natural ability to reduce nitrous oxide (N_2O) emission by virtue of their biological nitrification inhibition activity. N_2O emitted by the addition of synthetic nitrogenous fertilizers to soil is an important constituent of greenhouse gases causing global warming. It accounts for about 5% of all greenhouse gases emitted from human activities in the United States (EPA, 2015). Roots of sorghum and pearl millet exude a biological nitrification inhibitor that can reduce the rate of nitrification and thus prevent the loss of nitrogenous fertilizer from the soil (Subbarao et al., 2007).

Small millets are highly tolerant to drought and high temperature. They are suitable crops for poor soils under hotter and dryer climates, a growing condition where other crops fail to perform. Foxtail millet is reported to have high water use efficiency since it requires about 257 g of water to produce 1 g dry biomass compared to 470 g of maize and 510 g of wheat (Diao, 2007; Li and Brutnell, 2001). Similarly, proso millet also requires very little water, which it can convert to dry matter and grain very efficiently. Millets are the most drought-tolerant cereals that require little input during growth and with the decreasing water supplies and increasing population; they represent important crops for the future. Millets are, therefore, truly called "smart crops" or "climate resilient crops."

1.3 MAJOR PRODUCTION CONSTRAINTS

Agriculture, being a biological production system, faces a host of constraints that have the potential to affect production and the productivity of the system. In the case of crop production, weather or climate plays the most crucial role. Climate of a particular region influences the land topography, soil type, soil fertility, water availability, pest and disease incidence, and even socioeconomic conditions of the farming community. All these factors are intricately related and impact the production and productivity of an agricultural production system. This section discusses the major constraints related to millet-based agricultural production systems in the world.

1.3.1 BIOTIC CONSTRAINTS

Biotic stresses caused by living organisms, such as fungi, oomycetes, bacteria, mycoplasma, nematodes, insects, birds, weeds, and parasitic plants are the most important constraints of agricultural production worldwide. Apart from the true living organisms, entities like virus and viroids also cause considerable yield loss in agriculture. The intensity of biotic stress varies depending on the weather, cropping system, cultivation practices, type of crops, crop varieties, and their resistance levels. Generally, hot and humid weather, input-rich intensive cultivation, and poor crop-management practices make the crop vulnerable to these stresses. As millets are mostly grown in dry climates, the adverse effect of biotic stresses in millets is less compared to other crops. There are huge areas under millet cultivation in sub-Saharan Africa, where traditional landraces of millets are grown in poor soil with no or negligible input, and pest and disease problems are relatively lower. In spite of that, the total losses in millets due to biotic stresses are enormous since the acreage under millet cultivation across the globe is high. The estimated loss of grain sorghum production due to biotic stresses (diseases, pests, striga, weeds, and birds) in nine countries of eastern and southern Africa is about 5.88 million tonnes year^{-1} compared to 2.11 million tonnes year^{-1} due to water deficits or drought (Wortmann et al., 2009). Pests and disease problems in millets are relatively higher in areas where intensive cultivation with high yielding varieties are the common practice.

In general, millets suffer more from fungal disease than bacterial, viral, and nematode diseases. The important diseases of millets are grain mold (sorghum), downy mildew (pearl millet, sorghum, foxtail millet, proso millet), blast (finger millet, foxtail millet, pearl millet), leaf blight (sorghum, finger millet), smut (sorghum, foxtail millet, tef), rust (sorghum, foxtail millet, tef), ergot (sorghum, pearl millet), anthracnose (sorghum), and charcoal rot (sorghum) (Strange and Scott, 2005; Das, 2013). A few viral diseases (maize stripe virus, maize mosaic virus, etc.) are important in sorghum but not in other millets. Among the bacterial diseases, leaf streak, leaf stripe, and leaf spot are observed on sorghum in tropical or temperate humid environments. Recently, the bacterial soft rot caused by *Erwinia chrysanthemi* has been reported to occur in destructive forms in the Tarai region of India (Kharayat and Singh, 2013). A few species of plant parasitic nematodes have been reported to cause disease in sorghum and pearl millet, especially under poor soil and water environments. However, their damaging potential, supported by actual yield loss data are lacking. Important insect pests of millets include stem borer, shootfly, aphids, midges, headbug, etc. Besides these, grubs, armyworms, cutworms, locust, termites, black ant, and rodents also assume the dimension of important pests in some parts of the world. *Striga*, a parasitic weed, is one of the most serious constraints to cereal production in Africa (Ethiopia, Zimbabwe, Uganda, Rwanda, and Kenya), causing extensive yield losses on millets. It attacks the roots of young crops and starves them of nutrients, leading to low grain yields. Yield losses due to *Striga* are higher on pearl millet and sorghum than other millets. An estimated 100 million ha of the African savannah zones is infested with *Striga* (Ejeta, 2007).

The weeds are a global problem in agriculture and they are a major deterrent to increasing the productivity of millets, especially during the rainy season due to the weather conditions being congenial for their growth. Since millets are grown under a rainfed condition, soil moisture and nutrients are the most limiting factors. Weeds compete with millets for light, soil moisture, and nutrients and reduce their grain yield. The reduction of grain yield in sorghum may vary from 15% to 83%, depending on the crop, nature and intensity of weeds, duration of weed infestation, and environmental conditions (Stahlman and Wicks, 2000). The bird damage of millet-grains is now considered a potential threat to millet growers. Generally, the small-grain crops are more prone to severe damage by birds than large-grain crops. The extent of damage varies depending on growing conditions and the crop and the loss may go up to 100% in isolated and unprotected conditions. Although it is a common problem in many countries, it is severe in some African countries like Ethiopia, Kenya, and Rwanda. It is estimated that *Quelea* and other birds can cause yield loss of about 1.6 million tonnes year^{-1} in eastern and southern Africa (Wortmann et al., 2009). Details of various biotic stresses of millets, their significance, and management with emphasis on host-plant resistance will be discussed in detail in the following sections of this book.

1.3.2 ABIOTIC CONSTRAINTS

Abiotic constraints acting against the production and productivity of millets are related to soil and environments. They include the prevalence of drought conditions due to insufficient soil moisture and rainfall, heat and light stress (high irradiation), atmospheric drought (dry weather), poor status of soil fertility, and high soil salinity. Among these, drought is the most important constraint in the millet producing areas of the world. Since millets are mostly grown in regions that receive scarce rainfall with an erratic distribution, the standing crop often does not receive any rain and is exposed to drought. Drought or moisture stress may occur during the seedling stage, vegetative stage, or at the time of grain filling. Drought and water stress are considered to be one of the most important stresses of millet production in many African countries, such as Eritrea, Ethiopia, Ghana, Kenya, Namibia, Sudan, Tanzania, Uganda, and Zimbabwe (Matanyaire, 1996; Gebretsadik et al., 2014). Soil-water deficits during millet-crop establishment and grain fill stages are common in Ethiopia, while midseason deficits are found in Uganda and Kenya. In India, unlike in the other parts of the world, sorghum is cultivated in both rainy and postrainy season. The postrainy sorghum production environment is characterized by increasing moisture stress as the season progresses. Frequent dry spells with varying intensities, periods, and timing are the causes for the low productivity of sorghum in India during rainy and postrainy seasons (Patil, 2007).

Low soil fertility is another factor that limits the productivity of millet, which are mostly grown in sandy soils having poor nutrient status and low water holding capacity. The soils are predominantly low in organic carbon and deficient in nitrogen and phosphorus. Farmers hardly practice crop rotation or apply any fertilizer. These stresses in combination with drought, severely limit the production potential of the millets in the semiarid tropics. Soil salinity is another important constraint that impacts the millet production. However, this problem is not as widespread as drought and soil fertility. Inadequate drainage facility and poor quality water for irrigation aggravates the salinity problem. Saline soils have higher levels of soluble salts, such as sulfates, carbonates, and chlorides, which interfere with the seed germination process and cause a reduction in germination, plant growth, and yields of sorghum and other millets. Damage is greater in the seedling emergence stage than at any other stage in sorghum (Macharia et al., 1994).

1.3.3 SOCIOECONOMIC FACTORS

Socioeconomic conditions of the farming community and the people living in that region influence the productivity of an agricultural production system. Low levels of socioeconomic development, limited infrastructure, lack of input availability, higher dependency on natural resources, lack of institutional capacity, traditional mindset, lack of proper marketing and credit system, and lack of demand for the produce are some of the factors that adversely impact agriculture. These factors differ depending on the country and cropping system. The above factors assume greater significance in the millet production system in Africa than other countries since millets are mostly grown as subsistence agriculture on marginal land by the resource poor farmers of Africa. Recurrent drought and crop failure, small land holdings (around $0.5-2.0$ ha household^{-1}), population growth, and joblessness are considered the key reasons for leaving agriculture and the migration of people from rural to urban areas in the semiarid tropics. In these regions, the farmers use seeds of traditional varieties, which are low yielding landraces. Because of the fear of crop failure due to drought, the farmer uses hardly any fertilizer or pesticide and often undertakes mixed cultivation of many crops that might meet their domestic needs. Women are primarily responsible for the production, postharvest handling, and marketing. Lack of the adequate technology backup needed for improvement of productivity of millets in rainfed agro-ecosystems also contributes to low production. Although the improved production technologies, such as supply of improved cultivar, quality seed, and up-to-date information on crop management practices, may be available at the institutional level, they often do not reach the farmers on time. The other problems include poor prices and inadequacy of demand for the processed products, which compels the small-scale agribusiness entrepreneurs to sell their products on loan, a debt which later turns out to be a bad debt (Ja'afar-furo et al., 2011).

The situations in other millet growing nations like India, China, United States, and Australia are different since these countries have widespread communications networks, well-developed agricultural research and training networks, well-developed marketing systems for crops, credit facilities for rural people, and large scale fertilizer and pesticide manufacturing and distribution facilities. These factors mostly have a positive impact on the millet production and productivity. However, the developed countries have a different set of production constraints for millets and the shifting of millet-lands to more remunerative crops is the major factor limiting the millet production. For example, the production of small millets in India decreased to around 2.5 million tonnes during $2011-12$ compared to around 4 million tonnes during late 1940s. During the same period, the area under these crops was reduced to 2.3 million ha from 8 million ha (Seetharam, 2015). As the millets are relatively more intensively cultivated in these countries, the biotic stresses like pests, diseases, and weed problems play a greater role in decreasing the productivity under such conditions.

1.4 CONCLUSIONS

Millets are hardy cereals that can be grown with meager input under adverse agricultural situations. These crops originated mostly in Africa and Eurasia and were later domesticated in many countries of the world. In order of global crop importance, millets occupy the fifth position after rice, wheat, maize, and barley and constitute an important staple in the semiarid tropics. They ensure food and nutritional security for poor people in the drier parts of the world, where the cultivation of other major food crops is limited due to low rainfall and poor soil fertility. Presently, millet, particularly sweet sorghum, is being researched intensively for its utilization as a source of bioenergy production. Like other agricultural crops, the millets also face many obstacles toward achieving increased production and productivity. One of the major concerns is the diversion of millet growing lands to the more remunerative crops, resulting in the shrinking of global areas under millet production. The world is continuously experiencing the ominous effects of accumulation of greenhouse gases in the form of global warming and climatic changes. This situation needs crops that can mitigate the effects of these changes and millets offer the best alternative since they can help in the reduction of N_2O emission.

ACKNOWLEDGMENTS

We thank A.V. Umakanth for his support while preparing the manuscript and P. Rajendrakumar for critically reviewing the final draft.

REFERENCES

Berenji, J., Dahlberg, J., 2004. Perspectives of sorghum in Europe. J. Agron. Crop Sci. 1905, 332−338.

Bidinger, F.R., Mahalakshmi, V., Talukdar, B.S., Alagarswamy, G., 1982. Improvement of drought resistance in pearl millet. Drought Resistance in Crop With Emphasis on Rice. International Rice Research Institute (IRRI), Los Banos, pp. 357−375.

Crawford, G.W., Lee, G.A., 2003. Agricultural origins in the Korean peninsula. Antiquity 77 (295), 87−95.

Dahlberg, J.A., Wasylikowa, K., 1996. Image and statistical analyses of early sorghum remains (8000 B.P.) from the Nabta Playa archaeological site in the Western Desert, southern Egypt. Veg. Hist. Archaeobot 5, 293−299.

Dahlberg, J., Berenji, J., Sikora, V., Latković, D., 2011. Assessing Sorghum [*Sorghum bicolor* (L) Moench] Germplasm for New Traits: Food, Fuels & Unique Uses. Advance access publication. Maydica, 56−1750, 85−92.

D'Andrea, A.C., Lyons, D.E., Haile, M., Butler, E.A., 1999. Ethnoarchaeological approaches to the study of prehistoric agriculture in the Ethiopian Highlands. In: Van der Veen (Ed.), The Exploitation of Plant Resources in Ancient Africa. Kluwer Academic: Plenum Publishers, New York, NY.

Das, I.K., 2013. Disease management in grain, forage and sweet sorghum. In: Chapke, R.R., Bhagwat, V.R., Patil, J.V. (Eds.), Sorghum Cultivation for Value-added Diversified Products and Sweet Sorghum Perspectives. Directorate of Sorghum Research, Hyderabad, pp. 99−104.

Diao, X., 2007. Foxtail millet production in China and its future development tendency. In: Chai, Y., Wan, F.S. (Eds.), The Industrial Development of China Special Crops. Chinese Agricultural Science and Technology Press, Beijing, pp. 32−43.

Ejeta, G., 2007. The *Striga* scourge in Africa—a growing pandemic. In: Ejeta, G., Gressel, J. (Eds.), Integrating New Technologies for *Striga* Control Towards Ending the Witch-hunt. World Scientific Publishing Co., Singapore, pp. 3−16.

EPA, 2015. Overview of greenhouse gases. United States Environmental Protection Agency. Accessed on 15 July 2015. <http://epa.gov/climatechange/ghgemissions/gases/n2o.html>.

FAOSTAT, 2013. Download data-production. Accessed on 18 August 2014. <http://faostat3.fao.org/faostat-gateway/go/to/download/Q/QC/E>.

Fuller, D.Q., 2003. African crops in prehistoric South Asia: a critical review. In: Neumann, K., Butler, A., Kahlheber, S. (Eds.), Food, Fuel and Fields. Progress in Africa Archaeobotany. Africa Praehistorica 15 Series. Heinrich-Barth-Institute, Cologne, pp. 239−271.

Gebretsadik, R., Shimelis, H., Laing, M.D., Tongoona, P., Mandefro, N., 2014. A diagnostic appraisal of the sorghum farming system and breeding priorities in *Striga* infested agro-ecologies of Ethiopia. Agr. Syst. 123, 54−61.

Gopalan, C., Rama Sastry, B.V., Balasubramanian, S.C., 1989. Nutritive Value of Indian Foods. National Institute of Nutrition, Indian Council of Medical Research, Hyderabad, p. 156.

Harlan, J.R., 1975. Crops and Man. American Society of Agronomy and Crop Science Society of America, Madison, WI.

Harlan, J.R., de Wet, J.M.J., 1972. A simplified classification of cultivated sorghum. Crop Sci. 12, 172−176.

Hedge, P.S., Chandra, T.S., 2005. ESR spectroscopic study reveals higher free radical quenching potential in kodo millet (*Paspalum scrobiculatum*) compared to other millets. Food Chem. 92, 177−182.

Ja'afar-Furo, M.R., Bello, K., Sulaiman, A., 2011. Assessment of the prospects of value addition among small-scale rural enterprises in Nigeria: evidence from North-Eastern Adamawa. State J. Dev. Agric. Econ. 3 (3), 144−149.

Katie, M., Pelling, R., Higham, T., Schwenniger, J.L., Fuller, D.Q., 2010. 4500-year old domesticated pearl millet (*Pennisetum glaucum*) from the Tilemsi Valley, Mali: new insights into an alternative cereal domestication pathway. J. Archaeol. Sci. 38 (2), 312−322.

Khairwal, I.S., Rai, K.N., Diwakar, B., Sharma, Y.K., Rajpurohit, B.S., Bindu, N., Ranjana, B., 2007. Pearl Millet: Crop Management and Seed Production Manual. ICRISAT, Patancheru, p. 104.

Kharayat, B.S., Singh, Y., 2013. Characterization of *Erwinia chrysanthemi* isolates inciting stalk rot disease of sorghum. Afr. J. Agric. Res. 10 (22), 2309−2314.

Kowal, J.M., Kassam, A.H., 1978. Agricultural Ecology of Savanna. Oxford University Press, Oxford.

Li, P., Brutnell, T.P., 2001. *Setaria viridis* and *Setaria italic*, model genetic systems for the Panicoid grasses. J. Exp. Bot. 62, 3031−3037.

Macharia, J.M., Kamau, J., Gituanja, J.N., Matu, E.W., 1994. Effects of sodium salinity on seed germination and seedling root and shoot extension of four sorghum [*Sorghum bicolor* (L.) Moench] cultivars. Int. Sorghum Millet Newslett. 35, 124−125.

Mann, J.A., Kimber, C.T., Miller, F.R., 1983. The Origin and Early Cultivation of Sorghum in Africa. Texas A&M University, College Station, TX, pp. 1−21.

Marchais, L., Tostain, S., 1993. Evaluation de la diversité génétique des Mils (Pennisetum glaucum, (L.) R. BR.) au moyen de marqueurs enzymatiques et relation entre formes sauvages et cultivées. In: Hamon, S. (Ed.), Le mil en Afrique diversité génétique et agrophysiologique: Potentialités et contraintes pour l'amelioration génétique et l'agriculture. Actes de la réunion thématique sur le mil (Pennisetum glaucum, L.), Montpellier du 24−26 Novembre, 1992.

Matanyaire, C.M., 1996. Pearl millet production system(s) in the communal areas of northern Namibia: priority research foci arising from a diagnostic study. In: Leuschner, K., Manthe, C.S. (Eds.), Drought-tolerant Crops for Southern Africa: Proceedings of the SADC/ICRISAT Regional Sorghum and Pearl Millet Workshop, 25−29 July 1994, Gaborone, Botswana. ICRISAT, Patancheru, Andhra Pradesh, India, pp. 43−58.

McDonough, C.M., Rooney, L.W., Serna-Saldivar, Sergio, O., 2000. The Millets. Food Science and Technology: Handbook of Cereal Science and Technology, second ed. CRC Press, Boca Raton, FL, pp. 177−210.

Millennium Ecosystem Assessment, 2005. Ecosystems and Human Well-Being: Desertification Synthesis. World Resources Institute, Washington, DC.

Obilana, A.B., 2003. Overview: importance of millets in Africa. Accessed on 10 July 2015. <http://www.afripro.org.uk/papers/Paper02Obilana.pdf>.

Oliver, A.L., Pedersen, J.F., Grant, R.J., Klopfenstein, T.J., Jose, H.D., 2005. Comparative effects of the sorghum BMR-6 and BMR-12 genes: II. Grain yield, stover yield, and stover quality in grain sorghum. Crop Sci. 45, 2240–2245.

Oumar, I., Maria, C., Pham, J.L., Vigouroux, Y., 2008. Phylogeny and origin of pearl millet (*Pennisetum glaucum* [L.] R. Br) as revealed by microsatellite loci. Theor. Appl. Genet. 117, 489–497.

Patil, J.V., Chapke, R.R., Mishra, J.S., Umakanth, A.V., Hariprasanna, K., 2013. Sorghum Cultivation—A Compendium of Improved Technologies, vol. I. Directorate of Sorghum Research, Hyderabad.

Patil, S.L., 2007. Performance of sorghum varieties and hybrids during post-rainy season under drought situations in Vertisols in Bellary, India. J. SAT Agric. Res. 5 (1), 1–3.

Peacock, J.M., Wilson, G.L., 1984. Sorghum. In: Goldsworth, P.R., Fisher, N.M. (Eds.), The Physiology of Tropical Field Crops. John Willey and Sons Ltd., New York, NY, pp. 249–279.

Porteres, R., 1976. African cereals: Eleusine, Fonio, black Fonio, Tef, *Brachiaria*, *Paspalum*, *Pennisetum* and African rice. In: Harlan, J.R., de Wet, J.M.J., Stemler, A. (Eds.), Origins of African Plant Domestication. Mouton Publishers, La Hague.

Prabhakar, Aruna, C., Bhat, B.V., Umakanth, A.V., 2015. Genetic improvement in sorghum. In: Patil, J.V., Tonapi, V.A. (Eds.), Millets: Ensuring Climate Resilience and Nutritional Security. Daya Publishing House, New Delhi, pp. 109–196.

Rakshit, S., Hariprasanna, K., Gomashe, S., Ganapathy, K.N., Das, I.K., Ramana, O.V., et al., 2014. Changes in area, yield gains and yield stability of sorghum in major sorghum-producing countries during 1970–2009. Crop Sci. 54, 1571–1584.

Reddy, B.V.S., Ashok Kumar, A., Sanjana Reddy, P., Elangovan, M., 2008. Sorghum germplasm: diversity and utilization. In: Reddy, B.V.S., Ramesh, S., Ashok Kumar, A., Gowda, C.L.L. (Eds.), Sorghum Improvement in the New Millennium. ICRISAT, Patancheru, pp. 153–169.

Seetharam, A., 2015. Genetic improvement in small millets. In: Tonapi, V.A., Patil, J.V. (Eds.), Millets: Ensuring Climate Resilience and Nutritional Security. Daya Publishing House, New Delhi, p. 649.

Seetharam, A., Riley, K.W., Harinarayana, G., 1986. Small millets in global agriculture. Proceedings of the First International Small Millets Workshop. Oxford & IBH Publishing Co. Pvt. Ltd., New Delhi, pp. 19–30.

Stahlman, P.W., Wicks, G.A., 2000. Weeds and their control in sorghum. In: Smith, C.W., Fredricksen, R.A. (Eds.), Sorghum: Origin, History, Technology and Production. John Wiley & Sons, New York, NY, pp. 535–590.

Strange, R.N., Scott, P.R., 2005. Plant disease: a threat to global food security. Annu. Rev. Phytopathol. 43, 83–116.

Subbarao, G.V., Rondon, M., Ito, O., Ishikawa, T., Rao, I.M., Nakahara, K., et al., 2007. Biological nitrification inhibition (BNI)—is it a widespread phenomenon? Plant Soil. 294, 5–18.

Taylor, J.R.N., 2015. Overview: importance of sorghum in Africa. Department of Food Science, University of Pretoria, Pretoria 0002, South Africa. Accessed on 16 July 2015. <http://www.afripro.org.uk/papers/Paper01Taylor.pdf>.

University of Kentucky, 2013. Sweet sorghum for biofuel. Accessed on 16 July 2015. <http://www.uky.edu/Ag/CCD/introsheets/sorghumbiofuel.pdf>.

Vishnu-Mittre, 1968. Prehistoric records of agriculture in India. Trans. Bose Res. Inst. 31, 87–106.

Wendorf, F., Close, A.E., Schild, R., Wasylikowa, K., Housley, R.A., Harlan, J.R., et al., 1992. Saharan exploitation of plants 8,000 years BP. Nature 359, 721–724.

Wortmann, C.S., Mamo, M., Mburu, C., Letayo, E., Abebe, G., Kayuki, K.C., et al., 2009. Atlas of Sorghum Production in Eastern and Southern Africa. University of Nebraska, Lincoln, NE.

Zohary, D., Hopf, M., 2000. Domestication of Plants in the Old World, third ed. Oxford University Press, Oxford.

DISEASES AND INSECT PEST RESISTANCE

DISEASE RESISTANCE IN SORGHUM

2

I.K. Das and P. Rajendrakumar

ICAR-Indian Institute of Millets Research, Hyderabad, India

2.1 INTRODUCTION

Sorghum (*Sorghum bicolor* L. Moench) is one of the most important cereal crops grown extensively in the semiarid tropics for its nutritious grain and precious fodder. Grain yields are generally low in the resource-poor farmers' fields and incidence of diseases and pests further reduce the productivity. Diseases are one of the major constraints that come in the way of increasing productivity, production of quality grain and fodder in grain sorghum, and sugar yield in sweet sorghum. In grain sorghum, grain mold, downy mildew, anthracnose, and ergot are major diseases during the rainy season, whereas root and stalk rot and chlorotic stripe virus are common during the winter season. Other diseases such as leaf spots and smuts occur sporadically and assume economic significance under specific environments depending on relative humidity and temperature during the crop growth period in a particular year. A list of important diseases of sorghum in major sorghum growing countries is given in Table 2.1. Most of the diseases of grain sorghum also occur in sweet sorghum depending on the growing conditions and environment. Leaf anthracnose, red stalk rot, leaf blight, downy mildew, rust, sugary disease, head mold, and virus diseases are common in sweet sorghum. Leaf diseases cause destruction or reduction of active leaf area for photosynthesis, and thus adversely affect accumulation of sugar in stalk. As sweet sorghum is a crop having commercial value, cost-intensive management practices can be advocated. Diseases contribute negatively to fodder yield and quality in forage sorghum. Foliar diseases, that is, leaf spots, sooty stripes, leaf blight, downy mildew, anthracnose, and rust, are of economic significance for forage. There is report that infection of foliar diseases results in the reduction of the protein, zinc, and in vitro dry matter digestibility thereby affecting the fodder quality (Rana et al., 1999).

Utilization of host-plant resistance and development of disease-resistant hybrids and varieties is the only economic and practicable way of disease management in sorghum. To achieve this, a comprehensive and current knowledge on sorghum diseases, their distribution, importance, biology, epidemiology, and all aspects of the host-plant resistance is prerequisite. This chapter deals with all the above aspects of sorghum diseases.

Biotic Stress Resistance in Millets. DOI: http://dx.doi.org/10.1016/B978-0-12-804549-7.00002-0

Table 2.1 Important Diseases of Sorghum in the Top-10 Sorghum Growing Countries

Country	Share in World Acrage (%)	Major Uses (Grain)	Most Important Diseases	Intermediate Important Diseases
Sudan	17	Food	Grain mold, anthracnose, charcoal rot & ergot	Downy mildew, ergot, leaf blight, charcoal rot, loose & covered smuts, rust, gray leaf spot, sooty stripe, bacterial stripe
India	15	Food, feed	Grain mold, anthracnose, ergot, downy mildew & charcoal rot	Leaf blight, rust, zonate leaf spot, sooty stripe, viral diseases
Nigeria	13	Food	Grain mold, anthracnose, loose & covered smuts	Gray leaf spot, zonate leaf spot, sooty stripe, leaf blight
Niger	7	Food	Grain mold, anthracnose, long smut	Leaf spots, leaf blight
United States	6	Animal feed	Head blight/grain mold, ergot, anthracnose, *Fusarium* stalk rot	Charcoal rot, head smut, leaf spots
Mexico	4	Animal feed	Head blight/grain mold, leaf blight, rust, head smut	Charcoal rot, Anthracnose, zonate leaf spot, gray leaf spot, ergot, bacterial leaf streak, maize dwarf mosaic virus
Ethiopia	4	Food	Grain mold & anthracnose	Downy mildew, ergot, leaf blight, charcoal rot, loose & covered smuts, rust, gray leaf spot, zonate leaf spot, sooty stripe, bacterial stripe
Burkina Faso	4	Food	Head blight, anthracnose, head smut	Zonate leaf spot, gray leaf spot, sooty stripe, grain mold
Argentina	2	Animal feed	Grain mold & *Fusarium* stalk rot	Downy mildew, bacterial leaf stripe, sugarcane mosaic virus
Mali	2	Food	Anthracnose, loose, covered & long smuts, sooty stripe	Zonate leaf spot, gray leaf spot, ergot, charcoal rot, grain mold

2.2 DISEASE, BIOLOGY, AND EPIDEMIOLOGY

2.2.1 FUNGAL DISEASES

2.2.1.1 Panicle diseases

Panicle diseases refer to all the diseases of sorghum that are primarily caused by the infection of panicle tissues including peduncle, rachis, floret, or grain. Major diseases in this group include grain mold, head blight, ergot or sugary disease, and smuts.

2.2.1.1.1 Grain mold

Grain mold is a major disease of rainy season sorghum and is common in many countries in Asia, Africa, North America, and South America. The disease is severe in Asia and Africa where white grain sorghum are more widely grown. Improved short- and medium-duration sorghum cultivars that mature during the rainy season in humid, tropical, and subtropical climates suffer more. Late-maturing photoperiod sensitive sorghums generally escape grain mold as they flower and fill grain during dry weather. Colored grain sorghum which is grown for feed purpose in the United States, Mexico, Argentina, and Australia suffers less from this disease. The disease is most severe in India where the high yielding white grain hybrids are grown during rainy season. The major sorghum growing states in India are Maharashtra, Karnataka, Andhra Pradesh, Tamil Nadu, and parts of Gujarat and Madhya Pradesh. Molded grains become discolored and lose their appeal in the market and fetch 20−40% lower prices. Production losses range from 30−100% depending on cultivar and prevailing weather (Singh and Bandyopadhyay, 2000). Economic losses are around US$ 130 million in Asia and Africa (ICRISAT, 1992) and US$ 50−80 million in India (Das and Patil, 2013). Grain mold reduces the seed value of grain, the nutritive value of food and feed, and the cooking quality of the grain. Molded grains often contain mycotoxins some of which are harmful to human, animal, and poultry birds.

Symptoms of grain mold vary with the severity of infection and grain development stages. The first visible symptom is pigmentation of spikelet tissues including sterile lemma, palea, lodicules, and glumes. In the case of severe infection anthers and filaments also develop symptoms of fungal colonization. Infection at anthesis results in the loss of caryopsis formation, blasted florets, poor seed set, and the production of small, shriveled grains. Under humid conditions the pathogen grows quickly and may cover the entire grain with fungal growth before physiological maturity. The severely infected grains become soft and disintegrate under slight pressure and such a disease condition is often termed as "premature seed rotting" (Das et al., 2013). Depending on the fungal species the color of fungal bloom on grain varies from whitish, pinkish, grayish, to shiny black. Internal colonization of grain sometimes induces sprouting in grains under wet conditions. Sprouted grains become soft due to digestion of parts of the endosperm by α-amylase and are predisposed to extensive colonization by mold fungi, primarily species of *Fusarium* and *Curvularia* (Steinbach et al., 1995). The most obvious sign of mold infection on mature grain is the appearance of pink, orange, gray, white, or black mycelium on the grain surface. Discoloration of grain is more prominent on white grain than on brown/red grain sorghum. Fungal growth first occurs at the hilar end of the grain, and subsequently extends on the pericarp surface.

Many fungi are involved in the development of grain mold. Based on their nutritional behavior they may be pathogenic, parasitic, or saprophytic. Among grain mold pathogens *Fusarium moniliforme*, *Curvularia lunata*, *Alternaria alternata*, and *Phoma sorghina* are common and have worldwide distribution. They also have a saprophytic mode of nutrition and can infect and colonize mature grain. In addition to the above four genera, many saprophytic fungi may colonize mature sorghum grain under the condition of high humidity and rainfall. Frequently encountered fungi are *Bipolaris* spp., *Colletotrichum* spp., *Aspergillus* spp., *Cladosporium* spp., *Exserohilum* spp., etc. Among the *Fusarium* spp., the predominant one is *F. moniliforme* (now split into many different genera). As per present nomenclature the *Fusarium* causing grain mold in sorghum comes under *F. andiyazi*, *F. nygamai*, *F. proliferatum*, *F. thapsinum*, and *F. verticillioides* (Das et al., 2012). Among *Curvularia* spp., *C. lunata* is the most frequently occurring species on molded sorghum grain. Infection occurs at flowering by air-borne inocula of the causal fungi. Moderate temperature ($25-35°C$) and high humidity (>90% RH) favors infection and subsequent disease development. Cool and dry weather are not favorable for grain mold. The pathogen can be soil-borne, air-borne, or carried in plant residue. Plant residues and soil debris containing fungal hyphae and conidia seem to be the primary sources of inocula in the field. Senescing lower leaves also produce abundant spores, which are readily disseminated by wind and rain splash. Special fungal structures may not be essential for winter survival. Conidia and hyphae of *F. moniliforme* in sorghum stalks could survive two winters without any loss of viability. Microconidia of this fungus can survive up to 900 days at different levels of humidity and temperature under laboratory conditions.

Use of a cultivar that matures during a period of no rain is the best option to escape grain mold. But this is hardly practicable under an erratic rainfall situation. Use of a mold tolerant cultivar and harvesting the crop at physiological maturity followed by drying of grain is the second best option to avoid grain deterioration due to weathering. Fungicidal spray reduces grain mold damage but it is hardly economical because of the low price of sorghum grain and high cost of the fungicide and its application. Two sprays of propiconazole 25% EC @ 0.2% starting at flowering and another spray after 15 days, significantly reduce mold incidence.

2.2.1.1.2 Head blight

Head blight, as the term indicates, is a disease of sorghum panicle. It is mainly observed in North America, Mexico, and in a few countries in South America, such as Argentina and Venezuela. In these countries it occurs in a moderate to severe form and can cause a significant reduction of grain yield especially in susceptible hybrids or varieties. However, in the major sorghum regions of the world (ie, Africa and Asia) it is considered to be a minor disease with sporadic incidence.

Reddening of the pith tissues of the peduncle and the rachis or rachis branches are the predominant internal symptom of the disease. This may lead to externally visible symptoms of necrosis of the peduncle and rachis or rachis branches. In many cases external symptoms may be overlooked unless favorable weather conditions help expresses the necrosis phase. In severe infection, part or whole of the panicle may collapse inducing premature death of the part or whole of the panicle. Grains in an infected panicle may look drier than normal grains.

The disease is caused by the fungus *F. moniliforme*. Other grain mold pathogens like *Alternaria* spp., and *Curvularia* spp. have also been found associated with this disease. This disease may have some relationship with grain mold. However, considering the plant parts affected it is dealt with as a separate disease. The initial infection takes place at the lower part of the panicle near the base of the rachis and spreads upward. Continuous high humidity at the time of flowering favors infection and colonization. Use of a resistant cultivar is the best way to manage the disease.

2.2.1.1.3 Ergot

Ergot or sugary disease is a serious limiting factor in production of hybrid seed, particularly if seed-set in male sterile line is delayed due to lack of viable pollen. It happens if there is a lack of synchrony in flowering between the male and female parents. Initially the disease was observed in Kenya (in 1915) and India (in 1917). Subsequently the disease has been reported from most of the sorghum producing countries in Asia, Africa, South America, North America, and Australia. Ergot causes twofold damages—directly through loss in seed yield, seed quality, germination, seedling emergence and indirectly through rejection of a seed lot due to contamination with the ergot sclerotia.

The hybrid seed production suffers from the risk of ergot infection. Estimated production loss is about 10−80% in India and South Africa (Bandyopadbyay et al., 1996). Apart from the hybrid, the variety also gets ergot infection when environmental conditions are favorable. Sclerotia of sorghum ergot contain dihydroergotamine, a toxic alkaloid which can affect milk production in cows and pigs and weight-gain in cattle. The disease has quarantine implications and a seed lot containing sclerotia more than the acceptable limit is often rejected in trading (acceptable limit: 0.3% sclerotia (w/w)).

The first visible symptom is exudation of a honeydew-like droplet from an infected floret. The droplets are thin, sticky or viscous, pinkish to brownish in color, and sweet in test. This gives the name sugary or honeydew disease, which can be seen on a single, few, or all florets in a panicle depending on severity of infection. Honeydews fall on leaf underneath the panicle and often attract the growth of saprophytic fungi (eg, *Cerebella* spp.). Saprophytic growth of fungi on leaf surface gives it a black coloration. The initial structure produced after infection of a floret is called sphacelium (plural, sphacelia). Gradually a wart-like fungal structure called sclerotium (plural, sclerotia) evolves in place of grain. The sclerotia of *Claviceps africana* are always found in association with the sphacelia and it is impossible to distinguish between them using the naked eye. Sclerotia are purple black to black in color, elongated in shape, hard in texture, and bigger than sorghum seed.

Three species of *Claviceps* (*C. africana*, *C. sorghi*, and *C. sorghicola*) cause sorghum ergot in different parts of the world. *Claviceps africana* is widespread across continents, that is, southern and eastern Africa, South America, Australia, and South East Asia, while *C. sorghi* is limited to India (Kulkarni et al., 1976) and South East Asia and *C. sorghicola* to Japan (Tsukiboshi et al., 1999). There is a difference in an amount of inoculum produced by *C. africana* and *C. sorghi*. *Claviceps africana* produces large numbers of secondary conidia compared to few or no secondary conidia by *C. sorghi*. Several conidial cycles are completed in a season and conidia are spread by wind, rain, and insects. Sclerotia germinate to produce pigmented stipe containing capitulum inside which perithecia are developed. Perithecia contain asci each of which produce ascospores on maturity. Primary infection takes place either by ascospores or by conidia. Low night temperatures (<12°C) during the period between 3 and 4 weeks before flowering to 5 days after flowering, high RH (>90%), and cloudy weather or panicle wetness after stigma emergence favor ergot disease development. The pathogen survives off-seasons via infected panicles left in the field or via sclerotia that are mixed with the seed during threshing and processing. Collateral hosts also play a role in survival.

Removal of collateral host plants from the field bunds helps to reduce pathogen inoculum and disease. Mechanical removal of sclerotia from seeds, by washing in 30% salt water followed by rinsing in plain water before sowing reduces seed contaminated infection. In seed production plots, ensuring synchrony of flowering between A and R lines avoids the occurrence of disease. Rapid pollination and fertilization of an ovary prevents ergot infection.

Recently, improvement of pollen fertility by introgressing effective restorer genes are being experimented in a few crop including sorghum (Miedaner and Geiger, 2015). Use of tolerant cultivar and spraying panicles with fungicides (0.1% Bavistin or 0.2% Propiconazole or 0.2% Mancozeb) at flowering minimizes disease incidence and its subsequent spread.

2.2.1.1.4 Smuts

Smuts is a fungal disease that converts parts or whole of an earhead into black powdery mass. They are seen in almost all sorghum growing regions in the world. However, incidences are more in the African countries (Nigeria, Niger, Burkina Faso, Mali) and in North America (Mexico) than other countries. In India they are found here and there sporadically and are of minor importance. Their occurrence has been reduced especially after the advent of seed treatment fungicides.

Four different types of smuts namely, head smut (*Sporisorium reilianum*), covered smut (*Sporisorium sorghi*), loose smut (*Sporisorium cruenta*), and long smut (*Tolyposporium ehrenbergii*) are observed in sorghum. They produce distinctly different symptoms. In the covered smut, a fungal sorus is formed in the place of the grain. Most of the grains of an infected earhead are replaced by the sori. A membrane-like fungal structure covers the spore masses in the spikelet and such a membrane generally persist unless it is broken by the mechanical force applied during threshing. The loose smut infected plants become stunted, produce thinner stalks, and more tillers. They flower earlier than the other healthy plants in the field. All the spikelets of an infected earhead get malformed and hypertrophied. The membrane-like structure covering the spore masses generally ruptures soon after the head emergence and the smut spores are blown away leaving behind the empty spikelet. In the case of head smut, a sorus fully covered with a grayish-white membrane emerges from the boot leaf in place of normal inflorescence. When fully emerged the fungal membrane ruptures releasing spore masses in the air leaving filamentous vascular tissues of the host. In the case of long smut, the sorus is covered by a whitish to dull yellow, fairly thick membrane. The sori are much longer (about 4.0 cm long) than those of the covered and loose smuts.

Head smut is soil-borne. Teliospores of *Sporisorium reilianum* fall on the soil from the smutted head and survive until the next season. They germinate and infect the shoot apex of the germinated seed. The fungal mycelia continue to grow inter- and intracellularly inside the growing plant and multiply vigorously on the panicle tissues at the time of flowering developing head smut/teliospores. Loose and covered smuts are primarily carried externally on the seed surface. Spores get adhered to the seed and germinate in the soil along with the seed. Mycelia infect the juvenile plant and grow internally until they infect the floret and convert them to smut sori. In case of long smut the spore balls deposited on the soil surface get air-borne and fall on the flag leaf where they infect the developing florets.

Efficient management of smut diseases requires awareness about the disease among the farmers. The practice of clean cultivation, such as collecting smutted heads in cloth bags and dipping in boiling water to kill the pathogen, will reduce the inoculum for the next year and minimize incidence. Adjusting sowing dates seems to help in avoiding the disease. Loose and covered smuts are externally seed-borne and easily controlled by seed dressing with sulfur (@ 4 g kg^{-1} seed).

2.2.1.2 Foliar diseases

2.2.1.2.1 Anthracnose

Anthracnose is an important disease causing substantial economic losses to grain, forage, and sweet sorghum. The disease was first reported from Togo (in 1902) and subsequently from Texas (in 1912) and West Africa (in 1980). Since then the disease has been observed in most of the regions of the world where sorghum is grown. It occurs in moderate to severe form on sorghum in Sudan, Nigeria, Niger, Ethiopia, Burkina Faso, Mali, India, and United States. In India the disease is common on forage sorghum that is widely grown in Rajasthan, Gujarat, Haryana, Uttarakhand, Uttar Pradesh, and Madhya Pradesh. Anthracnose can affect several plant parts causing symptoms such as seedling blight, leaf blight, stalk rot, and head blight. Leaf anthracnose affects photosynthetic areas and thus reduces the amount of current photosynthesis. Red rot phase, on the other hand, damages stalk tissue and affects the movement of water and nutrients to the earhead. Seedling blight phase of the disease may sometimes affect final plant stand in the field. Grain yield losses of 50% or more can occur under severe epidemics which may go up to 80% on highly susceptible cultivars. The disease has a profound effect on grain and stover yield and stover quality in forage sorghum and sugar accumulation in sweet sorghum. Reduction in forage yield is particularly important since it can significantly impact a farmer's income from a crop—livestock production system.

Initial symptoms of anthracnose on the leaf appear as small, elliptic to circular spots, with straw-color center, and wide margin. The lesion-margin may be red, orange, blackish purple, or tan, depending on the pigment present in the cultivar (purple or tan). Adjoining spots may coalesce to give a blighted appearance on the leaf. A black dot-like acervulus is often seen at the center of the necrotic spot, which is the characteristic diagnostic symptom for leaf anthracnose. Examination of lesions with a hand lens reveals small, black, hair-like structures (setae) protruding from the acervuli. In the case of severe infection, plants get defoliated and die before reaching maturity. Infected mature stalks may develop reddish internal lesions, which may be continuous or discontinuous giving the stem a ladder-like appearance. Nodal tissues in such a stalk are rarely discolored. If the infection is early and severe, preemergence damping-off may occur and the seedlings wilt and die.

Colletotrichum graminicola (Ces.) Wilson (Synonym: *C. sublineolum* (Henn.) Kabat & Bub; Perfect state: *Glomerella graminicola* Politis) causes anthracnose in sorghum. The fungus produces dark brown to black colored acervulus, which develop numerous erect, hyaline, nonseptate conidiophores on which conidia are borne terminally among the setae. The conidia are single-celled, hyaline, and cylindrical to obclavate but become sickle-shaped with age. The pathogen has many pathotypes or races in different geographical areas of the world. The first report of the existence of races in this fungus came from the United States during 1967. Since then some 11 races/pathotypes have been reported from the United States and Puerto Rico, 12 from Brazil, 9 from India, 15 from Nigeria, and 2 from West Africa (Nakamura, 1982; Pande et al., 1991; Prom et al., 2012). Conidia produced on wild sorghum species or crop residues serve as the primary inoculum, which are spread by wind or rain. Conidium germinates through a germ tube, which is subsequently developed into an appressorium that penetrates directly through the epidermis or stomata. The disease is most severe during extended periods of cloudy, warm, humid, and wet weather, especially when these conditions occur during the early grain-filling period. The production of conidia is favored by high humidity and the spore requires about 14 h to mature at 22°C. The fungus can survive as mycelium in host residue, wild sorghum species, and some weeds and as conidia or mycelium on seed. It can persist up to 18 months in diseased residues on the soil surface. The fungal mycelia survive for only a few days in the absence of residues. The fungus has some host specialization on maize, sorghum, sugarcane, and a group of grasses. A specialized form that attacks one host species does not necessarily attack others.

The best control for anthracnose is the use of resistant cultivars. Cultural practices like 1-year crop rotation with a species other than a host offers reasonable control. Clean cultivation, elimination of crop residues and grasses on which the fungus can survive, and enhancement of the conditions that hasten decomposition of host residues have also been used to control the disease.

2.2.1.2.2 Downy mildew

Sorghum downy mildew (SDM) has been reported from many countries in the tropical and subtropical zones. It was initially reported from Africa and Asia and subsequently spread to Americas in the late 1950s. It occurs in most sorghum growing areas and often appears in a potentially destructive form in the peninsular India, Sudan, Ethiopia, and Argentina. The pathogen infects both the sorghum and maize and causes a significant loss in grain and fodder yield. Systemic infection of the plant results in a barren inflorescence. Crop losses due to systemic infection vary from 2−20% depending on the time of infection, cultivars, and weather conditions. Local infection during grain development stages results in poor grain filling and yield loss. Several epidemics of SDM have occurred in the past. An annual yield loss was reported to be at least 0.1 million tons in parts of India. In a single season in the United States, an SDM epidemic in Texas caused an estimated loss of US\$ 2.5 million (Frederiksen et al., 1969). Loss of stover has an adverse effect on livestock production systems in the peninsular India since sorghum fodder is an important feed for the cattle in this region.

SDM is visible as either a systemic or localized infection. The systemic infection occurs when the pathogen colonizes the apical meristematic tissues. Systemically infected seedlings are pale yellow or have light-color streaking on the leaf, are chlorotic and stunted, and may die prematurely. Usually, the first symptoms are visible on the lower part of the leaf blade, which later progress upward. In cool, humid weather, the lower surfaces of chlorotic leaves become covered by a white, downy growth consisting of conidia and conidiophores of the pathogen. The leaves emerging from the whorl subsequently exhibit parallel stripes of green and white tissue. The infected striped areas die, turn brown, and disintegrate, resulting in a shredded appearance of the leaf. Systemically infected plants do not produce an earhead. Conidia produced in the infected plants become airborne and cause rectangular shaped local lesions on the leaf.

SDM is caused by *Peronosclerospora sorghi* (Weston & Uppal.) Shaw (Synonym: *Sclerospora sorghi* (Kulk.) Weston & Uppal). It is an obligate parasite and can infect both the sorghum and the maize. However, it has been successfully grown in dual culture with host tissue on a modified White's medium (Kaveriappa et al., 1980). Conidophores are erect, hyaline, and dichotomously branched and emerge through stomata in the lower surface of leaf. Conidia are hyaline, obovate, nonpapillate, and nonporoid. Oospores are produced in the mesophyll tissue between the vascular bundles. Nine pathotypes/races of *Peronosclerospora sorghi* have been reported in United States, Brazil, and Honduras, but there is no such report from India. *Peronosclerospora sorghi* produces sexual (oospores) and asexual (conidia & zoospores) spores, which serve as a source of inoculums for infection. Systemic infection of young seedlings occurs either from oospores in soil or from conidia produced on early-infected plants. Such infection normally occurs a week after emergence. Low soil temperature (10°C) and soil moisture favor infection by oospores. A cool environment and high humidity favor the production of conidia and the infection process. The optimum temperature for conidia production is 18°C. *Peronosclerospora sorghi* survives as oospore in soil and plant debris. It spreads through oospores in the glumes of seeds and in plant debris mixed with the seed. Conidia are fragile and play hardly any role in the long distance dissemination of the fungus. The pathogen may be present as mycelium in infected seeds, but mycelium is inactivated when seeds are dried to a moisture level of 12% and thus is seemed to be of negligible importance in spreading the disease. *Peronosclerospora sorghi* can infect grasses such as *Euchlaena*, *Panicum*, *Pennisetum*, and *Zea* spp. on which it can produce conidia and oospores. The collateral hosts serve as source of inoculums for the sorghum crop.

Use of disease-resistant cultivars is the best practice for management of SDM. Cultural practices like deep summer plowing, crop rotation with nonhost plant, rouging of infected plants, and practice of clean cultivation help reduce inoculum and disease incidence. Seed treatment with Ridomyl-MZ @ 6 g Kg^{-1} seed followed by one spray of Ridomyl-MZ @ 3 g L^{-1} reduces disease incidence. Ridomil MZ-72 (metalaxyl and mancozeb @ 0.25%) also found to be effective in controlling this disease.

2.2.1.2.3 Rust

Rust is observed in almost all sorghum growing regions in Africa, Asia, North and South America and is particularly common in Mexico, India, Sudan, and Ethiopia. Though the disease occurs regularly in many places it hardly ever becomes severe. Early sown crops often escape rust injury and the disease is mainly observed on late sown crops. Sorghum lines containing purple-pigment are generally susceptible to rust (Singh et al., 1994). The cultivars especially tan type varieties and hybrids possess a good amount of tolerance to this disease. The disease may become severe under cool humid weather especially on forage sorghum. The disease weakens the plant and thus may make it vulnerable to other diseases. Severely rusted plants may fail to produce earhead. Yield loss vary from 29% to 65% under certain conditions (Bandyopadhyay, 1986; Hepperly, 1990).

Reddish brown rust pustules appear first on both the surfaces of the lower leaves. Generally the upper half of the leaf gets more severe infection than the lower half. As the disease advances, the infection spreads to the younger leaves. Several adjoining pustules may coalesce to form large patches on the leaves and the infected leaves die prematurely giving the plants an unhealthy appearance which becomes visible from a distance. The pustules may appear in any parts of the plant including leaf, mid-rib, peduncle, and stem.

Puccinia purpurea Cook. causes rust in sorghum on which it produces two types of spores—urediniospores and teleutospores. The pathogen survives as urediniospores produced on ratoon or volunteer sorghum or on successively grown sorghum in the field and perennial and collateral hosts. Urediniospores become air-borne and cause primary infection in the field. Urediniospores produced on freshly infected plants serve as a source of secondary infection. *Oxalis corniculata* serves as alternate host for *Puccinia purpurea*.

Use of resistant cultivars is the best known control measure for sorghum rust. Practices of clean cultivation, destruction of weeds, volunteer, wild sorghum, and alternate hosts help to reduce primary inoculum. Foliar spray of Mancozeb @ 0.2% effectively controls the disease.

2.2.1.2.4 Leaf blight

Leaf blight is an important disease causing substantial economic losses to forage and grain sorghum in Mexico, Brazil, India, Sudan, Nigeria, Niger, Kenya, and Ethiopia. The first report of the disease was from India (Butler, 1918). The disease incidence is more on the purple or red-pigmented genotype than the nonpigmented or tan genotype. In India the disease occurs regularly in a moderate to severe form on the forage sorghum in the states of Haryana, Rajasthan, Uttarakhand, and Uttar Pradesh. The disease severity may assume an epidemic form occasionally and severely affects forage yield, quality, and grain yield due to extensive damage of photosynthetic machineries on the leaf. Grain yield loss of 50% or more can occur under severe epidemics (Frederiksen, 1986). The disease may cause reduction in sugar accumulation in the sweet sorghum. Financial loss, however, varies depending on the incidence, prevailing environment, and resistance in the cultivar grown.

Leaf blight symptoms are characterized by the presence of long, elliptical, and necrotic lesions on the leaf lamina. The center of the lesion is straw in color and the margin is usually dark brown. The dark color of the margin is, however, not conspicuous in nonpigmented (tan type) cultivar. The size and shape of the lesions vary depending on the level of host resistance. The lesions, in a susceptible genotype, enlarge and coalesce to form purplish gray or tan color necrotic areas on the leaf. The symptoms first appear on the lower or older leaves and then progresses to the upper or younger leaves. The surface of the necrotic lesions appears dark-gray, olive, or black in color due to production of spores by the pathogen, especially under damp weather. A severe disease gives the crop a distinctly burnt appearance.

The disease is caused by *Exserohilum turcicum* (Pass.) Leonard & Suggs (Synonym: *Helminthosporium turcicum* (Pass.); *Bipolaris turcica* (Pass.) Shoemaker; and *Drechslera turcica* (Pass.) Subramanian & Jain), whose perfect state is *Trichometasphaeria turcica* Luttrell. The fungus produces light gray, straight or curved, and septate conidium singly at the tip of a conidiophore. The conidium has a protuberant hilum at its basal cell and germinates by a polar germ tube. Although the pathogen is highly variable in its pathogenicity, there are no reports about the existence of races of *E. turcicum* in sorghum (Mathur et al., 2011). *Exserohilum turcicum* persists as mycelia and conidia in the infected crop residues or in the soil. It can infect hosts, that is, maize, Johnson grass, teosinte, paspalum, and other cereals. Secondary spread of the inoculum within and between fields takes place by air-borne conidia. Foliar pathogens favor moderate temperature, extended periods of cloudy weather, high humidity, dew, and warm and humid weather. A moderate temperature (18−25°C) and high humidity favors infection by *E. turcicum* and subsequent disease development (Thakur et al., 2007c).

Use of good-quality healthy seeds, crop rotation or intercropping with nonhost crops, clean cultivation before and after planting, cultural practices like adjusting dates of sowing, and proper tillage reduce leaf blight incidence. Destruction of weeds, volunteer, wild sorghum, and alternate hosts help to reduce primary inoculum. Need-based use of fungicides with the right dosage and at the right time is beneficial. However, use of disease-resistant cultivar is thought to be the best option.

2.2.1.2.5 Leaf spots

Sorghum is the host of many fungal leaf spot diseases. A list of such diseases, their distribution, and key symptoms is given in the Table 2.2. Although they may occur in almost all the sorghum growing countries, their distribution given here represents countries where they have been important in sorghum. Generally these diseases occur sporadically and are less damaging, unless they are favored by congenial weather and cultivar susceptibility. They are more visible on purple or red pigment lines and relatively less on tan cultivars. A few of them have economic importance in a few countries, while others assume significance under specific conditions. For example, zonate leaf spot is serious disease on forage sorghum in northern India. It can damage up to 85% of photosynthetic area under humid and cloudy weather conditions (Agnihotri and Pandey, 1977).

Like leaf blight, leaf spot diseases are favored by moderate temperature, high humidity, extended periods of cloudy weather, and heavy dews. Intermittent rains and cool winds favor secondary spread of conidia within fields and the wind helps in their long distance dissemination. Sooty stripe pathogen survives as sclerotia in leaf residue on or below the soil surface. When conditions become favorable sclerotia germinate and produce abundant conidia. The fungus can also survive on few perennial hosts, such as *S. bicolor* subsp. *bicolor*, *S. halepense*, and

Table 2.2 Common Fungal Leaf Spot Diseases of Sorghum

Disease	Pathogen	Key Symptoms	Distribution
Gray leaf spot	*Cercospora sorghi* Ellis & Everhart	Narrow rectangular lesions delimited by veins; longitudinal spot enlargement to develop irregular blotches; lesions turn gray with age	Mexico, Sudan, Nigeria, Burkina Faso, Mali, Ethiopia, Venezuela, Philippines, India
Oval leaf spot	*Ramulispora sorghicola* Harris	Small, water-soaked spots with tan or red border and straw center	Nigeria, Sudan, Ghana, Uganda, Niger, Senegal, India, Pakistan
Rough leaf spot	*Ascochyta sorghina* Sacc.	Oval lesions with well-defined dark margins or halos; hard, black, raised pycnidia giving a sandpaper roughness to the lesion	India, United States, Sudan, Nigeria, Mali
Sooty stripe	*Ramulispora sorghi* (Ellis & Everhart) Olive & Lefebvre (syn. *R. andropogonis* Miura)	Circular to elongated, reddish brown spots on leaf with distinct yellow haloes; necrotic lesions with blackish or sooty center; sclerotia on the lesion surface giving it a rough appearance	Nigeria, Niger, Sudan, Mali, Tanzania, Senegal, Burkina Faso, China, India, United States, Argentina, Australia
Target leaf spot	*Bipolaris sorghicola* (Lefebvre & Sherwin) Alcorn	Oval to cylindrical, purple to red spots with irregular margin and straw-colored center; coalescing of spots to form large lesion	United States, India, Thailand, Sudan, Philippines, Japan, Venezuela
Tar spot	*Phyllachora sacchari* Henn. (syn. *P. sorghi* Hahn)	Black spot with hard stromatic mass sometimes surrounded by yellowish ring	India, Thailand, Malaysia, Philippines
Zonate leaf spot	*Gleocercospora sorghi* Bain & Edgerton	Circular, purplish bands alternating with straw-colored zones resembling a bull's-eye target; spots along the leaf margin in semicircular patterns	India, Nigeria, Mexico, Ethiopia, Burkina Faso, Mali, Korea

S. purpureosericeum (Frederiksen, 1986). Zonate leaf spot pathogen survives as sclerotia formed within the dead tissue of old lesion or on other millets and grasses. During wet weather, conidia are produced on the new lesions and cause further spread of the disease. The fungus may also be carried on the seed.

Use of healthy seeds of a disease-resistant cultivar is the best option for management of foliar diseases. Practice of clean cultivation before and after planting, cultural practices like adjusting sowing dates, proper tillage, and intercropping with nonhosts reduces foliar disease incidence. Destruction of weed, volunteer, wild sorghum, and alternate host help to reduce primary inoculums for the next season. Uses of fungicides (Tunwari et al., 2014) or bio-control agents (Purohit et al., 2013) are reported to be useful in reducing disease severity.

2.2.1.3 Root and stalk diseases
2.2.1.3.1 Charcoal rot

Charcoal rot is a widespread disease of sorghum in dry areas of the world. It is mostly a stress-associated fungal disease, which is prevalent in many sorghum growing countries, where postflowering moisture stress is a common occurrence. Soil moisture stress and drought act as the inducing factors for the disease. The disease causes a great deal of economic losses in many countries, that is, India, Sudan, United States, Mexico, Brazil, Australia, Ethiopia, and Mali. It is one of the most important impediments of sorghum production during postrainy season in India. Though it is predominant disease on postrainy sorghum, it may also occur on rainy sorghum if the crop is exposed to postflowering drought. The pathogen infects root, destroy cortical tissues, and may block water movement through vascular bundles and thus physiologically weaken the plants. Rotting and breaking of the basal internodes cause lodging of the crop, which in turn facilitates further loss of water from the cracks in the stalk. Yield losses vary, depending on weather and the growth stage of the cultivar at the time of infection. The earlier the infection the greater is the total loss under favorable conditions. The disease results in loss of yield and quality of grain and fodder, and postproduction loss of grain in the lodged plants due to destruction by termites and rodents. Crop lodging of 100% could cause 23−64% loss in grain yield under experimental conditions. Financial loss due to charcoal rot is approximately Rs. 450−650 per acre at current productivity in India. The diseased stalk may have an impact on the health of the farm animals as the fungus produces toxin (phaseolinone), which is reported to have anemic effects on the mice (LD_{50} 0.98 g kg^{-1} body wt) (Bhattacharya et al., 1994).

Macrophomina phaseolina can infect the sorghum anytime during the seedling to postflowering stages. The first visible symptom appears in the form of discoloration on the basal part of the stalk. Lodging of the crop and poor grain filling are the indications of charcoal rot infection. The premature lodging is, however, the most apparent symptom of charcoal rot. The infected root and stalk show water-soaked lesions that slowly turn brown or black. Such stalks become soft at the base and often lodge even due to moderate wind or by bending the plants. The pith and the cortical tissues in an infected stalk are disintegrated and the vascular bundles get separated from one another. The vascular tubes contain numerous minute, dark, charcoal-colored sclerotia of the pathogen, which gives the disease its name, charcoal rot. Normally the disease appears during postflowering stage, but in some cases seedlings can be infected causing seedling blight.

The disease is caused by *M. phaseolina* (Tassi) Goidanich (Perfect state: *Sclerotium bataticola* Taub). The mycelium is aerial, superficial or immersed, hyaline to brown, septate, and profusely branched or dendroid. The fungus produces microsclerotia, which are loose type, brown to black, irregular in shape, and highly variable in size. Pycnidia production by this fungus is quite uncommon in sorghum, maize, and soybean. Physiological specialization has been reported in this pathogen (Cloud and Rupe, 1991). Existence of biotype was not ruled out. It was observed that chlorate-sensitive isolates were distinct from chlorate-resistant isolates within a given host. Chlorate-sensitive isolates scored significantly higher disease than chlorate resistant ones (Das et al., 2008b). Recently, molecular techniques like amplified fragment length polymorphism (AFLP), restriction fragment length polymorphism (RFLP), random amplified polymorphic DNA (RAPD), and simple sequence repeats (SSR) have been used to unveil genetic variability in these soil-borne filamentous fungi. A wide range of variability was detected among isolated from different hosts. Isolates from a given host were, however, genetically similar to each. *Macrophomina phaseolina* is present in the most cultivated soils and can infect several plant species worldwide, including a wide range of agriculturally important crops. The sclerotia, which survive on the infected plant debris or in the soil, serve as primary source of inoculum. Germination of sclerotia is triggered by the root exudates of the sorghum seedlings. Germinated sclerotia infect the primary root of sorghum seedling and cause seedling blight. If infection occurs before the emergence of secondary roots, the plants die. Less severely infected seedlings, however, survive and establish secondary roots and grow to mature plants. Dry weather, high temperature (35−38°C), and low soil moisture are the important predisposing factors for the disease. Fungal mycelium colonizes the xylem vessels blocking the translocation of water and carbohydrate to the upper plant parts. Under the condition of stress (moisture, temperature, and photosynthesis) that often coincide with the onset of flowering, the host-defence system is weakened and activity of *M. phaseolina* increases many folds leading to a rapid and extensive rotting of roots and stalks that result in lodging of the crop. The pathogen produces at least six phytotoxins, which injure cell protoplast and often prime the infected plants toward more severe disease.

Deep sowing, conservation of soil moisture, optimum plant density, wheat straw mulching, and mixed cropping with pigeonpea are some of the practices recommended for management of charcoal rot. The early maturing varieties generally escape the disease. High level of genetic resistance is not available. Strong relation of the disease with yield and environmental stresses, particularly moisture and temperature, makes the task of evaluating host resistance more challenging. Drought tolerant, lodging resistant, and nonsenescing sorghum genotypes are supposed to have good tolerance to charcoal rot. However, finding such genotype with high grain yield under desirable agronomic background is often not easy. Seed treatment with talc formulation of *Pseudomonas chlororaphis* SRB127 is reported to reduce disease incidence and increase seed weight (Das et al., 2008a).

2.2.1.3.2 *Fusarium* stalk rot

Fusarium stalk rot is a disease mostly observed in the Americas (United States, Argentina) and Australia. In Africa and Asia, however, it is rare and considered as a minor and sporadic one. Damages caused by this disease are similar to that of charcoal rot. Widespread lodging of the crop and poor grain filling in the diseased plant make it an important disease of sorghum. Lodging causes a great problem for mechanical harvesting of the crop in the developed countries like, the United States and Australia.

Though it is assumed that *F. moniliforme* is the primary cause for the disease, many other fungi may also be isolated from the infected stalk. The disease is characterized by a reddish brown or purple discoloration of the basal part of the stem. Internal pith tissue of the infected stalk becomes reddish in color. Unlike charcoal rot, shredding of vascular stalk tissue is not common in *Fusarium* stalk rot. In case of severe infection many plants lodge giving a crushed view of the field. Stalk infection is invariably accompanied by root rot. *Fusarium* infection may start very early in the root causing seedling blight.

The pathogen survives as conidia or resting spores in the soil or crop residues in the field. Infection takes place in the root. Low temperature and high soil moisture favor the disease development. The disease becomes severe when these conditions suddenly coincide after a dry spell. Good crop management practices like clean cultivation, maintenance of soil moisture, and avoidance of stress reduce the chances of stalk rot development.

2.2.2 BACTERIAL DISEASES

Many bacterial diseases are reported on sorghum from several countries. However, they hardly appear in a severe form and by and large show less interferance to sorghum production and productivity. A list of the common bacterial diseases, their distribution and the key symptoms is given in Table 2.3. Apart from the diseases that are listed in Table 2.3, there are many other reports of bacterial diseases on sorghum, but they are mostly at the level of report and not much work has been done on them. Generally the bacterial diseases of sorghum occur sporadically and are less damaging, unless they are favored by congenial weather and susceptible cultivars. Yield loss estimation data, in general, are not available for such diseases. A few of them, however, have economic importance in few countries, while others assume significance in specific regions. For example, yield loss due to the bacterial stalk rot ranged from 34% to 56% during summer 1985 in Puerto Rico (Hepperly and Ramos-Dávila, 1987). Recently severe damage by this disease has been reported from Taiwan and India (Hseu et al., 2008; Kharayat and Singh, 2013).

Information on epidemiological studies of the bacterial diseases of sorghum is lacking. The bacteria enter into the plant through the natural openings like the stomata or through the injuries created on the leaf or the root. The disease development is favored by warm temperature and high humidity. The pathogen mostly survives in the crop residue left in the soil or in other living crop-hosts and grasses. Cereals like Johnson grass, Sudan grass, broomcorn, maize, and sugarcane, and legumes like soybean, chickpea, broad-bean, and common bean may serve as host for these bacteria. Occasionally, they may be seed-borne. Rain splash, the irrigation water, insects, and the wind help in dispersal of the bacteria within and other fields. Sprinkler irrigation facilitates the spread and establishment of the bacterial disease.

Table 2.3 Common Bacterial Diseases of Sorghum

Disease	Pathogen	Key Symptoms	Distribution
Bacterial leaf spot	*Pseudomonas syringae* pv. *syringae* van Hall	Small, elliptical or irregular shaped spots with straw-color center and dark margin; spots coalesce to form large bands	United States, Argentina, Mexico, Venezuela, Russia, India, China, West Africa
Bacterial leaf stripe	*Pseudomonas andropogoni* (Smith) Stapp	Long, narrow, interveinal red stripes on leaves; bacterial exudates on the lesions	Nigeria, Sudan, United States, Mexico, Argentina, Australia, India, China, Japan, Philippines
Bacterial leaf streak	*Xanthomonas campestris* pv. *holcicola* (Elliott) Starr & Burkholder	Narrow, interveinal, water-soaked streaks with tan centers and red margins; profuse bacterial exudates on the lesions	West Africa, Ethiopia, RSA, Mexico, Venezuela, Argentina, United States, Australia, Iran, Japan, India
Bacterial stalk rot	*Erwinia chrysanthemi* Burkholder, McFadden & Dimock	Sudden wilting, discoloration of top leaves; water-soaked lesions on basal stalk, pith disintegration, lodging; slimy soft-rot with foul-smell	India, United States, Brazil, Puerto Rico, Philippines, Taiwan

Scanty information is available on management of bacterial disease of sorghum since they caused little damage in the past. A few resistant sources are reported for various bacterial diseases of sorghum. Use of the healthy seeds of a disease-resistant cultivar is the best option for the management. Practice of clean cultivation, destruction of crop residues, weeds and other hosts, crop rotation with nonhost crop, and selection of the seeds from a disease-free plant are helpful to control the bacterial diseases. Pathogens may enter stalks or roots through wounds created by insects. Therefore, an insect-control measure helps in reducing losses. Sprinkler irrigation should be avoided in an infected field to prevent further spread of the disease to healthy plants.

2.2.3 VIRAL DISEASES

Members of three groups of the plant viruses, that is, potyvirus, tenuivirus, and rhabdovirus, infect sorghum (Table 2.4). The viruses that naturally infect sorghum in different countries are sugarcane mosaic virus (SCMV), maize dwarf mosaic virus (MDMV), Johnson grass mosaic virus (JGMV), sorghum mosaic virus (SrMV), maize stripe virus (MStV) (formerly MStpV), and maize mosaic virus (MMV).

MStV, MMV, and SRSV are economically important on sorghum in India. The sorghum isolates of chlorotic stripe virus were found to be the variants of MStV and designated as MStV-Sorg (or MStV-S) to distinguish them from MStV which readily infects maize. Sorghum mosaic was first observed on sorghum in peninsular India during 1988. In immuno-double diffusion tests, the virus reacted positively with antisera to MMV from Reunion (MMV-RN) and Hawaii (MMV-HI), and was designated as MMV-S isolate. One of the earliest reports of the occurrence of sorghum red stripe in India was from Maharashtra (Mali and Garud, 1977). A potyvirus naturally infecting sorghum grown in the proximity of sugarcane in Maharashtra was termed as sugarcane mosaic virus-Jg (SCMV-Jg) (Garud and Mali, 1985). Later the virus has been named as sorghum red stripe virus-Indian isolate (SRSV-Ind). SRSV was reported to be related to potyvirus SCMV but was distinct from MDMV-A, MDMV-B, and SrMV (Mali and Thakur, 1999).

Viral diseases occur in mild to moderate on sorghum. Disease incidence may vary from 4−14%. On average, a yield loss of 2−4% for grain and 2−8% for fodder is common particularly in

Virus Group	Virus	Shape	Size (nm)	Nucleic Acid	Genome Structure	Transmission
Table 2.4 Characteristics of Viruses Commonly Occurring on Sorghum						
Potyvirus	MDMV	Flexuous rod	770	ssRNA	Monopartite	Mechanical, Aphids (*Schizaphis graminus*)
	SCMV	Flexuous rod	755−770	ssRNA	Monopartite	Aphids (*Schizaphis graminus, Rhopalosiphum maidis*, etc.)
Tenuivirus	MStV	Filamentous	<10	ssRNA & dsRNA	Multipartite	Planthopper (*Peregrinus maidis*)
Rhabdovirus	MMV	Bacilliform	224 × 68	ssRNA	Monopartite	Planthopper (*Peregrinus maidis*)

peninsular India. Yield losses vary with the stages of infection. Infection at early stages results in higher grain and fodder losses in comparison to that at later stages. Reduction of plant height, earhead weight, and grain mass to the extent of 73%, 93%, and 25%, respectively, have been reported in the variety CSV15 (Revuru and Garud, 1998). Apart from effects on grain, viral diseases have severe adverse effect on forage quality. In forage sorghum, leaf protein content was reduced by 16.98% and 37.58% in MKV Chari-1 and SSG 59-3, while the total soluble solid content reduction was 3.25% and 3.19%, respectively (Rathod et al., 2004). Infection with SRSV significantly reduced plant height, leaf area, juice yield, and chlorophyll content at all stages of growth.

In general, losses to viral disease infection on cereals depend on the stage of infection. The earlier the infection occurs the greater is the damage to plant growth and grain yield. Upon artificial inoculation, under field conditions, significant reduction in grain yield up to 94.9% was observed when plants were inoculated at the 5th leaf stage in comparison to 70.2% loss at the $10-11^{th}$ leaf stage, respectively (Shukla et al., 2006).

Symptoms of viral diseases vary depending on virus. A plant may be infected by single or multiple viruses under natural field conditions. Chances of multiple infection increase when a single vector is able to transmit more than one virus. In such cases the symptoms will vary accordingly. Occurrence of MStV on sorghum (MStV-S) was first reported in India during the 1990s (Peterschmitt et al., 1991). The characteristic external symptoms on sorghum include the appearance of continuous chlorotic stripes/bands between the veins of the infected leaf. The width of chlorotic stripe varies depending of stages of disease development. Leaves on the infected plant appear as yellow with continuous stripes progressing from the base toward the tip of the leaves. The infection is systemic and subsequent leaves appear with yellow stripes on them. An early-infected plant dies sooner or later without emergence of earheads. Plants infected at later stages appear dwarf with short internodes, show partial exertion of earhead having few or no seed formations. The disease is also known as chlorotic stripe stunt or sorghum stripe disease, the name derived from its characteristic stripe symptoms on leaf. In an infected plant, expression of first chlorotic symptom can take place on any leaf starting from 4th to 11th leaf (Das and Prabhakar, 2002a). Sorghum mosaic (MMV-S) is characterized by fine discontinuous chlorotic streaks between the veins on leaf. The lesions become necrotic as the disease progress. Infected plants become stunted in growth with short internodes. An early-infected plant dies sooner or later without emergence of earhead. Plants infected at later stages may develop earhead with or without grain. SRSV is characterized by systemic symptoms of mosaic followed by necrotic red stripe and temperature dependent red leaf. Severely infected crops in a field may develop general necrosis and such a crop produces a burning appearance when viewed from a distance. Recently there was an outbreak of red stripe disease in Maharashtra, India (Narayana et al., 2011).

MStV-S and MMV-S are transmitted by insect vector *Peregrinus maidis*, the delphacid plant hopper on sorghum. *Peregrinus maidis* is known as shoot bug and is also a major pest on sorghum in India (Chellaiah and Basheer, 1965). Shoot bugs suck sap from the leaves, leaf sheaths, and stem during exploratory feeding and in the process transmit virus to healthy from diseased plants. *Peregrinus maidis* requires feeding on the infected plant for at least 4 h to acquire the virus. After the virus is acquired, the vector needs another 8−22 days to be ready for infection of other plant. The vector requires a minimum of 1 h feeding to transmit the virus in a plant. A viruliferous vector retains the virus until its death and transmits it from one generation to another through eggs (Nault and Gordon, 1988). The virus multiplies in the vector and is transmitted in a persistent manner. The nymphs and macropterous females are more efficient transmitters of MStV-S and MMV-S than the males. Neither of these two viruses are transmitted by seed or mechanically by plant sap. The mode of transmission of SRSV is not yet known. The potyvirus SCMV which is related to SRSV may be transmitted mechanically by plant sap and by various aphids in a nonpersistent manner. Plant growth stages between 36 and 65 days after emergence have been identified as highly susceptible for the development of chlorotic stripe virus in rabi sorghum. MStV-S and MMV-S are transmitted by a shoot bug (*P. maidis*) in a persistent manner. Both these viruses and the vector can infect wild graminaceous hosts like Johnson grass. Such wild grasses on the field bund and ratooned sorghum crop in the field serve as important sources for primary inoculums at the beginning of the season. The persistent and transovarial nature of virus transmission across insect generations contributes greatly to the perpetuation of the disease. An early sown crop develops more incidence compared to a late sown one.

Viral diseases can be managed or their incidence can be reduced by practices like clean cultivation, vector control, and adjustment of sowing time. The practice of uprooting and burning of the infected plants help to reduce source of inoculum for the vector and thus reduce spread of the disease in the field. There is strong relation between vector population and field incidence of MStV and MMV (Reynaud et al., 2009). Spraying of Imidachlorpid @ 1.5 mL/L water effectively reduces vector population and the disease. Disease incidence in India is reported to be reduced as sowing of postrainy sorghum is shifted from Sep. to Oct. (Das and Raut, 2002).

2.3 HOST-PLANT RESISTANCE

The application of naturally occurring resistance in plants against diseases through breeding has greatly protected the plants from diseases and saved tonnes of agricultural products from wastage. Host-plant resistance is the most economic and environment friendly way of managing plant diseases. Its application has no interference with other agricultural practices like fertilizer application or irrigation. For poor farmers host-plant resistance is the only viable practice, as they hardly use any other methods of disease control. Since sorghum is grown by resource-poor farmers of the semiarid tropic it has great relevance to this crop. As disease occurs naturally its antidote, that is, resistance is also available in nature. They need to be identified and applied correctly. The process starts with the screening of probable sources of resistance and ends with judicious deployment of the product (cultivar) in the field. In between there are many challenging steps that warrant application of adequate knowledge and expertise in the specified area.

2.3.1 SCREENING FOR RESISTANCE

In simple terms "screening" is a strategy or technique to identify or filter out a desired individual from a population. Screening for disease resistance is, therefore, a technique to identify disease-resistant sources. The basic requirement of a screening programme is availability of a large number of variable germplasm, in which a desired resistance can be searched. The plant materials can be screened either in a greenhouse or in the field, depending on the requirement and availability of the facility. Screening under field conditions can be performed using artificial inoculation (application of pathogen inoculums grown in the laboratory or use of infector row) or natural inoculation (without any artificial inoculation,

Table 2.5 Relation Among Disease Reactions, Rating Scale, and Percent Grain Infection in a Panicle

Disease Reactions	Rating Scale *vis-a-vis* Percent Grain Infection in a Panicle	
	1−5 Scale	**1−9 Scale**
Highly resistant	1 = 0 < 1	1 = 0 < 1
Resistant	2 = 1−10	2 = 1−5
		3 = 6−10
Moderately resistant	3 = 11−25	4 = 11−20
		5 = 21−30
Susceptible	4 = 26−50	6 = 31−40
		7 = 41−50
Highly susceptible	5 = >50	8 = 51−75
		9 = >75

usually in a "disease hot-spot"). Unlike field conditions, in a greenhouse, weather parameters (humidity, temperature, light intensity) can be regulated as per requirement for a particular disease and hence, the results are more repeatable and reliable. Screening should be carried out by a skilled-staff or by a plant pathologist having a sound knowledge of biology and epidemiology of the disease, culture or strain of the pathogen, resistant and susceptible lines, inoculation method, artificial disease development techniques, and disease rating scales. The basic principle is to provide adequate pathogen inoculum at the most susceptible growth stage of the crop under optimal environmental conditions for infection and disease development.

Visual scoring of disease severity is recorded following either a qualitative (immune, resistant, moderately resistant, susceptible, and highly susceptible) or a quantitative scale (1−9 scale, 1−5 scale, or percentage). The relationship among disease reactions, rating scale, and percent grain infection in a panicle is given in Table 2.5, which can be used for panicle diseases like grain mold and ergot. The scale 1−5 or 1−9 can also be used for diseases like anthracnose, rust, leaf blight, leaf spots, and charcoal rot, but the descriptor will differ. The percentage (%) scale is used for systemic diseases like downy mildew, viral diseases, and smut, where a whole plant is counted to be affected. The quantitative scores are useful for statistical analysis of resistance. They can also be converted into qualitative forms to describe disease reaction types. The resistant line score is used for comparison with the scores of other lines and the susceptible line score helps in deciding whether adequate disease pressure was prevalent. Screening techniques for most of the important diseases of sorghum have been developed and subsequently refined over the years. Recently, Thakur et al. (2007c) described screening techniques for major diseases of sorghum. Artificial inoculation methods, disease development techniques, disease scoring techniques, and plant material used for identification of resistance against major sorghum diseases are described in the following sections.

2.3.1.1 Grain mold

Field screening procedure is the most widely used screening technique for evaluation of resistant to grain mold in sorghum. In this method rows of test entries are grown in a field along with the known resistant and susceptible lines. As relative humidity plays crucial role in grain mold development it must be ensured that the crop is exposed to high relative humidity ($\sim$90%) during flowering to grain maturity period. Overhead sprinkler irrigation is generally used during rain-free days to maintain humid conditions in the field. Duration of wetness is an important factor, which is positively correlated with grain mold score (Navi et al., 2005). Bandyopadhyay and Mughogho (1988) evaluated three different variants of field screening techniques, that is, overhead sprinkler irrigation on rain-free days, inoculation of panicles with mold causing fungi, and bagging of panicles. They reported that if rainfall was frequent during the period between flowerings to maturity there was no difference between mold scores of the sprinkler-irrigated and nonsprinkler-irrigated plots. In such a situation artificial inoculation gave similar results as observed in noninoculated plots. Later they quantified the spores of the grain mold causal fungi in the air over a mold-susceptible sorghum crop and concluded that spores of mold causing fungi were naturally available in the air and their number was sufficient to initiate grain mold epidemics under suitable weather conditions (Bandyopadhyay et al., 1991). Major limitations of the field screening technique are: can be used only in rainy season, does not allow screening of photoperiod sensitive material constituting 75% of the total world germplasm, labor and cost-intensive, requires huge amounts of water, and there is chance of disease escape during a dry season. In order to overcome the limitations of a field screening technique Singh and Navi (2001) developed an in vitro screening technique involving dip inoculation of mature seed in a mixed spore suspension of major grain mold fungi, transferring inoculated seed to the sterilized Petri plate humid chamber, and incubating at $28 \pm 1°C$ with or without light for 5 days. Mold development on the seed was scored by visual observation using a 1$-$9 scale. Resistance in this method depends only on the surface characters of the grain, while many other plant characters that are known to confer mold resistance are overlooked. Therefore, resistance identified in this method needs to be confirmed by a field or greenhouse study. In a greenhouse screening procedure potted plants of the test entries are transferred into a greenhouse just before flowering. They are inoculated with spore suspension of the causal fungi of grain mold at 80% anthesis and kept in a dew chamber for 2 days. Mold scores are noted at physiological maturity after exposing the panicle in a dew chamber for 3 days (Thakur et al., 2007c). This method allows identification of mold resistance against individual grain mold fungus. Screening for resistance to individual fungus by artificial inoculation has been infrequent and only recently has got some attention (Prom et al., 2003; Nutsugah and Wilson, 2007; Das et al., 2010).

2.3.1.2 Ergot

Ergot development is highly influenced by a multitude of climatic and host factors. Cool nights ($\sim$10°C) and wet weather favor disease development. Even minute variations in temperature before flowering and during the first 4 days after pollen shed are reported to significantly affect ergot incidence. The factors that facilitate pollination and fertilization prevent infection. An efficient self-pollinating line shows greater escape from ergot than a nonefficient one. All these make reliable ergot screening a difficult task. A number of artificial inoculation techniques and disease evaluation methods have been used in the field and greenhouse (McLarcn and Wehner, 1992). These methods varied with respect to plant growth stage at the time of inoculation, number of inoculations, and panicle bagging. Still they need further refinement. Due to a lack of proper screening techniques many lines previously reported resistant have proved to be susceptible in subsequent tests. Tegegne et al. (1994) experimented with different combinations of trimming of panicle, inoculation, and bagging to develop a suitable screening technique and observed that a single inoculation of non-trimmed panicles (at the beginning of anthesis) was the most suitable method. Recently ICRISAT, Patancheru, has developed an effective field screening technique for pearl millet ergot, which is also used for sorghum ergot.

2.3.1.3 Smut

The head, covered, and kernel smuts of sorghum are predominantly soil-borne, while long smut is air-borne. Due to the lesser economic significance of these diseases, except head smut, systematic development of a screening technique has not taken place. A field screening technique of inoculating seedling with sporidial suspension with the help of a hypodermal syringe was developed by Edmunds (1963) for screening sorghum lines for head smut resistance. Later, Thakur et al. (2007c) modified the technique and developed a more refined greenhouse screening method. In this method sporidial suspension obtained from teliospores (collected previous year) are injected at the apical meristem of a 2−3-week-old seedling, and sorus development is recorded at the time of flowering. For the field screening for long smut resistance, the sporidial suspension can be injected at the boot just before flowering. The covered and kernel smut resistance can be screened under field conditions by sowing teliospore-coated sorghum seed followed by recording the percentage of the spikelet infection.

2.3.1.4 Anthracnose

Field as well as greenhouse screening procedures for identification of resistance to sorghum anthracnose has been available since long ago. Over time the procedures have been modified and refined to decide the correct stages of inoculation or susceptibility, weather parameters, and disease scoring time (Harris et al., 1964; Ferreira and Warren, 1982; Pande et al., 1994). In field screening, infector row technique is used to ensure supply of inoculums to the test entries at the correct stage. Infector rows are sown 20 days before the sowing of the test entries in about 1:8 ratio (one infector row followed by 8 rows of the test entries) and inoculated by applying pathogen-infested sorghum grain into the whorl of 3-week-old plants. High humidity is ensured by sprinkler irrigation. Disease scoring is taken on the top five leaves at the soft dough stage. In the greenhouse procedure, around 3-week-old potted plants are used for inoculation and disease scoring is done as described above.

2.3.1.5 Foliar diseases

Field screening techniques for foliar diseases like leaf blight, various leaf spots, and rust are more or less similar to those of anthracnose. The pathogens, susceptible and resistant sorghum lines, and optimum time for inoculation may vary depending on the disease. Weather factors for development of diseases are different and these should be taken into account based on the disease concerned. Infector row technique can be used for leaf blight and spots. For rust, generally direct spray-inoculation of seedling with rust urediniospores are practiced. Scoring of disease severity is done at the grain filling stage on the top four leaves using a 1−9 scale (Cobb's scale may be used for rust with modification).

2.3.1.6 Downy mildew

Downy mildew screening techniques are well standardized for field, greenhouse as well as laboratory methods. However, the field screening procedures are mostly used for SDM screening (Pande et al., 1997). This technique consists of a combination of the sick-plot and the infector row techniques, in which infector rows are grown in a downy mildew sick-plot. Infected plant tissues containing downy mildew oospores are mixed with the soil before sowing the infector row. Infector rows are sown 20 days before the test entries at about 1:8 ratio (one infector row after every 8 test rows) and inoculated by spraying conidial suspension of the pathogen at one leaf stage. Maintenance of high humidity (>90%) is essential for downy mildew development. The test entries are sown in-between the infector rows when sufficient disease has developed (around 20 days later) on them, allowed to grow and scored for percent disease incidence at seedling (20 days after emergence) and reproductive (60 days after emergence) stages. Greenhouse and laboratory screening procedures are equally important and can be used to save time and resources (Narayana et al., 1995; Thakur et al., 2007a).

2.3.1.7 Charcoal rot

Screening for resistance to sorghum charcoal rot can be completed by the sick-plot method, toothpick method, or a combination of both. Charcoal rot infected stover and stubbles are mixed into the soil every year to maintain inoculum density in the sick-plot. Soil moisture plays a very critical role in infection and subsequent development of charcoal rot in sorghum but creating optimum moisture stress in the soil at optimum growth stage of the plant is not always practicable. Therefore, toothpick inoculation of individual plants with pathogen-infested toothpicks are undertaken to ensure infection. Test entries are sown in rows along with the charcoal rot resistant and susceptible lines. About 10 plants per row in each replication are inoculated with toothpick at the 2nd internode at 1 week after flowering. The number of nodes crossed by the charcoal rot lesion is recorded by split-opening each plant at maturity of the crop. Each plant is rated for charcoal rot damage on a 1−5 scale (1, one internode invaded, but rot does not pass through any nodal area; 2, two; 3, three; 4, four; and 5, more than four internodes extensively invaded, shredding of stalk and death of plant) (Rosenow, 1984). Percent charcoal rot incidence, nodes crossed by the lesion, and lesion length are considered for assessing disease reactions of a particular line (Das et al., 2007).

2.3.1.8 Bacterial diseases

Bacterial diseases in sorghum hardly appear in a severe form and, therefore, literature on systematic studies on their screening techniques is not available. However, bacterial stalk rot is becoming important in recent times in a few countries. As the bacteria is soil-borne and causes systemic infection, inoculation of bacterial suspension into the stalk (at around 25−35 days after sowing) by a hypodermal syringe may work efficiently. For bacterial spot, streak, and stripe spraying of inoculum on the foliage may be practiced.

2.3.1.9 Viral diseases

A repeatable and efficient screening technique not only helps to identify stable resistant sources but also is essential to identify susceptible stages of the host, maintenance of virus inoculum, mass multiplication of insect vectors, and developing disease assessment scales. Field screening as well as greenhouse screening techniques have been reported for sorghum viruses. Group seedling inoculation method for greenhouse screening for resistant to MStV is a simple and efficient technique that can develop 98% virus infection. The young nymphs of the insect vector (*Peregrines maidis*) are allowed to feed on virus-infected sorghum leaves for 4 days and allowed to incubate for a week to get viruliferous. The viruliferous adults are used for inoculation of sorghum plants both in greenhouse and in field conditions.

In a field screening technique, a sorghum genotype highly susceptible to MStV is planted in the field as an infector row after every five blank rows. Each seedling at 3−4 leaf stages are infested with two viruliferous vectors per plant in the leaf-whorl. The test materials are planted in the following day in five consecutive rows in-between two infector rows. The viruliferous insects inoculate the virus to plants in infector rows and also multiply and the subsequent generations of insects from infected plants move to young seedlings to feed and at the same time inoculate the virus to fresh young seedlings. The observations on number of plants showing disease symptoms over total plants are expressed as percent disease.

2.3.2 SOURCES OF RESISTANCE

Availability of disease-resistance sources is the prerequisite for starting a resistant breeding programme. The resistant sources may be identified in a wild or cultivated species related to a crop. In cultivated sorghum, all of the variation in the primary gene pool is accounted by the five basic races (*bicolor, guinea, caudatum, kafir, durra*) and their ten intermediate combinations. Most of the disease-resistant sources also come from this primary gene pool and little attention is paid to interspecific or intergeneric crosses including wild species related to the crop. Numerous sorghum germplasm have been screened worldwide over the years for identification of resistant sources to various diseases. Emphasis has been on identification of sources of resistance to major diseases of global importance like, grain mold, anthracnose, ergot, downy mildew, smut, and charcoal rot. This section will discuss the resistance sources identified through conventional methods other than using molecular tools (discussed in a separate section).

During the 1960s and early 1970s the farmers grew traditional long-duration tall photoperiod-sensitive varieties, which matured after the rainy season, and had hardly any problems with grain mold in India. With the introduction of hybrids during the mid-1970s the crop phenology had been dramatically changed to photoperiod-insensitive, short- and medium-duration, and high-yielding sorghum. This invited the problem of grain mold, and the search for the solution started. Since the mid-1970s huge collections of sorghum germplasm has been screened for grain mold resistance and many resistant sources have been identified. These sources are mostly resistant to moderately resistant type and represent different gene pools from various countries (Table 2.6). Absolute resistance against grain mold is rare mainly because of nature of the disease, complexity of causal fungi, and involvement of multiple resistance mechanisms. Initially derivatives of Zera-zera germplasm from Sudan and Ethiopia were used extensively in breeding programmes in India to produce high yielding mold tolerant progenies. These progenies had better tolerance to grain mold than high yielding elite lines at low pressure of mold. Bandyopadhyay and Mughogho (1988) screened 7132 germplasm lines that represented almost all the sorghum growing regions of the world, and flowered and matured during the rainy season. They reported that 156 colored grain lines had high levels of mold resistance, of which 14 lacked testa but showed resistance. This indicated that grain mold resistance without testa is possible. Later on, many mold resistant sources have been identified in white grain background, which are known to be devoid of a testa layer (Singh et al., 1995; Audilakshmi et al., 2005; Sharma et al., 2010b). However, getting high levels of resistance in white grain has been always a challenge. Grain mold resistant sources in white grain guinea races generally have small-seed with hard grain and corneous endosperm. Resistance in them has been difficult to transfer into an agronomical superior background. Gradually, in addition to germplasm, grain

Table 2.6 Grain Mold Resistant Sources Reported in Sorghum

Screening Conditions	Type of Resistance	Name of the Resistant Sources	References
Field screening	Resistant to *Fusarium* & *Curvularia*	SC No. 0630, 0297, 0566	Castor and Frederiksen (1980)
Artificial inoculation, multiple years	Resistant to moderately resistant	IS No. 625, 2821, 2825, 2867, 3547, 8545, 8614, 8763, 8848, 9353, 9487, 9498, 10301, 10892, 11227, 14332, 14375, 14380, 14384, 14387, 14388, 17141, 18759, 20620, 21454	Bandyopadhyay and Mughogho (1988)
Field screening with sprinkler	Immune	Guinea lines IS No. 7173, 23773, 23783, 34219	Rao et al. (1995)
Field screening with sprinkler	Resistant to moderately resistant	Guinea lines IS No. 7326, 4963, 5726, 4011, 5292, 27761	Rao et al. (1995)
Laboratory screening	Resistant to moderately resistant	Photoperiod sensitive guinea lines IS No. 7326, 4963, 5726, 4011, 5292, 27761	Singh and Prasada Rao (1993)
Field screening, multiple years	Resistant	SC 103-11E, SC 650-11E, SC 748-5	Rooney et al. (2002)
Lab and field, multiple years	Resistant	Converted Zerazera selections IS No. 18758C-618-2, 18758C-618-3, 18758C-710-4, 18758C-710-5	Navi et al. (2002)
Field screening with sprinkler	Moderately resistant	B-lines ICSB 382, 384, 400, 321, IVSB 401	Ashok Kumar et al. (2008), Thakur et al. (2010)
Field screening with sprinkler	Resistant to moderately resistant	*R*-lines PVK 801, GD 65028, ICSR 93034, NSS 254, RSSV 106 resistant; ICSR 89058, ICSR 91011, IS 41675, GD 65055, ICSV 96105 moderately resistant	Ashok Kumar et al. (2008), Thakur et al. (2010)
With sprinkler & artificial inoculation	Resistant	Sorghum accessions from Sudan PI No. 570011, 570027, 569992, 569882, 571312, 570759	Prom and Erpelding (2009)
Field screening with sprinkler	Resistant	Fifty mini-core collections consisting of 5 basic races (*bicolor, guinea, caudatum, kafir, durra*) and 10 intermediate races	Sharma et al. (2010a,b)
Field screening	Resistant to moderately resistant	Indian land races E 12, E 5, EJ 15	Nageshwar Rao et al. (2010)
Field screening with artificial inoculation	Resistant	Sorghum accessions from Uganda PI534117, PI576395, SC719-11E	Prom et al. (2011)
Field screening with artificial inoculation	Resistant to moderately resistant	GM RIL No. 25, 92, 98, 124, 203, 170, 83, 169	Audilakshmi et al. (2011)

(Continued)

Table 2.6 Grain Mold Resistant Sources Reported in Sorghum *Continued*

Screening Conditions	Type of Resistance	Name of the Resistant Sources	References
Field screening	Resistant to moderately resistant	DSR-GMN No. 41, 42, 46, 52, 58, 59	Ambekar et al. (2011)
Field screening with artificial inoculation	Resistant to *Fusarium* induced seed rotting	RIL No. 4, 166, 92, 118, 161, 172, 30	Das et al. (2013)
Field screening with artificial inoculation		Sorghum accessions from Burkina Faso PI No. 586182, 586186, 647705, 647706, 647707, 647708, 647710, 647712	Cuevas et al. (2016)

mold resistance has been identified in many parental lines, recombinant inbred lines, and segregating materials (Reddy et al., 2005; Ashok Kumar et al., 2008; Thakur et al., 2010; Audilakshmi et al., 2011; Ambekar et al., 2011). A2-based cytoplasm has been used to increase genetic diversity (Stack and Pedersen, 2003).

Since ergot is a problem mainly in hybrid seed production plots, not much specific work on identification of host-plant resistance has been done. There are many reports of ergot resistance since 1964, but they lack confirmed sources of resistance (Bandyopadhyay, 1992; Bandyopadbyay et al., 1996). This might be due to lack of a reliable screening techniques or reports based on the results of unreplicated trials (Tegegne et al., 1994). Tegegne et al. (1994) developed a suitable screening technique and identified 6 ergot-resistant lines (ETS 1446, 2448, 2465, 3135, 4457, and 4927) from 213 Ethiopian sorghum accessions. Later Thakur and Rao (2007) reported the germplasm accessions IS 8525, IS 14131, and IS 14257 as highly resistant to ergot.

Anthracnose pathogen *Colletotrichum graminicola* has many races distributed all over the world (Nakamura, 1982; Pande et al., 1991; Prom et al., 2012). Therefore, resistance sources identified in one region may not be as useful to other regions. However, in the case of cultivars, which are meant for use in a specific region, they may be tested in that region. Ferreira and Warren (1982) tested sorghum cultivars in Purdue and reported 932027, 152319, 932062, 159198, 159569, 954206, and Br64 as highly resistant to foliar anthracnose. Marley and Ajayi (2002) evaluated 21 elite sorghum lines for resistance to foliar, peduncle, rachis, grain, and overall panicle anthracnose under natural infection in Samaru and Bagauda in Nigeria and found three genotypes (R 6078, IS 14384, and CCGM 1/19-1-1) were completely resistant to all forms of the disease. Many world germplasms have been evaluated simultaneously in multiple hot spot locations to identify stable sources of resistance. Pande et al. (1994) evaluated about 13,000 sorghum germplasms in multilocation hot spots in Burkina Faso, India, Nigeria, and Zimbabwe for 1−10 years and reported stable sources of resistance across these locations over the years in 11 lines (A 2267-2, IS 3547, IS 8283, IS 9146, IS 9249, IS 18758, SPV 386, PB 18601-3, PM 20873-1-3, M 35610). Thakur et al. (2007b) tested 15 sorghum lines at 14 anthracnose hot spots in India, Thailand, Ethiopia, Kenya, Zambia, Nigeria, and Mali for 4−7 years and identified 3 IS lines (IS No. 6928, 18758, and 12467) as highly resistant to anthracnose across locations. Later, Thakur and Mathur (2007) reported IS 3547, IS 6958, IS 6928, IS 8283, IS 9146, IS 9249, IS 18758, M 35610, A 2267-2, SPV 386, and ICSV 247 as stable sources of resistance to anthracnose. Some of these lines with white grain have been used to develop resistant lines and hybrid parents. Recently many sorghum accessions from Uganda (PI No. 534117, 534144, 576337, 297199, 533833, 297210, and SC748-5) (Prom et al., 2011) and Burkina Faso (PI No. 644285, 656070, 576422, 533976, 644302, 644291, 644328, 533954, 533961) (Cuevas et al., 2016) were reported to be resistant to anthracnose when tested in Texas and Puerto Rico, respectively.

For identification of stable sources of resistance to SDM a large number of sorghum germplasm has been screened in national and international laboratories by exposing the genotypes under greenhouse screening conditions. Out of 13,101 accessions from 72 countries that were evaluated initially under field exposure and subsequently under greenhouse conditions, 46 entries were found resistant under field exposure (Narayana et al., 2002). Whereas in seedlings spray inoculated under greenhouse conditions, 17 entries showed <10% downy mildew incidence and 20 were free from downy mildew. Out of 800 late flowering (>80 days to flower) sorghum accessions evaluated only 10 entries (IS 18512, 18552, 18713, 18714, 19018, 19096, 19239, 20049, and 20205) remained free from downy mildew and 4 entries (IS 18716, 19019, 19506, and 19971) showed <5% downy mildew incidence (Navi and Singh, 1994). Germplasm accessions consisting of 308 wild and weedy sorghum were evaluated under greenhouse conditions, 29 entries found free from downy mildew where as 8 entries showed high levels of resistance (1−10%). Of the downy mildew free accessions, 8 belonged to para sorghum and 21 to other sorghum (Karunakar et al., 1994). At AICSIP Centre, Dharwad, 18 germplasm lines expressed stable resistance to downy mildew over 6 year (IS No. 2204, 2473, 2474, 2482, I3443, 3546, 3547, 5628, 7179, 8185, 8203, 8607, 10710, 14332, 14375, 1875, 18757, and 27042). In addition to the above, 18 germplasm lines (IS No. 473, 602, 1004, 1233, 3151, 4515, 7131, 13549, 20625, 21645, 22986, 27887, 28449, 28451, 28614, 29606, 29689, and 33023) were identified as resistant to downy mildew. Out of 18 parental lines, 7 lines (AKMS 14 A&B, CK60-A&B, DKMS-9101 A&B, DKMS 9106-A & B, DKMS-9108 A & B DNA-2 A & B, and DNA 5 A & B) were found resistant to downy mildew. Out of 26 varieties and hybrids evaluated 11 entries (DSV-1, DSV-2, DSV-6, SPV1633, JKSH-2, Pro-Agro-8340, SPV1632, CS 354, GK4010, and CSV-15) showed <10% downy mildew reactions.

Development of resistant cultivars against charcoal rot, especially in high-yielding background, is still a challenge. Early maturing varieties generally escape disease as they are not exposed to severe moisture stress that is common during the later part of the cropping season. The strong relation of the disease with yield and environmental stresses, particularly moisture and temperature, makes the task of evaluation and identification of host resistance more challenging. Moreover, nonuniformity of soil inoculum in the screening field and the unreliability of the toothpick inoculation method make the task of identification of resistance more difficult. Drought tolerant, lodging resistant, and nonsenescing (stay-green) sorghum genotypes are reported to have good tolerance to charcoal rot. E36-1 and B35 are resistant because of their earliness and stay-green properties. However, such genotypes have yield limitation and lack agronomic superiority. Among the hybrid seed parents, ICSA/B 307, -351,-371, -373, -375, -376, -405, -589, -675, -678, and 702, and among male parents/varieties ICSV 21001 through 21025 are quite promising for stay-green traits (Das et al., 2007). Some other lines, such as SLB 7, SLB 8, SLR 17, and SLR 35, are also reported to be tolerant to charcoal rot.

Systematic studies to identify resistance to other sorghum disease like smuts, rust, leaf blight, leaf spots, and viral and bacterial diseases are limited. Among the smuts some resistance sources have been identified against head smut (Cao et al., 1988a,b; Shao-jie, 2006) and covered smuts

(Mirza et al., 1982). For leaf diseases there are reports of leaf blight resistance in germplasm accessions (IS No. 13868, 13869, 13870, 13872, 18729, 18758, 19669, and 19670), which were developed in a trait-specific breeding program at ICRISAT. Two sorghum accessions IS 26866 and IS 13996 showed high to moderate levels of resistance against eight isolates of *E. turcicum* from five different states in India (Bunker and Mathur, 2010), which are now used as stable source in breeding program (Mathur et al., 2011). Recently some leaf blight tolerant hybrid seed parents (ICSA/B 296 TO ICSA/B328) have been developed (Thakur et al., 2007c). Singh et al. (1994) screened 5218 sorghum accessions from Africa and India for rust resistance and shortlisted 15 lines, which on retesting identified 6 lines (IS No. 2300, 3443, 31446, 18758, 18758, 7023 and C40, C242, and C603), that were highly resistant to rust. Regarding viral diseases some preliminary works have been reported particularly on MStV in India and a few tentative sources resistance (IS 9600, Q 104, ICSB 15, NIC 21222, NIC 21298, IS 40279, IS 19444, IS 40191) have been identified (Narayana and Muniyappa, 1996; Das and Prabhakar, 2002b).

2.3.3 MECHANISMS OF RESISTANCE

For any living being to remain healthy is normal and disease is an exception, in spite of the omnipresence of numerous disease-causing factors. Every plant has inbuilt resistance mechanisms or natural barriers against diseases. By nature the barrier may be morphological, physiological, or biochemical in character. Unveiling the mechanism of disease resistance is important for the planning of a sound management strategy. Sometimes disease does not occur in spite of inherent susceptibility in the plant. This may be due to lack of cooccurrence of host, virulent pathogen, and favorable environment, leading to disease escape. Escape mechanism is an important tool for disease avoidance, which is mostly a crop management issue and not a part of genetic control in the plant. Among the diseases in sorghum, fungal diseases comprising of foliar diseases, grain molds, and downy mildew cause extensive damage to the crop resulting in severe losses in grain as well as forage yields. Several mechanisms of resistance have been unveiled in these diseases and efforts have been made to incorporate them in a cultivar and enforce resistance.

The resistance to grain mold is a complex trait manifested by panicle structural features as well as constitutive and inducible physiological mechanisms that work in collaboration with each other in preventing fungal infection and avoiding grain damage. It has been reported that open panicles, increased glume coverage, and endosperm hardness are some of the structural features that are associated with grain mold resistance (Mansuetus et al., 1988; Audilakshmi et al., 1999; Chandrashekar et al., 2000). Even though Sharma et al. (2010a,b) reported a negative correlation between glume coverage and grain mold, Menkir et al. (1996) and Audilakshmi et al. (1999) did not observe any significant relationship between them. Hard endosperm derived from α-, β-, and γ-kafarins and having numerous antifungal proteins (Waniska et al., 2002) are associated with grain mold resistance, which was also confirmed by the presence of a direct and positive relationship between these two traits reported by Ghorade et al. (1997) and Audilakshmi et al. (1999). Colored pericarp exhibited better resistance to grain mold than white or lighter pericarp, which was supported by the observation of a strong association of purple or black glumes with grain mold resistance (Audilakshmi et al., 1999; Sharma et al., 2010a,b). Resistance is greater in colored grain with tannins, followed by colored grain without tannin, and less in white grained sorghums. The accumulation of flavon-4-ols was observed in sorghum genotypes with a red pericarp thereby contributing to grain mold resistance (Jambunathan et al., 1990). On the contrary, Forbes (1986) observed little accumulation of gallic acid, which is associated with fungistatic activity. With respect to defense response genes, the mRNA of PR-10 was induced to significantly higher levels at 48 h after inoculation (*Fusarium thapsinum* or *Curvularia lunata*) in glumes of resistant cultivars as compared to susceptible cultivars (Katile et al., 2010). Similarly, higher levels of antifungal proteins such as glucanase, chitinase, and sormatin, which inhibit the fungal growth, were noticed in the resistant cultivars than in the susceptible cultivars (Rodriguez-Herrera et al., 1999; Bueso et al., 2000).

Ergot resistance is manifested by two physiological mechanisms, that is, pollen-based and nonpollen-based. In pollen-based mechanism the fungal infection is restricted by efficient pollination as well as fertilization (Bandyopadhyay et al., 1998). In a number of ergot-resistant sources reported earlier, the resistance could be due to pollen-mediated disease escape or genetic effects overshadowing the environmental effects on the phenotype (Sundaram, 1971; Tegegne et al., 1994). Nonpollen-based mechanisms involve a combination of floral characters. The floral features such as small stigma, minimum exposure of stigma to fungal inoculums before pollination, and quick drying of stigma after pollination were reported to be associated with ergot resistance (Dahlberg et al., 2001). Several new sources of ergot resistance were identified to be effective in male sterile backgrounds (Dahlberg et al., 2001), which supports the fact that ergot resistance is not totally pollen-mediated.

Sorghum is known to produce defense molecule phytoalexins in response to fungal attack. On attack by anthracnose pathogen, epidermal cells of sorghum produce phytoalexins, which interact with the fungal appressoria and kill them (Snyder and Nicholson, 1990). Aanthracnose pathogen exhibits an intracellular type of hemibiotrophy. Several cell wall-associated defense reactions such as accumulation of reactive oxygen species and cell wall barrier formation are activated during the biotrophic stage, leading to the prevention of fungal development (Basavaraju, 2011). Even though several sources of anthracnose resistance are identified, a clear understanding of the mechanism of resistance is still lacking.

Mechanisms of resistance against SDM are poorly understood and not much work has been done on this topic in sorghum. Like pearl millet downy mildew, recovery of symptom is occasionally observed in systemically infected sorghum plants. Unraveling host characters linked to this phenomenon might be helpful in throwing light on downy mildew resistance mechanism.

Charcoal rot is a stress related disease and no clear-cut resistance mechanism is available. Dodd (1977) developed a photosynthetic stress-translocation balance concept to explain the predisposition of sorghum to charcoal rot. The hypothesis implies that the interaction of drought stress and pathogens causes charcoal rot. Due to the strong association between drought and charcoal rot infection, indirect selection for drought tolerance has resulted in the improvement of charcoal rot resistance. Most of the resistant sources identified for charcoal rot are reported to be stay-green genotypes (Woodfin et al., 1988). Therefore, stay-green can be used as an important trait for indirect selection of genotypes possessing resistance to charcoal rot.

2.3.4 GENETICS OF RESISTANCE

Disease resistance in plant has generally two basic genetic mechanisms—monogenic or qualitative resistance and polygenic or quantitative resistance. In monogenic resistance the disease resistance is controlled by a single major gene, whereas in polygenic resistance multiple genes are involved.

Monogenic resistance shows specific gene-for-gene interactions and imparts complete resistance, but is specific to a particular race of a pathogen and is often overcome by the evolution of a new race of the pathogen. Durability of resistance is a big concern in monogenic resistance. Polygenic resistance, on the other hand, shows no obvious genetic interaction with the pathogen and controls disease mostly by slowing down the disease development process rather than giving complete resistance. Genetics of disease resistance in sorghum has been worked out for few important diseases. Both qualitative and quantitative resistance is in operation against sorghum diseases. Information in literature suggests that resistance against anthracnose, downy mildew, rust, and leaf blight is qualitative in nature and controlled by dominant genes, whereas resistance against grain mold, head smut, and zonate leaf spot are quantitative in nature and controlled by many genes (Rooney et al., 2002), though there are contradictions. A few examples are given in the following sections.

2.3.4.1 Qualitative resistance

Among sorghum diseases, qualitative resistance is reported to be in operation against anthracnose, downy mildew, rust, and leaf blight. Mehta et al. (2000) worked on 13 germplasm lines and suggested that at least 6 different sources of genetic resistance are present in those lines. Inheritance in resistance groups 2, 3, and 6 was controlled by a single dominant gene, while that in groups 4 and 5 by a single recessive gene. Later it was noted that resistance may be controlled by a dominant or recessive gene depending on the sources of resistance. In sorghum line SC748-5, a single dominant gene *Cg1*, confers resistance to anthracnose (Mehta et al., 2005). Even it was reported to vary depending on the type of symptoms produced—the leaf anthracnose gene was dominant, the leaf mid-rib anthracnose gene was recessive (Erpelding, 2007). Operation of a gene-for-gene hypothesis was suggested by de Costa et al. (2011). They observed that for most crosses, resistance was dominant, and the frequencies of resistant and susceptible plants conformed to the hypothesis that one gene with two alleles controls host resistance. A recessive gene conferring anthracnose resistance was mapped with a RAPD marker (Boora et al., 1998) followed by another recessive gene with RAPD marker OPJ_{011437} at the same loci at 3.26 cM (Singh et al., 2006). A dominant gene linked to anthracnose resistance (*Cg1*) present on the distal region of SBI-05 linkage group was mapped using four AFLP markers (Ramasamy et al., 2009). Eight single nucleotide polymorphism (SNP) loci associated with stable resistance to anthracnose were detected by Upadhyaya et al. (2013b) through a genome-wide association mapping approach, two of which were colocated with the major quantitative trait loci (QTL), *QAnt3*, and *QAnt2*, reported by Murali Mohan et al. (2010). Recently, Cuevas et al. (2014) studied anthracnose resistance response with respect to specific pathotype. They observed that a series of nearest single loci at the distal region of chromosome 5 control the resistance responses against pathotypes from Puerto Rico, Texas, and Arkansas and the region associated with these three loci was not associated with the anthracnose resistance locus Cg1, located in the same chromosome.

Resistance against downy mildew is reported to be governed by one or two dominant or partially dominant genes (Craig and Schertz, 1985; Sifuentes and Frederiksen, 1988; Reddy et al., 1992). However, in maize resistance against the same pathogen is probably controlled by one, two, or many genes (Geetha and Jayaraman, 2002; Nallathambi et al., 2010). Regarding inheritance of rust resistance only a few old literature are available, which suggested that inheritance of resistance was complex and was governed by one to three major dominant genes and there might be predominant additive effects (Patil-Kulkarni et al., 1972; Dabholkar et al., 1980; Tao et al., 1998). The "tan" plant type and resistance were strongly linked (Rana et al., 1976).

2.3.4.2 Quantitative resistance

Among sorghum diseases, quantitative resistance is reported to be in operation against grain mold, head smut, and zonate leaf spot. Grain mold development is governed by many plant traits, few of which may be qualitatively inherited, but their individual contribution to total grain mold resistance generally account for only a small portion of the total variation. Therefore, many genes might be involved in grain mold resistance. Analyzing grain mold resistance in colored sorghum genotypes Audilakshmi et al. (2000) suggested that resistance to grain mold was dominant and mold reaction in colored grain sorghum was generally controlled by two or three major genes. Around the same time Rodriguez-Herrera et al. (2000) reported that at least 4−10 genes contributed to grain mold resistance in grain sorghum without a pigmented testa. Later, Audilakshmi et al. (2005) reported that grain mold resistance in the white grained resistance sources was polygenic and the additive gene action and additive × additive gene interaction played a significant role in inheritance. The grain hardness, one of the major contributors of grain mold resistance, was reported to be polygenic in nature (Aruna and Audilakshmi, 2004). Now it is well accepted that grain mold resistance is mostly quantitative with additive, dominance, and some epistatic effects (Dabholkar and Baghel, 1980; Murty and House, 1984; Kataria et al., 1990; Rodriguez-Herrera et al., 2000; Audilakshmi et al., 2005). Quantitative resistance against head smut in sorghum depends on the cultivars. Jianhua and Chenghu (1988) observed that different sorghum varieties had different inheritance of resistance to head smut. Some varieties possess characteristic of quantitative inheritance, while others qualitative inheritance. Resistance in line with quantitative inheritance was mainly controlled by additive genes. Even though the resistance to zonate leaf spot is governed by a duplicate epistasis (Grewal, 1988), resistance to anthracnose is by a single dominant gene and resistance to leaf mid-rib anthracnose is by a single recessive gene (Erpelding and Wang, 2007); a polygenic control of resistance was proposed for these diseases by Murali Mohan et al. (2009).

2.3.5 UTILIZATION OF HOST RESISTANCE

Once a stable and heritable source of disease resistance is identified, the next step is to use the source in breeding programs for development of disease-resistant hybrid parents and varieties. Conventional breeding methods have been followed predominantly for transfer of the resistance to a desired background. Huge amounts of work have been done for the development of disease-resistant sorghum hybrid parental lines and varieties worldwide. During the last two decades, however, due to advancements in molecular biology tools, the molecular methods like marker-assisted breeding and transgenics are being used increasingly. Over the years, a number of resistant lines for each of the major diseases of sorghum have been identified and many of those have been effectively utilized in developing hybrid parental lines and varieties.

2.3.5.1 Conventional breeding

Development of a disease-resistant cultivar is a dynamic process as new forms of pathogens evolve in nature continuously. For monogenic resistance the process of new pathogen evolution is faster than the polygenic resistance. Therefore, continuous efforts through a focused disease resistance breeding programme are necessary to develop new cultivars keeping in pace with the development of new forms of pathogen. In the past, most of the disease resistance breeding programmes gave more focus on major gene (R gene) resistance than polygene. Identification of major gene resistance was probably easier than detection of minor effects of the multiple genes. Disease screening techniques with high inoculum pressure, in many cases, might have facilitated identification of major genes. Once a variety or hybrid is released durability of resistance often becomes an issue. Because of frequent breakdown of major-gene resistance in the released hybrid or varieties, focus has been shifted to QTL. Conventional or classical breeding techniques, including emasculation, pollination/crossing, backcrossing, selfing, and selection, have been used to develop disease-resistant cultivars. However, hand emasculation was a tedious job restricting the speed of a breeding programme. Discovery of a cytoplasmic male sterility system with nuclear suppressor genes (Stephens and Holland, 1954) helped making crosses rapidly and soon led to commercial production of high yielding hybrids. Developments of various methods in the classical resistance breeding schemes with respect to sorghum have been reviewed by Magill (2013). Most of the disease-resistant hybrid parental lines and varieties that are in use across the globe now are developed using the classical breeding techniques. Molecular breeding in sorghum being in its infancy, conventional breeding still dominates most of the resistance breeding programme. The developed materials once stabilized (generally after F_6) are tested in multiple hotspot locations within the target regions for diseases resistance over two to three seasons, and only a few superior lines are selected and released for cultivation (Rakshit et al., 2014). Like other countries, in India many disease-resistant hybrids and varieties have been developed using conventional breeding techniques. All the centrally released rainy season, white grain sorghum hybrids and varieties in India since 1980 (eg, CSH Nos. 9, 10, 11, 13, 14, 16, 27, 30; and CSV Nos. 9, 10, 11, 13, 15, 17, 20, 25) possess a good amount of tolerance to grain mold and foliar diseases. All these grain hybrids and varieties are in tan background, which make them tolerant to leaf diseases.

2.3.5.2 Molecular breeding

Development of molecular markers facilitated accurate tagging and identification of disease resistance genes in the host genome, thus permitting marker-assisted selection and gene pyramiding for durable disease resistance. The availability of mapping populations, molecular markers, and genetic maps in sorghum over the last two decades offered ample opportunities for the identification of genomic regions/QTL associated with traits of interest, especially those associated with biotic and abiotic stresses. In the current era of genomics and molecular breeding, it has become imperative for the identification and mapping of major QTL associated with disease resistance in sorghum along with tightly linked DNA markers thereby paving way for the transfer of those effective QTL to the elite variety or parental line through marker-assisted introgression. Specific biparental mapping population have been developed by different research groups across the globe and employed for the identification and mapping of QTL associated with various diseases in sorghum (Table 2.7). The first linkage map in sorghum was reported by Hulbert et al. (1990) with 36 RFLPs representing 8 of the 10 linkage groups. Later, Subudhi and Nguyen (2000) aligned 5 major RFLP-based linkage maps constructed by Boivin et al. (1999), Chittenden et al. (1994), Pereira et al. (1994), Ragab et al. (1994), and Xu et al. (1994), integrating with 10 linkage groups. Subsequently, the linkage maps were saturated with AFLP markers by various research groups (McIntyre et al., 2005; Murray et al., 2008; Ramu et al., 2009; Ritter et al., 2008; Shiringani and Friedt, 2011). High density linkage maps were constructed by Bhattramakki et al. (2000) by using 323 RFLPs and 143 SSR and by

Table 2.7 Important Major Effect Genes and QTL Mapped in Sorghum

Character	QTL/Marker	LG	R^2	References
Anthracnose resistance	*QAnt*	6	40	Murali Mohan et al. (2010)
Zonate leaf spot resistance	*QZls*	6	17	
Target leaf spot resistance	*QTls*	6	45	Murali Mohan et al. (2010)
Rust resistance				
	BNL5.09	2	25	Tao et al. (1998)
	RZ323-ISU102	4	24	
	PSB47-TXS422	10	43	
	QRust	6	24	Murali Mohan et al. (2010)
Ergot resistance				
Infection %	*Sb4-32*	9	20	Parh et al. (2008)
Pollen viability	*AAGCCT6*	6	20	Parh et al. (2008)
Charcoal rot resistance				
Internodes crossed	*Xtxp297*	2	19	Reddy et al. (2008)
Lodging %	*Xtxp343*	4	15	Reddy et al. (2008)

QTL, quantitative trait loci; LG, linkage group; R^2, % contribution to phenotypic variance.

Bowers et al. (2003) using 2512 RFLP loci, the former map was further saturated by Menz et al. (2002) with 2926 markers comprising of AFLPs, SSRs, and RFLPs. Linkage maps were also constructed using EST-SSR (Ramu et al., 2009; Srinivas et al., 2009b) and unigene-based SSR markers (Srinivas et al., 2009a; Nagaraja Reddy et al., 2012). Employing high-quality SNPs comprising of 3418 bin markers, an ultrahigh density linkage map was constructed by Zou et al. (2012). The genomic tools along with mapping populations have led to the identification and mapping of important QTL for disease resistance in sorghum.

Resistance to grain mold, caused by a complex of fungi, primarily by *F. moniliforme* and *Curvularia lunata*, is governed by major and minor genes with additive and epistatic effects along with significant $G \times E$ interactions (Stenhouse et al., 1998; Rodriguez-Herrera et al., 2000). Five significant and environment-dependent QTL for grain mold resistance contributing a phenotypic variance of 10−23% were identified by Klein et al. (2001). By employing association mapping and SNP, Upadhyaya et al. (2013a) detected two SNP loci associated with grain mold resistance; one of the loci possessed a NB-ARCLRR class of R-gene (Sb02g004900) with 57% similarity to a resistance gene in maize, *Rxo1*. However, the map positions did not match with the QTL reported earlier (Klein et al., 2001), indicating the possibility of differential expression of resistance due to differences in the pathogen, and environmental factors over diverse environments (Audilakshmi et al., 2005; Little et al., 2012).

About 18 QTL were identified for percentage ergot infection, pollen quantity, and pollen viability, which are the component traits of ergot resistance (Parh et al., 2008). Among these, four major QTL on SBI-01, SBI-06, SBI-08, and SBI-09 were associated with percent ergot infection while one QTL each were associated with pollen quantity (SBI-06) and pollen viability (SBI-07). Interestingly, the major QTL for percent ergot infection (SBI-06) was colocated with QTL associated with grain mold identified by Klein et al. (2001) and those identified by Murali Mohan et al. (2010) for anthracnose, zonate leaf spot, and bacterial leaf spot resistance. Moreover, the genomic regions reported earlier by Tao et al. (1998) and Klein et al. (2001) that contained QTL for resistance to rust and grain mold (SBI-07, SBI-10, and SBI-08) also possessed an ergot resistance QTL.

Charcoal rot resistance has a complex inheritance with both additive and nonadditive gene actions (Indira et al., 1983; Garud and Borikar, 1985; Rao et al., 1993). Very limited reports are available on the identification of QTL associated with this trait. Nine QTL were identified for three morphological traits (number of internodes crossed by the rot, length of infection, and crop lodging) and two biochemical traits (lignin and total phenols) associated with charcoal rot resistance (Reddy et al., 2008; Patil et al., 2012). The peritoxin produced by the saprophytic fungus *Periconia circinata* causes the root and crown rot disease in sorghum. The *Pc* locus associated with dominant sensitivity to a host-selective peritoxin cloned by a map-based approach harbors three tandemly repeated genes resembling nucleotide-binding site-leucine-rich repeat (NBS-LRR) class of disease resistance genes (Nagy et al., 2007).

Resistance to rust also shows complex inheritance (Patil-Kulkarni et al., 1972; Rana et al., 1976) governed by at least four loci (Tao et al., 1998). Four major QTL on SBI-01, SBI-02, SBI-03, and SBI-08 contributing 16–42% to the phenotypic variation were identified by Tao et al. (1998). Another major QTL was identified by McIntyre et al. (2005) on SBI-08, hosting a key R-gene homolog of rust resistance gene, *Rp1-D*, from maize and sugarcane. In the recent past, five SNP loci were identified by Upadhyaya et al. (2013a) that are associated with rust resistance; two of them contained genes that are homologous to *Rp1-D* and *Lr1*, the rust resistance genes in maize and wheat, respectively.

A major QTL was identified between the marker interval, Xtxp95-Xtxp57 on SBI-06 for resistance to bacterial leaf blight, zonate leaf spot, and anthracnose (Klein et al., 2001). Moreover, using 168 F_7 RILs from the cross 296 B × IS18551, QTL linked to resistance to oval leaf spot and those associated with resistance to several foliar diseases were also colocated in this region (Murali Mohan et al., 2010). Regarding viral and bacterial diseases, there is no report on the identification and mapping of QTL for them. The availability of effective QTL for resistance to diseases such as grain mold, ergot, charcoal rot, and foliar diseases offers the breeders an excellent opportunity for the genetic improvement by the integration of a consistent major or a set of minor QTL associated with disease resistance through molecular breeding.

2.3.5.3 Transgenic

Even though direct gene delivery systems (Battraw and Hall, 1991) as well as *Agrobacterium*-mediated transformation systems (Gao et al., 2005a,b; Howe et al., 2006; Nguyen et al., 2007; Gurel et al., 2009) are successful for the recovery of stable transformants in sorghum, several research groups prefer the latter to get lower copy number insertions and/or have a higher frequency of coexpression of the nonselected transgenic cassette (Dai et al., 2001). In spite of employing a wide range of explants, such as mature and immature embryos, immature inflorescences, seedlings, leaf fragments, and anthers, to develop a dependable and efficient tissue culture system for regeneration through somatic embryogenesis or organogenesis, consistent success has been achieved with embryogenic cultures initiated from immature embryos or inflorescences.

The development of transgenic crops for resistance to fungal diseases followed by the plant genetic engineers can be categorized into the following approaches involving the expression of five different gene products (reviewed by Punja, 2001): (1) pathogenesis-related proteins (PR proteins) for reducing fungal growth (Mauch et al., 1988); (2) polygalacturonase, oxalic acid, lipase, polyphenols, and phytoalexins for neutralizing a pathogen component; (3) elevating the levels of peroxidase and lignin for enhancing the structural defenses in the plant; (4) production of specific elicitors, hydrogen peroxide, salicylic acid, and ethylene for regulating the plant defences; and (5) overexpression of resistance gene (R) products involved in the hypersensitive response (necrosis of infected area).

Identification of effective antifungal molecules that impart resistance to the fungi infecting sorghum is essential for the control of fungal diseases that include grain molds and foliar diseases. Transgenic plants have been successfully developed in many crops by introducing antifungal genes, especially those associated with cell wall-degrading enzymes, such as chitinase and glucanase (reviewed by Ceasar and Ignacimuthu, 2012). In addition to these genes, many antimicrobial proteins were also effective in imparting fungal disease resistance. Three different antifungal genes were introduced in rice by various groups: the trichosanthin gene by Xiaotian et al. (2000), an antifungal protein (AFP) gene of *Aspergillus giganteus* by Coca et al. (2004), and synthetically-prepared antifungal genes Ap-CecA and ER-CecA by Coca et al. (2006). A synthetic AFP gene of prawn (PIN) was introduced to develop transgenic finger millet (Latha et al., 2005) and pearl millet (Latha et al., 2006) conferring resistance to fungal diseases. These studies indicate that antifungal genes can be successfully employed for the development of fungal resistance in crop plants, in addition to the popularly used genes like chitinase and glucanase.

Introduction of a chitinase gene of rice into sorghum for imparting resistance against stalk rot was demonstrated by Zhu et al. (1998) and Krishnaveni et al. (2000). Biolistic transformation of embryogenic calli derived from scutellum was transformed by Muthukrishnan et al. (2001) with a plasmid DNA containing a selectable marker (*bar* gene) and the Class I rice chitinase gene under the control of CaMV35S promoter to produce transgenic sorghum plants. Progeny analysis showed that the chitinase gene segregated in a Mendelian fashion with the T_2 and T_3 plants exhibiting increased level of stalk rot resistance. Ayoo et al. (2011) attempted to develop transgenic sorghum plants with resistance to anthracnose disease by introducing genes encoding proteins such as chitinases and chitosanases that hydrolyze fungal cell wall. They demonstrated the effect of genetic background on resistance to anthracnose in transgenics. They suggested that these transgenes could be utilized to pyramid genes for higher tolerance to anthracnose in sorghum.

The viral diseases caused by MStV and MDMV result in substantial yield losses in sorghum. Development of transgenic sorghum possessing resistance to these viral diseases is the best way in combating the yield losses caused by them since very few germplasm lines were identified by Henzell et al. (1982) exhibiting limited resistance against these viruses. The demonstration of transgenics possessing viral coat protein genes in conferring resistance to plant viruses by Stark and Beachy (1989) has opened up the possibility of developing sorghum transgenics conferring resistance against major viruses. Various genetic constructs targeting the silencing of the specific gene products vital to viral replication were tried in many crops (Beachy et al., 2003; Prins, 2003), including sorghum (Gilbert et al., 2005). These strategies offer great promise for the development of sorghum transgenics conferring durable resistance against viral diseases caused by maize dwarf mosaic and MStVs.

2.4 CONCLUSIONS

Many diseases occur on sorghum worldwide and some are economically significant. Fungal diseases are of major importance followed by viral and bacterial diseases. Production, productivity, and grain and forage quality of sorghum is adversely affected by diseases across the globe. Several disease epidemics have occurred in the past in various parts of the world. The magnitude of losses varied with the type of disease, the severity, and agro-ecological conditions. Host-plant resistance has been the mainstay of sorghum disease management, as farmers thriving on this crop in drier parts of the world have hardly any other resources for disease management. Over the decades, a huge amount of research work has been carried out for development and use of disease-resistant cultivars. Greenhouse and field screening techniques for many important diseases have been developed and modified time to time for correct identification of resistance. A large number of sorghum germplasms representing major sorghum regions of the world have been screened for disease resistance in different national and international nurseries and many useful and stable sources of resistance have been identified. Genetics and inheritance of disease resistance and mechanism of resistance of many diseases like grain mold, ergot, anthracnose, downy mildew, smut, ergot, and foliar disease have been worked out for a better understanding and utilization of host resistance. Continuous evolution of new pathogen forms (race, pathotypes) especially in downy mildew, rust, leaf blight, and anthracnose have challenged the durability of resistance in released cultivar, and emphasized the need for gene pyramiding. Classical plant breeding procedures have been mostly used for incorporation of resistance to desirable background. Of late, molecular tools are being used increasingly in sorghum especially for identification of QTLs, attempting gene pyramiding and the development of transgenics. A lot more needs to be done in the future in the field of host resistance to generate versatile sources of disease resistance that can counteract the need of the future sorghum cultivation efficiently under changing climatic situations.

2.5 FUTURE RESEARCH NEED

Breeding for disease resistance in sorghum is an important objective in almost every breeding program. Application of host resistance is undoubtedly the cheapest and the safest way of managing sorghum diseases. Continuous development of new cultivars with new sources of resistance against all economically important diseases should be targeted with knowledge resources from multidisciplinary experts and all required facilities for such programmes should be strengthened to full capacity. However, there are a few sorghum diseases, for example, grain mold and charcoal rot, which need more than just host-plant resistance for their effective management. For these diseases, integration of other management practices such as cultural, biological, and even disease escape mechanisms should be encouraged, along with an emphasis on more basic studies on pathogen biology, epidemiology, and resistance mechanisms to find new avenues that can be targeted for formulating new control strategies. Since many fungi are involved in grain mold development, there is a need to understand the interaction of these fungi so as to ascertain if there is a synergistic or antagonistic effect. Also an understanding of the factors responsible for resistance to grain infection at different stages of development right from initiation is overdue. Many fungi produce mycotoxins and reduce the quality of grain for human consumption as well as for poultry and animal feed. And this problem is greater on white grain sorghum which are less tolerant than colored grain sorghum. Colored grain sorghum with red pericarp should be focused for use in biscuit industries, animal, and poultry feed. Hybrid seed companies face the problem of sorghum ergot, for which an accurate and reliable screening technique is required so as to identify sources of resistance and incorporate these into hybrid parents. Durability of resistance is a problem for SDM. In some cases molecular markers have been linked to downy mildew resistance but none of these markers has yet been placed on a map of the sorghum genome. Concerted efforts are required for gene pyramiding either through conventional breeding or by molecular methods. Leaf diseases, particularly anthracnose, leaf blight, zonate leaf spot, and rust, are major constraints for forage sorghum, especially that grown under high humidity conditions. Enhancement of foliar disease resistance in forage lines with tan plants needs more emphasis. Viral diseases cause considerable damage to sorghum, but few efforts have been put into developing a sound screening technique and identifying host resistance that can be incorporated into a desired cultivar through molecular tools. Similar is the case of bacterial diseases, which are assuming greater significance in some regions. Soft stalk is becoming severe in India and there is a need to develop a reliable screening technique for identification resistance and use.

ACKNOWLEDGMENTS

Authors are thankful to the scientists at IIMR who have actively contributed to the improvement of the manuscript.

REFERENCES

Agnihotri, V.P., Pandey, S., 1977. Zonate leaf spot of jowar caused by *Gloeocercospora sorghi* and its control through fungitoxicants. Ind. Phytopath. 29, 401–406.

Ambekar, S.S., Kamatar, M.Y., Ganesamurthy, K., Ghorade, R.B., Usha, S., Pooran, C., et al., 2011. Genetic enhancement of Sorghum (Sorghum bicolor (L) Moench) for grain mold resistance: II. Breeding for grain mold resistance. Crop Prot. 30, 759–764.

Aruna, C., Audilakshmi, S., 2004. Genetic architecture of grain hardness—a durable resistance mechanism for grain moulds in sorghum [*Sorghum bicolor* (L.) Moench]. Indian J. Genet. Plant Breed. 64 (1), 35–38.

Ashok Kumar, A., Reddy, B.V.S., Thakur, R.P., Ramaiah, B., 2008. Improved sorghum hybrids with grain mold resistance. J. SAT Agric. Res. 6, 1–4.

Audilakshmi, S., Stenhouse, J.W., Reddy, T.P., Prasad, M.V.R., 1999. Grain mould resistance and associated characters of sorghum genotypes. Euphytica 107 (2), 91–103.

Audilakshmi, S., Stenhouse, J.W., Reddy, T.P., 2000. Genetic analysis of grain mould resistance in coloured sorghum genotypes. Euphytica 116 (2), 95–103.

Audilakshmi, S., Stenhouse, J.W., Reddy, T.P., 2005. Genetic analysis of grain mould resistance in white seed sorghum genotypes. Euphytica 145, 95–101.

Audilakshmi, S., Das, I.K., Ghorade, R.B., Mane, P.N., Kamatar, M.Y., Narayana, Y.D., et al., 2011. Genetic improvement of sorghum for grain mould resistance: I. Performance of sorghum recombinant inbred lines for grain mold reactions across environments. Crop Prot. 30, 753–758.

Ayoo, L.M.K., Bader, M., Loerz, H., Becker, D., 2011. Transgenic sorghum (*Sorghum bicolor* L. Moench) developed by transformation with chitinase and chitosanase genes from *Trichoderma harzianum* expresses tolerance to anthracnose. Afr. J. Biotechnol. 10 (19), 3659–3670.

Bandyopadhyay, R., 1986. Grain mold. In: Frederiksen, R.A. (Ed.), Compendium of Sorghum Diseases. The American Phytopathological Society, St. Paul, MN, pp. 36−38.

Bandyopadhyay, R., 1992. Sorghum ergot. In: de Milliano, W.A.J., Frederiksen, R.A., Bengston, G.D. (Eds.), Sorghum and Millets Diseases: A Second World Review. International Crops Research Institute for the Semi-Arid Tropics, Patancheru, pp. 235−244.

Bandyopadhyay, R., Mughogho, L.K., 1988. Evaluation of field screening techniques for resistance to sorghum grain molds. Plant Dis. 72, 500−503.

Bandyopadhyay, R., Mughogho, L.K., Satyanarayana, M.V., Kalisz, M.E., 1991. Occurrence of airborne spores of fungi causing grain mould over a sorghum crop. Mycol. Res. 95, 1315−1320.

Bandyopadbyay, R., Frederickson, D.E., McLaren, N.W., Odvody, G.N., 1996. Ergot: a global threat to sorghum. Int. Sorghum Millet Newsl. 37, 1−32.

Bandyopadhyay, R., Fredricksen, D.E., McLaren, N.W., Odvody, G.N., Ryley, M.J., 1998. Ergot: a new disease threat to sorghum in the Americas and Australia. Plant Dis. 82, 356−367.

Basavaraju, 2011. Mechanism of Disease Resistance in Sorghum Infection Induced Defense Responses in Sorghum: Accumulation of Reactive Oxygen Species and Cell Wall Modifications, Saarbrücken: LAP Lambert Academic Publishing. 978-3-8443-8399-7, p. 156 .

Battraw, M., Hall, T.C., 1991. Stable transformation of *Sorghum bicolor* protoplasts with chimeric neomycin phosphotransferase II and β-glucuronidase genes. Theor. Appl. Genet. 82, 161−168.

Beachy, R.N., Fraley, R.T., Rogers, S.G., 2003. Protection of Plants Against Viral Infection. Monsanto Technology LLC and Washington University, United States patent no. 6,608, 241, 19 Aug 2003.

Bhattacharya, G., Siddiqui, K.A.I., Chakraborty, S., 1994. The toxicity of phaseolinone to mice. Indian J. Pharmacol. 26, 121−125.

Bhattramakki, D., Dong, J., Chhabra, A.K., Hart, G.E., 2000. An integrated SSR and RFLP linkage map of *Sorghum bicolor* (L.) Moench. Genome 43, 988−1002.

Boivin, K., Deu, M., Rami, J.F., Trouche, G., Hamon, P., 1999. Towards a saturated sorghum map using RFLP and AFLP markers. Theor. Appl. Genet. 98, 320−328.

Boora, K.S., Frederiksen, R., Magill, C., 1998. DNA-based markers for a recessive gene conferring anthracnose resistance in sorghum. Crop Sci. 38, 1708−1709.

Bowers, J.E., Abbey, C., Anderson, S., Chang, C., Draye, X., Hoppe, A.H., et al., 2003. A high-density genetic recombination map of sequence-tagged sites for sorghum, as a framework for comparative structural and evolutionary genomics of tropical grains and grasses. Genetics 165, 367−386.

Bueso, F.J., Waniska, R.D., Rooney, W.L., Bejosano, F.P., 2000. Activity of antifungal proteins against mold in sorghum caryopsis in the field. J. Agric. Food Chem. 48, 810−816.

Bunker, R.N., Mathur, K., 2010. Pathogenic and morphological variability in *Exserohilum turcicum* isolates causing leaf blight of sorghum (*Sorghum bicolor*). Indian J. Agric. Sci. 80 (10), 888−892.

Butler, E.J., 1918. Fungi and Disease in Plants. Calcutta and Simla, India Thacker, Spink & Co., Calcutta, p. 547.

Cao, Y.H., Wang, X.L., Ren, J.H., Nan, C.H., 1988a. Study of resistance to head smut in sorghum and its inheritance. Acta Genetica Sinica 15 (3), 170−173.

Cao, R.H., Wang, X.L., Ren, J.H., Nan, C.H., 1988b. The Resistance of Sorghum to Head Smut and Its Inheritance. In: Suzuki S., (Ed.), Crop Genetic Resources of East Asia, International Board for Plant Genetic Resources, Rome, pp. 121−124.

Castor, L.L., Frederiksen, R.A., 1980. *Fusarium and Curvularia* grain molds in Texas. In: Sorghum Diseases—A World Review. Proceedings of the International Workshop on Sorghum Diseases, sponsored jointly by Texas A&M University (USA) and ICRISAT, 11−15 December 1978, Hyderabad, India, pp. 93−102.

Ceasar, S.A., Ignacimuthu, S., 2012. Genetic engineering of crop plants for fungal resistance: role of antifungal genes. Biotechnol. Lett. 34 (6), 995–1002.

Chandrashekar, A., Bandyopadhyay, R., Hall, A.J., 2000. Technical and Institutional Options for Sorghum Grain Mold Management: Proceedings of an International Consultation. ICRISAT, Patancheru, p. 299. 18–19 May 2000.

Chellaiah, S., Basheer, M., 1965. Biological studies of *Peregrinus maidis* (Ashmead) (Araccopidae: Homoptera) on sorghum. Indian J. Entomol. 27, 466–471.

Chittenden, L., Schertz, K., Lin, Y., Wing, R., Paterson, A., 1994. A detailed RFLP map of *Sorghum bicolor* × *S. propinquum*, suitable for high-density mapping, suggests ancestral duplication of sorghum chromosomes or chromosomal segments. Theor. Appl. Genet. 87, 925–933.

Cloud, G.L., Rupe, J.C., 1991. Morphological instability in a chlorate medium of isolates of *Macrophomina phaseolina* from soybean and sorghum. Phytopathology 81, 892–895.

Coca, M., Bortolotti, C., Rufat, M., Peñas, G., Eritja, R., Tharreau, D., et al., 2004. Transgenic rice plants expressing the antifungal AFP protein from *Aspergillus giganteus* show enhanced resistance to the rice blast fungus *Magnaporthe grisea*. Plant Mol. Biol. 54, 245–259.

Coca, M., Peñas, G., Gómez, J., Campo, S., Bortolotti, C., Messeguer, J., et al., 2006. Enhanced resistance to the rice blast fungus *Magnaporthe grisea* conferred by expression of a cecropin A gene in transgenic rice. Planta 223, 392–406.

Craig, J., Schertz, K.F., 1985. Inheritance of resistance in sorghum to three pathotypes of *Peronosclerospora sorghi*. Phytopathology 75, 1077–1078.

Cuevas, H.E., Prom, L.K., Erpelding, J.E., 2014. Inheritance and molecular mapping of anthracnose resistance genes present in sorghum line SC112-14. Mol. Breed. 34 (4), 1943–1953.

Cuevas, H.E., Prom, L.K., Isakeit, T., Radwan, G., 2016. Assessment of sorghum germplasm from Burkina Faso and South Africa to identify new sources of resistance to grain mold and anthracnose. Crop Prot. 79, 43–50.

Dabholkar, A.R., Baghel, S.S., 1980. Inheritance of resistance to grain mould of sorghum. Indian J. Genet. Plant Breed. 40 (2), 472–475.

Dabholkar, A.R., Sharma, H.C., Baghel, S.S., 1980. Inheritance of rust resistance in *Sorghum bicolor* (L.) Moench. JNKVV Res. J. 14 (1/2), 12–14.

da Costa, R.V., Zambolim, L., Cota, L.V., da Silva, D.D., Rodrigues, J.A.S., Tardin, F.D., et al., 2011. Genetic control of sorghum resistance to leaf anthracnose. Plant Pathol. 60 (6), 1162–1168.

Dodd, J.L., 1977. A photosynthetic stress-translocation balance concept of corn stalk rot. Proc. 32nd Ann. Corn Sorghum Res. Conf. 32, 122–130.

Dahlberg, J., Bandyopadhyay, R., Rooney, W., Odvody, G.N., Madera-Toms, P., 2001. Evaluation of sorghum germplasm used in US breeding programmes for sources of sugary disease resistance. Plant Pathol. 50, 681–689.

Dai, S., Zhenh, P., Marmey, P., Zhang, S., Tian, W., Chen, S., et al., 2001. Comparative analysis of transgenic rice plants obtained by *Agrobacterium*-mediated transformation and particle bombardment. Mol. Breed. 7 (1), 25–33.

Das, I.K., Patil, J.V., 2013. Assessment of economic loss due to grain mold of sorghum in India. In: Rakshit, S., Das, I.K., Shyamprasad, S., Mishra, J.S., Ratnavathi, C.V., Chapke, R.R., Tonapi, V.A., Rao, B.D., Patil, J.V. (Eds.), Compendium of Papers and Abstracts: Global Consultation on Millets Promotion for Health and Nutritional Security, 18–20 December 2013 Society for Millet Research, Directorate of Sorghum Research, Hyderabad, pp. 59–63.

Das, I.K., Prabhakar, 2002a. Effect of growth stages of winter sorghum on susceptibility of stripe disease. Indian Phytopath. 55, 313–314.

Das, I.K., Prabhakar, 2002b. Resistance sources for sorghum stripe disease. J. Maha. Agric. Univ. 27, 221−223.

Das, I.K., Raut, M.S., 2002. Effect of sowing dates and weather parameters on the occurrence of stripe disease in winter sorghum. Indian J. Mycol. Plant Path. 32, 21−24.

Das, I.K., Rao, V.P., Reddy, P.S., 2007. Charcoal rot. In: Thakur, R.P., Reddy, B.V.S., Mathur, K. (Eds.), Screening Techniques for Sorghum Diseases. Information Bulletin No. 76. ICRISAT, Patancheru, pp. 41−46.

Das, I.K., Indira, S., Annapurna, A., Prabhakar, Seetharama, N., 2008a. Biocontrol of charcoal rot in sorghum by fluorescent pseudomonads associated with rhizosphere. Crop Prot. 27 (11), 1407−1414.

Das, I.K., Fakrudin, B., Arora, D.K., 2008b. RAPD cluster analysis and chlorate sensitivity of some Indian isolates of *Macrophomina phaseolina* from sorghum and their relationships with pathogenicity. Microbiol. Res. 163 (2), 215−224.

Das, I.K., Vijay Kumar, B.S., Ratnavathi, C.V., Komala, V.V., Annapurna, A., Seetharama, N., 2010. Toxigenicity of *Fusarium* isolates and fumonisin B1 contamination in rainy season sorghum [*Sorghum bicolor* L. (Moench)]. Indian J. Agric. Sci. 80, 67−72.

Das, I.K., Audilakshmi, S., Patil, J.V., 2012. *Fusarium* grain mold: the major component of grain mold in sorghum (*Sorghum bicolor* L. Moench). The Eur. J. Plant Sci. Biotechnol. 6, 45−55.

Das, I.K., Annapurna, A., Patil, J.V., 2013. Effect of panicle characters and plant height on premature seed rot caused by *Fusarium* grain mold in sorghum. Indian J. Plant Prot. 41 (3), 238−243.

Edmunds, L.K., 1963. Use of sporidial hypodermic injection to test sorghum for head smut resistance. Plant Dis. Rep. 47, 909−913.

Erpelding, J.E., 2007. Inheritance of anthracnose resistance for the sorghum cultivar redlan. Plant Pathol. J. 6 (2), 187−190.

Erpelding, J.E., Wang, M.L., 2007. Response to anthracnose infection for a random selection of sorghum germplasm. Plant Path. J. 6, 127−133.

Ferreira, A.S., Warren, H.L., 1982. Resistance of sorghum to *Colletotrichum graminicola*. Plant Dis. 66, 773−775.

Forbes, G.A., 1986. Characterization of Grain Mold Resistance in Sorghum (*Sorghum bicolor* L. Moench) (Ph.D. thesis). Texas A&M University, College Station, TX, 75 pp.

Frederiksen, R.A., 1986. Compendium of Sorghum Diseases. The American Phytopathological Society and Department of Plant Pathology and Microbiology. Texas A&M University, Texas, p. 82.

Frederiksen, R.A., Amador, J., Jones, B.L., Reyes, L., 1969. Distribution, symptoms and economic loss from downy mildew caused by *Sclerospora sorghi* in grain sorghum in Texas. Plant Dis. Rep. 53, 995−998.

Gao, Z., Jayaraj, J., Muthukrishnan, S., Claflin, L., Liang, G.H., 2005a. Efficient genetic transformation of sorghum using a visual screening marker. Genome 48, 321−333.

Gao, Z., Xie, X., Ling, Y., Muthukrishnan, S., Liang, G.H., 2005b. *Agrobacterium tumefaciens* mediated sorghum transformation using a mannose selection system. Plant Biotechnol. J. 3, 591−599.

Garud, T.B., Borikar, S.T., 1985. Genetics of charcoal rot resistance in sorghum. Sorghum Newsl. 28, 87.

Garud, T.B., Mali, V.R., 1985. A red stripe virus disease of sorghum in India. Indian Phytopath. 36, 545−546.

Geetha, K., Jayaraman, N., 2002. Inheritance of sorghum downy mildew resistance in maize. Indian J. Agric. Res. 36 (4), 234−240.

Ghorade, R.B., Shekar, V.B., Gite, B.D., Sakhare, B.A., 1997. Some general combiners for grain mould resistance in sorghum. J. Soils Crops 7, 8−11.

Gilbert, R.A., Gallo-Meagher, M., Comstock, J.C., Miller, J.D., Jain, M., Abouzid, A., 2005. Agronomic evaluation of sugarcane lines transformed for resistance to sugarcane mosaic virus strain E. Crop Sci. 45, 2060−2067.

Grewal, R.P.S., 1988. Genetic basis of resistance to zonate leaf spot disease in forage sorghum. Theor. Appl. Genet. 76, 550–554.

Gurel, S., Gurel, E., Kaur, R., Wong, J., Meng, L., Tan, H.Q., et al., 2009. Efficient reproducible Agrobacterium-mediated transformation of sorghum using heat treatment of immature embryos. Plant Cell Rep. 28 (3), 429–444.

Harris, H.B., Johnson, B.J., Dobson Jr.J.W., Lutuell, E.S., 1964. Evaluation of anthracnose on grain sorghum. Crop Sci. 4, 460–462.

Henzell, R.G., Persley, D.M., Greber, R.S., Fletcher, D.S., Van Slobbe, L., 1982. Development of grain sorghum lines with resistance to sugarcane mosaic and other sorghum diseases. Plant Dis. 6, 900–901.

Hepperly, P.P., 1990. Sorghum rust II. Control and losses. J. Agric. Univ. Puerto Rico 74, 37–44.

Hepperly, P.R., Ramos-Dávila, E., 1987. *Erwinia chrysanthemi* Burk., McFaddan & Dimock: a bacterial whorl and stalk rot pathogen of sorghum [*Sorghum bicolor* (L.) Moench.]. J. Agric. Univ. Puerto Rico 71 (3), 265–275.

Howe, A., Sato, S., Dweikat, I., Fromm, M., Clemente, T., 2006. Rapid and reproducible Agrobacterium-mediated transformation of sorghum. Plant Cell Rep. 25 (8), 784–791.

Hseu, S.H., Kuo, K.C., Lin, H.F., Lin, C.Y., 2008. Bacterial stalk rot of sorghum occurred in Kimmen area caused by *Erwinia chrysanthemi*. J. Plant Pathol. Bull. 17 (3), 255–260.

Hulbert, S.H., Richter, T.E., Axtell, J.D., Bennetzen, J.L., 1990. Genetic mapping and characterization of sorghum and related crops by means of maize DNA probes. Proc. Natl. Acad. Sci. 87, 4251–4255.

ICRISAT, 1992. The Medium Term Plan, Vol. 2. International Crops Research Institute for Semi-Arid Tropics, Patancheru.

Indira, S., Rana, B.S., Rao, N.G.P., 1983. Genetics of host plant resistance to charcoal rot in sorghum. Indian J. Genet. 43, 472–477.

Jambunathan, R., Kherdekar, M.S., Bandyopadhyay, R., 1990. Flavan-4-ols concentration in mold-susceptible and mold-resistant sorghum at different stages of grain development. J. Agric. Food Chem. 38, 545–548.

Jianhua, C.R.W.X.R., Chenghu, N., 1988. Studies on the resistance to sorghum varieties head smut and its inheritance. Acta Genetica Sinica 3, 002.

Karunakar, R.I., Narayana, Y.D., Pande, S., Mughogho, L.K., 1994. Evaluation of wild and weedy sorghum for downy mildew resistance. Int. Sorghum Millet Newsl. 35, 104–106.

Kataria, S.K., Singh, R., Shrotria, P.K., 1990. Inheritance of resistance to grain mould fungi in three sorghum *Sorghum bicolor* crosses. Environ. Ecol. 1990 (8), 1111–1113.

Katile, S.O., Perumal, R., Rooney, W.L., Prom, L.K., Magill, C.W., 2010. Expression of pathogenesis-related protein PR-10 in sorghum floral tissues in response to inoculation with *Fusarium thapsinum* and *Curvularia lunata*. Mol. Plant Path. 11, 93–103.

Kaveriappa, K.M., Safeeulla, K.M., Shaw, C.G., 1980. Culturing *Sclerospora sorghi* in callus tissue of sorghum. Proc. Indian Acad. Sci. Plant Sci. 89, 31–38.

Kharayat, B.S., Singh, Y., 2013. Unusual occurrence of *Erwinia stalk* rot of sorghum in tarai region of Uttarakhand. Int. J. Agric. Sci. 9, 809–813.

Klein, R., Rodriguez-Herrera, R., Schlueter, J., Klein, P., Yu, Z., Rooney, W.L., 2001. Identification of genomic regions that affect grain-mould incidence and other traits of agronomic importance in sorghum. Theor. Appl. Genet. 102, 307–319.

Krishnaveni, S., Jeoung, J.M., Muthukrishnan, S., Liang, G.H., 2000. Transgenic sorghum plants constitutively expressing a rice chitinase gene show improved resistance to stalk rot. J. Genet. Breed. 55, 151–158.

Kulkarni, B.G.P., Seshadri, V.S., Hegde, R.K., 1976. The perfect stage of *Sphacelia sorghi* McRae. Mysore J. Agric. Sci. 10 (2), 286–289.

Latha, M.A., Reddy, V.D., Latha, M.A., Rao, K.V., 2005. Production of transgenic plants resistant to leaf blast disease in finger millet (*Eleusine coracana* (L.) Gaertn.). Plant Sci. 169, 657–667.

Latha, M.A., Rao, K.V., Reddy, T.P., Reddy, V.D., 2006. Development of transgenic pearl millet (*Pennisetum glaucum* (L.) R. Br.) plants resistant to downy mildew. Plant Cell Rep. 25, 927–935.

Little, C.R., Perumal, R., Tesso, T., Prom, L.K., Odvody, G.N., Magill, C.W., 2012. Sorghum pathology and biotechnology—a fungal disease perspective: Part I. Grain mold, head smut, and ergot. Eur. J. Plant Sci. Biotechnol. 6, 10–30.

Magill, C.W., 2013. Bridging classical and molecular genetics of sorghum disease resistance. In: Paterson, A.H. (Ed.), Genomics of the Saccharinae, Vol. II – Plant Genetics and Genomics: Crops and Models. Springer, New York, NY, pp. 347–366.

Mali, V.R., Garud, T.B., 1977. Studies on sorghum red stripe virus disease in Maharashtra. Indian J. Mycol. Plant Pathol. 7, 201–203.

Mali, V.R., Thakur, R.P., 1999. Reactions and virus titres of differential sorghum inbred lines mechanically inoculated with an Indian isolate of sorghum red stripe potyvirus (SRSV-Ind). Sugar Tech. 1, 13–18.

Mansuetus, A.S.B., Frederiksen, R.A., Waniska, R.D., Odvody, G.N., Craig, J., Rosenow, D.T., 1988. The effects of glume and caryopses characteristics of sorghum on infection by *Fusarium moniliforme* Sheldon. Sorghum Newsl. 31, 100.

Marley, P.S., Ajayi, O., 2002. Assessment of anthracnose resistance (*Colletotrichum graminicola*) in sorghum (*Sorghum bicolor*) germplasm under field conditions in Nigeria. J. Agric. Sci. 138, 201–208.

Mathur, K., Thakur, R.P., Rao, V.P., Jadone, K., Rathore, S., Velazhahan, R., 2011. Pathogenic variability in *Exserohilum turcicum* and resistance to leaf blight in sorghum. Indian Phytopath. 64, 32–36.

Mauch, F., Mauch-Mani, B., Boller, T., 1988. Antifungal hydrolases in pea tissue. II. Inhibition of fungal growth by combinations of chitinase and β-1,3-glucanase. Plant Physiol. 88, 936–942.

McIntyre, C.L., Casu, R., Drenth, J., Knight, D., Whan, V., Croft, B., et al., 2005. Resistance gene analogues in sugarcane and sorghum and their association with quantitative trait loci for rust resistance. Genome 48, 391–400.

McLaren, N.W., Wehner, F.C., 1992. Pre-flowering low temperature pre-disposition of sorghum to sugary disease (*Claviceps* at Ticana). J. Phytopath. 135, 328–334.

Mehta, P.J., Collins, S.D., Rooney, W.L., Frederiksen, R.A., Klein, R.R., 2000. Identification of different sources of genetic resistance to anthracnose in sorghum. Intl. Sorg. Millets Newsl. 41 (5), 1–54.

Mehta, P.J., Wiltse, C.C., Rooney, W.L., Collins, S.D., Frederiksen, R.A., Hess, D.E., et al., 2005. Classification and inheritance of genetic resistance to anthracnose in sorghum. Field Crops Res. 93 (1), 1–9.

Menkir, A., Ejeta, G., Butler, L., Melake-Berhan, A., 1996. Physical and chemical kernel properties associated with resistance to grain mold in sorghum. Cereal Chem. 73, 613–617.

Menz, M., Klein, R., Mullet, J., Obert, J., Unruh, N., Klein, P., 2002. A high-density genetic map of *Sorghum bicolor* (L.) Moench based on 2926 AFLP®, RFLP and SSR markers. Plant Mol. Biol. 48, 483–499.

Miedaner, T., Geiger, H.H., 2015. Biology, genetics and management of ergot (*Claviceps* spp.) in rye, sorghum, and pearl millet. Toxins 7, 659–678.

Mirza, M.S., Hamid, S.J., Hassan, S.F., 1982. Resistance of sorghum varieties to covered kernel smut. Pak. J. Agric. Res. 3, 31–33.

Murali Mohan, S., Madhusudhana, R., Mathur, K., Howarth, C.J., Srinivas, G., Satish, K., et al., 2009. Co-localization of quantitative trait loci for foliar disease resistance in sorghum. Plant Breed. 128, 532–535.

Murali Mohan, S., Madhusudhana, R., Mathur, K., Chakravarthi, D.V.N., Rathore, S., Nagaraja Reddy, R., et al., 2010. Identification of quantitative trait loci associated with resistance to foliar diseases in sorghum [*Sorghum bicolor* (L.) Moench]. Euphytica 176, 199–211.

Murray, S.C., Sharma, A., Rooney, W.L., Klein, P.E., Mullet, J.E., Mitchell, S.E., et al., 2008. Genetic improvement of sorghum as a biofuel feedstock: I. QTL for stem sugar and grain non-structural carbohydrates. Crop Sci. 48, 2165–2179.

Murty, D.S., House, L.R., 1984. Components of generation means for resistance to grain mold-causing fungi *Curvularia* and *Fusarium* in sorghum. Cer. Res. Comm. 12, 237−244.

Muthukrishnan, S., Liang, C.H., Trick, H.N., Gill, B.S., 2001. Pathogenesis-related proteins and their genes in cereals. Plant Cell Tiss. Org. Cult. 64, 93−114.

Nagaraja Reddy, R., Madhusudhana, R., Murali Mohan, S., Chakravarthi, D.V.N., Seetharama, N., 2012. Characterization, development and mapping of unigene-derived microsatellite markers in sorghum [*Sorghum bicolor* (L.) Moench]. Mol. Breed. 29, 543−564.

Nageshwar Rao, T.G., Elangovan, M., Kaul, S.L., 2010. New sources of resistance to grain molds of sorghum (Sorghum bicolor (L. Moench.). Indian Phytopath 63 (1), 98−99.

Nagy, E., Lee, T.C., Ramakrishna, W., Xu, Z., Klein, P., SanMiguel, P., et al., 2007. Fine mapping of the Pc locus of *Sorghum bicolor*, a gene controlling the reaction to a fungal pathogen and its host-selective toxin. TAG Theor. Appl. Genet. 114, 961−970.

Nakamura, K., 1982. Especializacao Fisiologic em *Colletotrichum graminicola* (Ces.) Wil. (Sensu Arx., 1957) Agente Causal da Antracnose em Sorgho (Ph.D. thesis). University of Estadual Paulista, Jaboticabal.

Nallathambi, P., Sundaram, K.M., Arumugachamy, S., 2010. Inheritance of resistance to sorghum downy mildew (*Peronosclerospora sorghi*) in Maize (*Zea mays* L.). Int. J. Agric. Environ. Biotechnol. 3 (3), 285−293.

Narayana, Y.D., Muniyappa, V., 1996. Virus-vector relationships of a plant hopper (*Peregrinus maidis*)-borne sorghum stripe tenuivirus. Int. J. Pest Manage. 42 (3), 165−170.

Narayana, Y.D., Mughogho, L.K., Bandyopadhyay, R., 1995. Evaluation of greenhouse inoculation techniques to screen sorghum for resistance to downy mildew. Euphytica 86, 49−53.

Narayana, Y.D., Bandypadhyay, R., Pande, S., 2002. Sources of resistance to sorghum downy mildew. J. Mycol. Pl. Pathol. 32 (2), 213−218.

Narayana, Y.D., Das, I.K., Bhagwat, V.R., Tonapi, V.A., Patil, J.V., 2011. Viral Disease of Sorghum in India. Directorate of Sorghum Research, Hyderabad, p. 30.

Nault, L.R., Gordon, D.T., 1988. Multiplication of maize stripe virus in *Peregrinus maidis*. Phytopathology 78, 991−995.

Navi, S.S., Bandyopadhyay, R., Reddy, R.K., Thakur, R.P., Yang, X.B., 2005. Effects of wetness duration and grain development stages on sorghum grain mold infection. Plant Dis. 89, 872−878.

Navi, S.S., Singh, S.D., 1994. Identification of sources of resistance to sorghum downy mildew in late flowering sorghum germplasm. Int. Sorghum Millet Newsl. 35, 104.

Navi, S.S., Singh, S.D., Gopal Reddy, V., Kameswara Rao, N., Bramel, P.J., 2002. New sources of resistance to grain mold in converted zerazera sorghum. ISMN 43, 77−80.

Nguyen, T.V., Thu, T.T., Claeys, M., Angenon, G., 2007. Agrobacterium-mediated transformation of sorghum (*Sorghum bicolor* (L.) Moench) using an improved in vitro regeneration system. Plant Cell Tiss. Org. Cult. 91, 155−164.

Nutsugah, S.K., Wilson, J.P., 2007. Development of a reliable inoculation technique to assess resistance in pearl millet to *Fusarium* grain mold. J. SAT Agric. Res. 5 (1), 1−3.

Pande, S., Bock, C.H., Bandyopadhyay, R., Narayana, Y.D., Reddy, B.V.S., Lenné, J.M., et al., Downy Mildew of Sorghum. Information Bulletin No. 51. International Crops Research Institute for the Semi-Arid Tropics, Patancheru, p. 32.

Pande, S., Mughogho, L.K., Bandyopadhyay, R.B., Karunakar, R.I., 1991. Variation in pathogenicity and cultural characteristics of sorghum isolates of *Colletotrichum graminicola* in India. Plant Dis. 75, 778−783.

Pande, S., Thakur, R.P., Karunakar, R.I., Bandyopadhyay, R., Reddy, B.V.S., 1994. Development of screening methods and identification of resistance to anthracnose in sorghum. Field Crops Res. 38, 157−166.

Parh, D.K., Jordan, D.R., Aitken, E.A., Mace, E.S., Jun-Ai, P., McIntyre, C.L., et al., 2008. QTL analysis of ergot resistance in sorghum. Theor. Appl. Genet. 117, 369−382.

Patil, A., Fakrudin, B., Salimath, P.M., Rajkumar, 2012. Genome-wide molecular mapping and QTL analysis, validated across locations and years for charcoal rot disease incidence traits in *Sorghum bicolor* (L.) Moench. Indian J. Genet. 72, 296–302.

Patil-Kulkarni, B.G., Puttarudrappa, A., Kajjari, N.B., Goud, J.V., 1972. Breeding for rust resistance in sorghum. Indian Phytopath. 25, 166–168.

Pereira, M., Lee, M., Bramel-Cox, P., Woodman, W., Doebley, J., Whitkus, R., 1994. Construction of an RFLP map in sorghum and comparative mapping in maize. Genome 37, 236–243.

Peterschmitt, M., Ratna, A.S., Sacks, W.R., Reddy, D.V.R., Mughogho, L.K., 1991. Occurrence of an isolate of maize stripe virus in India. Ann. Appl. Biol. 118 (1), 57–70.

Prins, M., 2003. Broad virus resistance in transgenic plants. Trends Biotechnol. 21, 373–375.

Prom, L.K., Erpelding, J.E., 2009. New sources of grain mold resistance among sorghum accessions from Sudan. Trop. Subtrop. Agroecosys. 10, 457–463.

Prom, L.K., Waniska, R.D., Kollo, A.I., Rooney, W.L., 2003. Response of eight sorghum cultivars inoculated with *Fusarium thapsinum*, *Curvularia lunata*, and a mixture of the two fungi. Crop Prot. 22, 623–628.

Prom, L.K., Isakeit, T., Perumal, R., Erpelding, J.E., Rooney, W.L., Magill, C.W., 2011. Evaluation of the Ugandan sorghum accessions for grain mold and anthracnose resistance. Crop Prot. 30, 566–571.

Prom, L.K., Perumal, R., Erattaimuthu, S.R., Little, C.R., No, E.G., Erpelding, J.E., et al., 2012. Genetic diversity and pathotype determination of *Colletotrichum sublineolum* isolates causing anthracnose in sorghum. Eur. J. Plant Path. 133, 671–685.

Punja, Z.K., 2001. Genetic engineering of plants to enhance resistance to fungal pathogens—a review of progress and future prospects. Can. J. Plant Pathol. 23 (3), 216–235.

Purohit, J., Singh, Y., Bisht, S., Srinivasaraghvan, A., 2013. Evaluation of antagonistic potential of *Trichoderma harzianum* and *Pseudomonas fluorescens* isolates against *Gloeocercospora sorghi* causing zonate leaf spot of sorghum. Bioscan 8 (4), 1327–1330.

Ragab, R., Dronavalli, S., Maroof, M.S., Yu, Y., 1994. Construction of a sorghum RFLP linkage map using sorghum and maize DNA probes. Genome 37, 590–594.

Rakshit, S., Hariprasanna, K., Gomashe, S., Ganapathy, K.N., Das, I.K., Ramana, O.V., et al., 2014. Changes in area, yield gains and yield stability of sorghum in major sorghum-producing countries during 1970–2009. Crop Sci. 54, 1571–1584.

Ramasamy, P., Menz, M., Mehta, P., Katilé, S., Gutierrez-Rojas, L., Klein, R.R., 2009. Molecular mapping of *Cg1*, a gene for resistance to anthracnose (*Colletotrichum sublineolum*) in sorghum. Euphytica 165, 597–606.

Ramu, P., Kassahun, B., Senthilvel, S., Kumar, C.A., Jayashree, B., Folkertsma, R., et al., 2009. Exploiting rice-sorghum synteny for targeted development of EST-SSRs to enrich the sorghum genetic linkage map. Theor. Appl. Genet. 119, 1193–1204.

Rana, B.S., Tripathi, D.P., Rao, N.G.P., 1976. Genetic analysis of some exotic × Indian crosses in sorghum. XV. Inheritance of resistance to sorghum rust. Indian J. Genet. 36, 244–249.

Rana, B.S., Rao, M.H., Indira, S., Singh, B.U., Appaji, C., Tonapi, V.A., 1999. Technology for Increasing Sorghum Production and Value Addition. Director and Project Coordinator (AICSIP) National Research Centre for Sorghum, Hyderabad.

Rao, G.M.R., Patil, S.J., Anahosur, K.H., 1993. Genetics of charcoal rot resistance in Rabi sorghum. Kar. J. Agric. Sci. 6, 113–116.

Rao, K.E.P., Singh, S.D., Stenhouse, J.W., 1995. Grain mold resistance in guinea sorghum germplasm. ISMN 36, 94–95.

Rathod, K.S., Telang, S.M., Rathod, R.M., 2004. Effect of chlorotic stripe stunt disease on fodder quality of forage cultivars. J. Soils Crops 14 (2), 351–353.

Reddy, P.S., Fakrudin, B., Rajkumar, Punnuri, S.M., Arun, S.S., Kuruvinashetti, M.S., et al., 2008. Molecular mapping of genomic regions harboring QTLs for stalk rot resistance in sorghum. Euphytica 159, 191–198.

Reddy, B.V.S., Reddy, S., Mughogho, L.K., Narayma, Y.D., Nicodemus, K.D., Stenhouse, J.W., 1992. Inheritance patterns of downy mildew resistance in advanced generations of sorghum. Ann. Appl. Biol. 121, 249–255.

Reddy, B.V.S., Ramesh, S., Ortiz, R., 2005. Genetic and cytoplasmic-nuclear male sterility in sorghum. Plant Breed. Rev. 25, 139–172.

Revuru, S.S., Garud, T.B., 1998. Effect of chlorotic stripe stunt disease on plant growth and grain yield of different sorghum cultivars. J. Maha. Agric. Univ. 23, 253–255.

Reynaud, B., Delattem, H., Peterschmittm, M., Fargettem, D., 2009. Effects of temperature increase on the epidemiology of three major vector-borne viruses. Eur. J. Plant Pathol. 123, 269–280.

Ritter, K.B., Jordan, D.R., Chapman, S.C., Godwin, I.D., Mace, E.S., McIntyre, C.L., 2008. Identification of QTL for sugar-related traits in a sweet × grain sorghum (*Sorghum bicolor* L. Moench) recombinant inbred population. Mol. Breed. 22, 367–384.

Rodriguez-Herrera, R., Waniska, R.D., Rooney, W.L., 1999. Antifungal proteins and grain mold resistance in sorghum with a non-pigmented testa. J. Agric. Food Chem. 47, 4802–4806.

Rodriguez-Herrera, R., Rooney, W.L., Rosenow, D.T., Frederiksen, R.A., 2000. Inheritance of grain mold resistance in grain sorghum without a pigmented testa. Crop Sci. 40, 1573–1578.

Rooney, W.L., Delroy Collins, S., Klein, R.R., Mehta, P.J., Frederiksen, R.A., Rodriguez-Herrera, R., 2002. Breeding sorghum for resistance to anthracnose, grain mold, downy mildew, and head smuts. In: Leslie, J.F. (Ed.), Sorghum and Millet Diseases. Iowa State Press, Ames, IA, pp. 273–279.

Rosenow, D.T., 1984. Breeding for resistance to root and stalk rots in Texas. In: Mughogo, L.K. (Ed.), Sorghum Root and Stalk Rots—A Critical Review: Proceedings of the Consultative Group Discussion on Research Needs and Strategies for Control of Sorghum Root and Stalk Rot Diseases. Bellagio, Italy, pp. 209–218.

Shao-jie, J.I.A.O., 2006. Identification of resistance to head smut, target leaf spot, sorghum aphid and corn borer of sorghum germplasm in Heilongjiang province. J. Plant Genet. Res. 3, 021.

Sharma, R., Rao, V.P., Varshney, R.K., Prasanth, V.P., Kannan, S., Thakur, R.P., 2010a. Characterisation of pathogenic and molecular diversity in *Sclerospora graminicola*, the causal agent of pearl millet downy mildew. Arch. Phytopath. Plant Prot. 43, 538–551.

Sharma, R., Rao, V.P., Upadhyaya, H.D., Reddy, V.G., Thakur, R.P., 2010b. Resistance to grain mold and downy mildew in a mini-core collection of sorghum germplasm. Plant Dis. 94, 439–444.

Shiringani, A.L., Friedt, W., 2011. QTL for fibre-related traits in grain × sweet sorghum as a tool for the enhancement of sorghum as a biomass crop. Theor. Appl. Genet. 123, 999–1011.

Shukla, S., Shukla, A., Pathak, M., Ansari, N.A., Srivastava, G.P., Tewari, J.P., 2006. Effect of sorghum mosaic virus (SrMV) on plant growth and grain yield of sorghum in eastern Uttar Pradesh. Vegetos 19, 59–62.

Sifuentes, J., Frederiksen, R.A., 1988. Inheritance of resistance to pathotypes 1, 2, and 3 of Peronosclerospora sorghi in sorghum. Plant Dis. 72, 332–333.

Singh, S.D., Prasada Rao, K.E., 1993. Sorghum grain molds – identification of resistance. Cereals Program Annual Report 1992. ICRISAT, Patancheru 502 324, Andhra Pradesh, India, p. 20.

Singh, M., Chaudhary, K., Singal, H., Magill, C., Boora, K., 2006. Identification and characterization of RAPD and SCAR markers linked to anthracnose resistance gene in sorghum [*Sorghum bicolor* (L.) Moench]. Euphytica 149, 179–187.

Singh, S.D., Bandyopadhyay, R., 2000. Grain mold. In: Frederiksen, R.A., Odvody, G.N. (Eds.), Compendium of Sorghum Diseases, second ed. The Am. Phytopathol. Soc., St. Paul, MN, pp. 38–40.

Singh, S.D., Navi, S.S., 2001. An in vitro screening technique for the identification of grain mould resistance in sorghum. Indian Phytopath. 54, 35–39.

Singh, S.,D., Sathiah, P., Rao, K.E.P., 1994. Sources of rust resistance in purple color sorghum. Int. Sorghum Millet Newsl. 35, 100–101.

Singh, S.D., Navi, S.S., Stenhouse, J.W., Rao, K.E.P., 1995. Grain mold resistance in white grain sorghum. Int. Sorghum Millets Newsl. 36, 95–96.

Snyder, B.A., Nicholson, L., 1990. Synthesis of phytoalexins in sorghum as a site specific response to fungal ingress. Science 248, 1637–1639.

Srinivas, G., Satish, K., Madhusudhana, R., Reddy, R.N., Murali Mohan, S., Seetharama, N., 2009a. Identification of quantitative trait loci for agronomically important traits and their association with genic-microsatellite markers in sorghum. Theor. Appl. Genet. 118, 1439–1454.

Srinivas, G., Satish, K., Madhusudhana, R., Seetharama, N., 2009b. Exploration and mapping of microsatellite markers from subtracted drought stress ESTs in *Sorghum bicolor* (L.) Moench. Theor. Appl. Genet. 118, 703–717.

Stack, J.P., Pedersen, J.F., 2003. Expression of susceptibility to *Fusarium* head blight and grain mold in A1 and A2 cytoplasms of *Sorghum bicolor*. Plant Dis. 87, 172–176.

Stark, D.M., Beachy, R.N., 1989. Protection against potyvirus infection in transgenic plants: evidence for broad spectrum resistance. Natl. Biotechnol. 7, 1257–1262.

Steinbach, H.S., Benech-Arnold, R.L., Kristof, G., Sanchez, R.A., Marcucci-Poltri, S., 1995. Physiologic basis of pre-harvest sprouting in *Sorghum bicolor* (L.) Moench. ABA levels and sensitivity in developing embryos of sprouting-resistant and -susceptible varieties. J. Exp. Bot. 46, 701–709.

Stenhouse, J.W., Bandyopadhyay, R., Singh, S.D., Subramanian, V., 1998. Breeding for grain mold resistance in sorghum. Genetic Improvement of Sorghum and Pearl Millet. Proceedings of the International Conference on Genetic Improvement of Sorghum and Pearl Millet. INSTORMILL and ICRISAT, Lubbock, TX, pp. 326–336, September 22–27.

Stephens, J.C., Holland, R.F., 1954. Cytoplasmic male-sterility for hybrid sorghum seed production. Agron. J. 46, 20–23.

Subudhi, P.K., Nguyen, H.T., 2000. Linkage group alignment of sorghum RFLP maps using a RIL mapping population. Genome 43, 240–249.

Sundaram, N.V., 1971. Possible resistance to sugary disease in sorghum. Indian J. Genet. Plant Breed. 31, 383–387.

Tao, Y.Z., Jordan, D.R., Henzell, R.G., McIntyre, C.L., 1998. Identification of genomic regions for rust resistance in sorghum. Euphytica 103, 287–292.

Tegegne, G., Bandyopadhyay, R., Mulatu, T., Kebede, Y., 1994. Screening for ergot resistance in sorghum. Plant Dis. 78 (9), 873–876.

Thakur, R.P., Mathur, K., 2007. Anthracnose. In: Thakur, R.P., Reddy, B.V.S., Mathur, K. (Eds.), Screening Techniques for Sorghum Diseases. Information Bulletin No. 76. ICRISAT, Patancheru, pp. 15–23.

Thakur, R.P., Rao, V.P., 2007. Ergot. In: Thakur, R.P., Reddy, B.V.S., Mathur, K. (Eds.), Screening Techniques for Sorghum Diseases. Information Bulletin No. 76. ICRISAT, Patancheru, pp. 47–52.

Thakur, R.P., Rao, V.P., Reddy, P.S., 2007a. Downy mildew. In: Thakur, R.P., Reddy, B.V.S., Mathur, K. (Eds.), Screening Techniques for Sorghum Diseases. Information Bulletin No. 76. ICRISAT, Patancheru, pp. 31–39.

Thakur, R.P., Rao, V.P., Wu, B.M., Subbarao, K.V., Mathur, K., Tailor, H.C., et al., 2007b. Genetic resistance to foliar anthracnose in sorghum and pathogenic variability in *Colletotrichum graminicola*. Indian Phytopath. 60, 13–23.

Thakur, R.P., Reddy, B.V.S., Mathur, K., 2007c. Screening Techniques for Sorghum Diseases. Information Bulletin No. 76. ICRISAT, Patancheru, p. 92.

Thakur, R.P., Sharma, R., Rao, P.S., Reddy, P.S., Rao, V.P., Reddy, B.V.S., 2010. Evaluation of sweet sorghum hybrid parents for resistance to grain mold, anthracnose, leaf blight and downy mildew. SAT eJ 8, 1–5.

Tunwari, B.A., Nahunnaro, H., Anaso, A.B., 2014. Eco-friendly management strategies for gray leaf spot disease of sorghum using cultivar selection and seed dressing fungicides in Maiduguri, Nigeria. J. Agric. Sustainab. 5 (1), 14−25.

Tsukiboshi, T., Shimanuki, T., Uematsu, T., 1999. *Claviceps sorghicola* sp. nov., a destructive ergot pathogen of sorghum in Japan. Mycol. Res. 103 (11), 1403−1408.

Upadhyaya, H.D., Wang, Y.H., Sharma, R., Sharma, S., 2013a. Identification of genetic markers linked to anthracnose resistance in sorghum using association analysis. Theor. Appl. Genet. 126, 1649−1657.

Upadhyaya, H.D., Wang, Y.H., Sharma, R., Sharma, S., 2013b. SNP markers linked to leaf rust and grain mold resistance in sorghum. Mol. Breed. 32, 451−462.

Waniska, R.D., Venkatesha, R.T., Chandrashekar, A., Krishnavenl, S., Bejosano, F.P., Jeoung, J., et al., 2002. Antifungal proteins and other mechanisms in the control of sorghum stalk rot and grain mold. In: Leslie, J.F. (Ed.), Sorghum and Millets Diseases. Iowa State Press, Ames, IA, pp. 287−297.

Woodfin, C.A., Rosenow, D.T., Clark, E., 1988. Association Between the Stay-Green Trait and Lodging Resistance in Sorghum. ASA, Madison, WI, p. 102. Agronomy Abstract.

Xiaotian, M., Lijiang, W., Chengcai, A.N., Huayi, Y., Zhangliang, C., 2000. Resistance to rice blast (*Pyricularia oryzae*) caused by the expression of trichosanthin gene in transgenic rice plants transferred through *Agrobacterium* method. Chin. Sci. Bull. 45, 1774−1778.

Xu, G.W., Magill, C., Schertz, K., Hart, G., 1994. A RFLP linkage map of *Sorghum bicolor* (L.) Moench. Theor. Appl. Genet. 89, 139−145.

Zhu, H., Muthukrishnan, S., Krishnaveni, S., Wilde, G., Jeoung, J.M., Liang, G.H., 1998. Biolistic transformation of sorghum using a rice chitinase gene. J. Genet. Breed. 52, 243−252.

Zou, G., Zhai, G., Feng, Q., Yan, S., Wang, A., Zhao, Q., et al., 2012. Identification of QTLs for eight agronomically important traits using an ultra-high-density map based on SNPs generated from high-throughput sequencing in sorghum under contrasting photoperiods. J. Exp. Bot. 63, 5451−5462.

DISEASE RESISTANCE IN PEARL MILLET AND SMALL MILLETS

3

A. Nagaraja[1] and I.K. Das[2]

[1]*UAS-Gandhi Krishi Vignana Kendra, Bengaluru, India*
[2]*ICAR-Indian Institute of Millets Research, Hyderabad, India*

3.1 INTRODUCTION

Pearl millet and small millets are important food and feed crops of the semiarid regions of the world. About 98% of the total pearl millet acreage of the world lies in Africa (64%) and Asia (34%) and the major producing countries are India, Niger, Nigeria, Sudan, Mali, Burkina Faso, Chad, China, and Senegal. India ranks first in the world, in terms of both harvested area (9.2 million ha) and production (10.9 million tonnes) and produces more than half the world's pearl millet. Rajasthan is the major pearl millet growing state in India encompassing 50% of total area, while the states of Maharashtra, Gujarat, Uttar Pradesh and Haryana encompass around 40% of the area. The small millets include around eight small-seeded grasses, such as finger millet or *ragi*, kodo millet, little millet, foxtail millet, barnyard millet, proso millet, tef and fonio. They are mostly used as human food in Asia (India, China, Nepal, Sri Lanka, Japan, Malaysia) and Africa (Nigeria, Niger, Kenya, Tanzania, Ethiopia, Uganda, Zaire, Somalia, Mali, Burkina Fasso) and as animal feed in some parts of Europe and America. Tef is mostly cultivated in the highlands of Ethiopia as a food and fodder crop. In India small millets are grown from the extreme southern tip at sea level to the northern Himalayan regions in Uttarakhand up to an altitude of 3000 m. The major growing states are Karnataka, Tamil Nadu, Odisha, Maharashtra, Andhra Pradesh, Uttarakhand, Uttar Pradesh and Bihar.

Millets constitute an extremely important group of crops. From a cultivation point of view, they are hardy crops and grow well in dry zones as rain-fed crops under marginal conditions of soil fertility and moisture, a condition where many other cereals fail to grow. They are highly resilient and ideal in adapting to different agro-ecological conditions, climate change, and contingency plantings. Long storability of the small millets under ordinary conditions has made them "famine reserves." Owing to their nutritive value small millets are now popularly called "nutricereals" and are a rich source of iron and zinc besides crude fiber. However, like all other crops, millets also are affected by many production constraints including biotic and abiotic factors. Among the yield-reducing biotic factors, diseases are extremely important and must be kept under control to have a good harvest. Fungal diseases are of a major concern as many epidemics especially downy mildew and blast have appeared in different parts of the world in the past. For management of the millet disease the use of a disease-resistant cultivars has been the cheapest and safest method of control. This is more so as millet farmers are basically poor and cannot bear the cost of expensive chemicals or other methods for

Biotic Stress Resistance in Millets. DOI: http://dx.doi.org/10.1016/B978-0-12-804549-7.00003-2

disease control. Disease-resistant cultivars of millets have been used widely across the world in the past for disease management and will continue to be used in the future. This requires continuous efforts to identify new sources of resistance and their adequate utilization for the benefit of the poor farmers. This paper gives a detailed account of all aspects of the millet diseases (pearl millet and small millets) and their management focusing host-plant resistance as a major way forward.

3.2 PEARL MILLET

Pearl millet is mainly grown under dry climate, yet it suffers from infection of many diseases. Compared to fungal diseases, the incidence of bacterial and viral diseases are negligible. King (1992) ranked the relative importance of pearl millet diseases as downy mildew 45%, *Striga* 32%, smut 9%, ergot 7%, rust 3%, viruses >1%, and other diseases 3%. Downy mildew is important in India and to some extent in western Africa, and *Striga* is important in western Africa. A list of common fungal diseases of pearl millet is given in Table 3.1. This chapter will restrict discussion to the major diseases causing significant economic loss to the crop worldwide.

Table 3.1 Common Fungal Diseases of Pearl Millet

Disease	Causal Fungus
Downy mildew	*Sclerospora graminicola* (Sacc.) Schroet.
Blast	*Pyricularia grisea* (Cke.) Sacc.
Ergot	*Claviceps fusiformis* Loveless; *Claviceps africana* Frederickson
Smut	*Moesziomyces penicillariae* (Bref.) Vanky
Rust	*Puccinia substriata* Ell. & Barth. *indica* Ramachar & Cumm.
Top rot	*Fusarium moniliforme* Sheldon
Head mold	*Various fungi*
Leaf spots	
Bipolaris leaf spot	*Bipolaris setariae* (Saw.) Shoem
Cercospora leaf spot	*Cercospora penniseti* (Chupp)
Curvularia leaf spot	*Curvularia penniseti* (Mitra) Boedijn
Dactuliophora leaf spot	*Dactuliophora elongate* Leakey
Drechslera leaf spot	*Drechslera dematioidea* (Bubak & Wroblewski) Subram. & Jain
Myrothecium leaf spot	*Myrothecium roridum* Tode ex. Fr.
Phyllachora leaf spot	*Phyllachora penniseti* Syd.
Zonate leaf spot	*Gloeocercospora sorghi* Bain & Edgerton
Leaf blights	
Exserohilum leaf blight	*Exserohilum rostratum* (Drechs.) Leonard & Suggs
Phyllosticta leaf blight	*Phyllosticta penicillariae* Speg.
Rhizoctonia blight	*Rhizoctonia solani* Kuhn; *Rhizoctonia zeae* Voorhees
Southern blight	*Sclerotium rolfsii* Sacc.

3.2.1 DISEASE, BIOLOGY AND EPIDEMIOLOGY

This part provides a brief description of the major diseases of pearl millet. This is intended to give the readers a general overview of pearl millet diseases including their importance, distribution, causal organism, and management practice, other than varietal resistance, which has been dealt separately. Fungal, bacterial, and viral diseases are discussed separately in the following sections.

3.2.1.1 Fungal diseases

Fungal diseases are of major importance in pearl millet compared to viral and bacterial diseases. Downy mildew or green ear disease is the most damaging disease of pearl millet worldwide followed by blast, while rust and leaf spots are of minor importance. Among panicle diseases smut is quite common in all the growing regions but its severity is not worrisome, while ergot is severe in some parts of the world, particularly India and is a concern for hybrid seed production.

3.2.1.1.1 Foliar diseases

Downy mildew. Downy mildew is the most important disease of pearl millet. Several epidemics of this disease have been reported in the past causing widespread damage to the crop. The disease is present in all pearl millet growing regions in India and Africa (Western and Central Africa). However, it has not yet been reported on pearl millet in the United States. The states of Gujarat, Rajasthan, Haryana in the North-West, and Tamil Nadu and Karnataka in the South are the major downy mildew prone areas in India. Disease incidence up to 90% is often recorded in farmers' field depending on location and specific cultivar (AICPMIP,2014). Newly developed hybrids during the 1970s and 1980s were the worst affected due to a severe outbreak of this disease (Singh et al., 1993). Many popular hybrids were then withdrawn from cultivation. Since the 1990s no major epidemic has occurred mainly due to diversification of the hybrid base (Thakur et al., 2011). Presently the downy mildew disease is somewhat under check. The disease affects grain yield and there is highly significant ($p = 0.05$) negative linear correlation between disease incidence and severity on yield (Gwary et al., 2009). Apart from yield the disease can adversely affect the fodder quality. There may be significant reduction in total soluble sugars in diseased plants compared to healthy plants of pearl millet hybrids (Upasana et al., 2010).

Symptoms are developed both on the leaves and on the earhead. Initial symptoms of the systemic infection are expressed as chlorosis or yellowing of the lower leaves that progressively spread to the upper leaves and the whole plant. Often the lower half of a leaf shows symptoms while its upper half remain symptomless. This is known as "half leaf" symptoms. Numerous sporangia are produced on the lower surface of an infected leaf when the environmental conditions are favorable. A severely infected plant becomes stunted and often fails to produce an earhead. Sometimes the infected plant produces symptoms only on the earhead in the form of the leafy structures known as "green ear" disease or "virescence." Local lesion symptoms, as seen in sorghum, are rare in pearl millet.

The disease is caused by *Sclerospora graminicola* (Sacc.) Schroet. It is an oomycete that reproduces asexually by producing sporangia and sexually by producing oospores. The fungus is largely heterothallic but homothallism has also been reported (Michelmore et al., 1982). Pushpavathi et al. (2006a,b) observed low frequency of homothallic isolates in India. Two mating types have been identified in *S. graminicola*. The existence of mating types and their frequency greatly contribute toward pathogen variability. Several pathotypes have been identified from pearl millet growing states in India. Extensive studies on pathogen variability have been carried out using various molecular markers (Sastry et al., 1995; Sudisha et al., 2009; Sharma et al., 2010). A Sequence characterized regions (SCAR) primer pair (UOM3-Sg-Path1-F/R) has been developed for rapid identification and specific detection of pathotype-1 (Sudisha et al., 2009). *Sclerospora graminicola* is reported to be highly cultivar-specific (Singh and Singh, 1987). The oospores in the soil germinate and cause primary infection at the underground part of the seedling. The leaves of the systemically infected plants produce abundant sporangia which become air-borne and cause secondary infection. Optimum temperature and humidity for production of sporangia is 20−25°C and RH 95−100%. Sporangia are short-lived, surviving for about 24 h at low temperature (5−15°C). Sporangia germinate through flagellate zoospores which require a film of water for germ tube production and infection. Sporangial infection takes place only up to the one- or two-leaf stage of the seedling. Thereafter, susceptibility of the seedling decreases sharply (Singh and Gopinath, 1985). The pathogen colonizes the leaf tissue inter- and intracellularly (Celia Chalam, 1996). Oospores are produced in infected leaves when compatible mating types are present in the same tissue, or when homothallism is operative. Available information is not clear about the role of alternate/collateral hosts on the survival of the pathogen. Many graminaceous species have been reported as hosts of *S. graminicola* but their role in disease epidemiology is doubtful. Although it is believed that the disease is seed-borne, internal seed transmission of the disease has been a subject of controversy (Safeeulla, 1976; Shetty et al., 1980).

Use of resistant cultivars has been the best and most economical method for management of pearl millet downy mildew (see "host-plant resistance" section). Considerable research has been done to control downy mildew by protective and systemic fungicides. The systemic fungicide metalaxyl has been used successfully to control the disease. Seed treatment with metalaxyl at 2 g a.i. kg^{-1} seed controls the disease for about a month after sowing. However, seed treatment coupled with a single foliar application was found to be superior to the seed treatment alone (Shankara Rao et al., 1987). Anaso and Anaso (2010) reported that Apron Star 42WS, a new formulation of metalaxyl, for seed dressing is cost-effective and sustainable in the downy mildew endemic areas of northern Nigeria. Seed priming with chitosan was reported to give systemic protection against this disease (Manjunatha et al., 2009).

Blast. *Pyricularia* leaf spot or blast is a common disease of pearl millet and found in many countries where the crop is grown. The disease was first recorded in Uganda in 1933 (Emchebe, 1975). Later it was reported from Kanpur, India (Mehta et al., 1953). The disease is important for grain and forage pearl millet in India, United States, and in many African countries. Blast was generally more severe in finger millet than pearl millet but recently the disease is increasing alarmingly in pearl millet (Lukose et al., 2007; AICPMIP, 2013). In the recent past, it has become severe on commercial hybrids in several states of India including Rajasthan, Gujarat, Maharashtra and Madhya Pradesh. The disease is severe both on seedlings and on grownup plants and results in premature drying of leaves and reduction in yield. Forage yield, dry matter yield, and digestive dry matter are negatively affected due to this disease. Mean disease severity ranges from 10−30% and sometimes may go up to 50% on hybrids and varieties, depending on the season. A field survey in Uganda estimated around 60% incidence and severity of leaf blast in pearl millet causing severe loss in grain yield (Lubadde et al., 2014).

The disease is caused by *Pyricularia grisea* (teleomorph: *Magnaporthe grisea* (Herbert) Barr). The initial symptoms appear as grayish, water-soaked lesions on the foliage that turn brown upon drying. The lesions are elliptical or diamond-shaped with a gray center and often surrounded by a chlorotic halo, which turns necrotic, giving the appearance of concentric rings. Extensive chlorosis causes premature drying of young leaves. The symptom may appear on leaves, leaf sheath and stem. Lesions produce abundant sporulation under high humid conditions.

Pyricularia grisea infects many cereals including rice, wheat, pearl millet, finger millet, foxtail millet and several grasses. It is highly specialized in its host range and the pathogen population that infects rice or any other host does not infect pearl millet and vice versa. High degree of pathogenic variation in this fungus is reported from rice, finger millet, foxtail millet, wheat and several weed hosts (Prabhu et al., 1992; Takan et al., 2012). Operation of mechanisms like sexual recombination, heterokaryosis and parasexual recombination in this pathogen help it in frequent race changes. Repeated race development is a big challenge for the development of durable resistance against this pathogen in other crops. In rice, resistance in most of the cultivars breaks down in a few years because of the rapid change in pathogenicity of this fungus (Suh et al., 2009). In pearl millet too this may be true, but until now only limited information has been available for this pathogen. The pathogen survives in the crop residues and on other cereals which act as collateral hosts. The initial inoculum most probably comes from weeds or some collateral hosts. The fungus spreads mainly by air-borne conidia. A temperature of 25−30°C, RH of 90% and above, and cloudy days with intermittent rainfall are favorable for disease development and rapid spread of the disease (Thakur et al., 2011). The disease become severe under humid weather and a densely populated crop gets more infection. Application of nitrogenous fertilizers increases the incidence of the disease.

Management practices like wide plant spacing and regulating the amount of nitrogenous fertilizer are important measures to minimize the occurrence of blast disease. In the absence of resistant varieties a vast number of chemicals are widely used for controlling blast. Two sprays of carbendazim (@ 0.05% a.i.) at 15 days interval was reported to reduce the blast intensity and increase the grain and fodder yields with the maximum net return (Lukose et al., 2007). There are reports of the use of biological control agents (Ramteke et al., 2011) but their efficacies under field conditions have been hardly testified.

Rust. Rust of pearl millet is observed in all the cropping regions including Asia, Africa, North America and South America. The disease was restricted to Africa and the western hemisphere but recently it has become damaging in central Brazil (de Carvalho et al., 2006). In India the disease is common in the states of Tamil Nadu, Karnataka, Rajasthan, Gujarat and Maharashtra. It is more severe on the postrainy season crop than on the rainy or summer season crop. Mean rust severity across India varies from 10% to 45% depending on season. The economic significance of the disease is small, especially when it appears late in the season and has little effect on grain yield. The disease is important for multi-cut forage hybrids in which it may cause losses in yield of digestible dry matter (Wilson et al., 1991).

Initial symptoms appear on the leaf as reddish-orange color pustules which are round to elliptical in shape. The distal half of the leaf is commonly infected first, then pustules (sori) spread over both the surfaces. Highly susceptible cultivars develop large pustules on the leaf blade and leaf sheath. The mature pustules rupture and release rusty spores. Symptoms may also appear on the stem and other plant parts. Severely rusted plants look reddish-brown in color.

Rust of pearl millet is caused by *Puccinia substriata* var. *indica* Ramachar & Cumm. (Synonym: *Puccinia substriata* Ell. & Barth. var. *penicillariae* de Carvalho et al.; *Puccinia penniseti* Zimm.). It is heteroecious rust and needs two different host plants to complete its life cycle. The fungus survives on the alternate host brinjal (*Solanum melongena*) on which it produces spermagonia, acea and aeciospores (de Carvalho et al., 2006). Air-borne aeciospores fall on pearl millet leaves, infect, and produce ureadia and then telia on them. Teleutospores produced in telia infect alternate host and thus complete the life cycle. Low night temperature favors rust development. Removal of alternate hosts and grassy weeds helps reduce rust in pearl millet. Use of resistant cultivars is the most cost-effective method for control of rust.

Leaf spots and blights. Many leaf spots and leaf blights occur on pearl millet (Table 3.1). They may occur in any growing conditions of pearl millet (Asia, Africa, and Americas) and are not generally location specific. In general they cause minor damages on leaves and are not considered economically important presently. However, they may become severe under favorable weather conditions limiting grain and fodder yield and quality substantially. The pathogen normally survives in crop residues, stray crops, or on collateral hosts and a few may be seed-borne.

3.2.1.1.2 Panicle diseases

Ergot. Ergot or sugary disease of pearl millet is common in India and many African countries, including Sudan, Ethiopia, Eritrea, Benin, Togo, Central African Republic, Ghana, Senegal, Tanzania, Cameroon and Nigeria (Marley et al., 2002). It is a major disease of pearl millet in India, especially during hybrid seed production. Pearl millet ergot was first reported from India (Thomas et al., 1945) and the first epiphytotic form of the disease was reported from Maharashtra (Bhide and Hegde, 1957). With the introduction of new hybrids HB1 and HB3 during the late sixties ergot became severe and occurred almost every year in India (Sundaram, 1975). Now the disease is common in many states of India including Rajasthan, Maharashtra, Karnataka, Haryana and Uttar Pradesh. Ergot infection causes loss in seed yield, seed quality, germination, and seedling emergence. Grain yield loss has been estimated to be as high as 58–70% in hybrids (Khairwal et al., 2007). Sclerotia of pearl millet ergot contain alkaloids that affect the health of human beings and animals. Alkaloid agroclavine is known to cause milklessness in female pigs and weakening of legs in chicks. The disease has quarantine implications and seeds harvested from infected fields may face rejection in trading.

The first visible symptom of ergot infection appears as exudates of viscous liquid from the florets. The honeydew-like exudates contain numerous micro- and macroconidia of the pathogen. The symptom can be seen on a single, few, or all florets in a panicle depending on severity of infection. Infection is spread to adjoining florets or panicles through mechanical contact. Infected florets produce fungal sclerotium in place of grain. Sclerotia are light pink to dark brown or black in color, round or elongated in shape, and brittle or hard in texture. They are larger than seed, and with a pointed apex which protrude from the florets in place of grain.

Ergot in pearl millet is caused by *Claviceps fusiformis*, an ascomycetous fungus. Sclerotia germinate by producing 1–16 fleshy, purplish stipes (up to 2.5 cm long) which bear a globular, dark brown capitulum at the head (Thakur et al., 1984). Asci are produced inside a pyriform perithecium. Each perithecium bears hyaline, nonseptate and filiform ascospores. Sclerotia mixed with the seed or left in the soil serve as source of primary inoculum for the next season. After a rain the sclerotia germinate and produce thread like ascospores which become air-borne and cause infection in the floret. Chances of infection increase when pollination does not happen after emergence of stigma. Ergot severity is inversely related to pollen shedding (Miedaner and Geiger, 2015). Infected florets produce honeydew symptoms under the conditions of high relative humidity (80–85%) and moderate temperature (20–30°C) (Thakur et al., 2011). Overcast skies and drizzling rain favor spread of the disease.

General disease management practices like cultural, chemical and biological control can be followed for management of ergot. In addition, the disease can be managed by pollen management that includes continuous supply of pollen by maintaining heterogenous plant populations of open-pollinated varieties. Deep plowing helps bury sclerotia at greater depth which prevents their germination and spore production. Removal of weeds from around pearl millet fields also helps reduce the inoculum. Healthy seed can be separated from ergot mixed or infected seed either manually or by soaking in salt solution. Among all, use of resistant cultivars is the most cost-effective method for the control of ergot.

Smut. Pearl millet smut is an important disease in India, countries in Western and Central Africa, and the United States. In India the disease commonly occurs in Rajasthan, Madhya Pradesh and Maharashtra and the incidence varies from 2−20% depending on season and cultivars. It causes direct loss of production as it replaces grain with smut sori. Grain yield losses of up to 30% have been reported. The disease is more severe in cytoplasmic male sterility-based single-cross hybrids than in open-pollinated varieties (Thakur et al., 1989). In recent years the disease has become more important on commercial F_1 hybrids in India and on early-maturing cultivars in countries of Africa.

Smut of pearl millet is caused by *Tolyposporium penicillariae* (synonym: *Moesziomyces penicillariae*), a basidiomycetes fungus. The fungus survives as teleutospores in infected seed or in soil. After getting optimum moisture and temperature, the teleutospore germinates to produce sporidia. A sporidium becomes air-borne, falls on stigma, and causes infection. In an infected floret the ovary is converted into a spore-bearing structure called sorus. Initially the sorus is bright or shiny green in color and later turns brown. The sorus is larger than grain and appears as an enlarged body in place of grain. The mature sorus ruptures to release spore balls containing teleutospores. The pathogen has a long latent period therefore secondary infection in the same aged crop is rare unless late-sown or late-flowering plants are available.

Cultural and chemical methods for smut management in pearl millet are not much practiced and they are not cost-effective. Use of resistant cultivars is the most sought after method for control of smut. Rapid pollination reduces the chance of smut infection following the same mechanisms as observed for ergot resistance. It is assumed that ergot resistant almost ensures smut resistance but the opposite is not true. Open-pollinated varieties in India are less vulnerable to smut.

3.2.1.2 Bacterial diseases

Bacterial diseases are not common in pearl millet mostly because of the dry agro-climatic conditions in which this crop is grown. A few bacterial diseases, that is, bacterial leaf streak (*Xanthomonas axonopodis* pv. *pennamericanum*), bacterial leaf stripe (*Pseudomonas avenae*), and bacterial spot (*Pseudomonas syringae*), are reported on this crop but they are not economically important. They cause minor and sporadic incidence and negligible economic damage. They may become important under specific climatic situations. Little information is available in literature about these diseases.

3.2.1.3 Viral diseases

More than half-a-dozen viral diseases have been reported on pearl millet. Most of them are minor in incidence and cause negligible or no economic damage to the crop. Two viruses, that is, maize dwarf mosaic virus and maize streak virus, have somewhat widespread distribution in many pearl millet growing countries in Africa, Asia and Americas. Other viruses (namely panicum mosaic virus, wheat streak mosaic virus, guinea grass mosaic virus and Indian peanut clump virus) are reported from a few countries. Because of their economic nonsignificance, little information is available in literature about the symptom, transmission, host range and economic damage caused by these viruses on pearl millet.

3.2.2 HOST-PLANT RESISTANCE

Host-plant resistance has been an important component of disease management in pearl millet. It is the most economic and environment friendly way of managing plant diseases. For economically poor farmers host-plant resistance is the only viable practice of disease control. However, the resistance needs to be identified and applied correctly. Availability of diverse sources of resistance and efficient disease screening techniques are prerequisite for identification of resistance. Genetic and heritability study of resistance helps to choose a suitable selection process and breeding method for incorporation of resistance to a desired background. Knowledge of the mechanisms of disease resistance, on the other hand, provides clues for making resistance stable and durable. All these aspects of host resistance in pearl millet is discussed in the following section.

3.2.2.1 Screening for resistance

Availability of an effective disease screening technique is important for initial identification of resistance sources followed by selection of the disease-resistant materials of a resistant breeding program. In pearl millet, field and greenhouse screening techniques have been developed for almost all the major diseases, including downy mildew, blast, ergot, smut and rust (Thakur and King, 1988a; Yadav et al., 1995; Singh et al., 1997). Recently Thakur et al. (2011) published an information bulletin describing new techniques and refinement of the earlier methodologies of screening for all these diseases.

For downy mildew the field-screening technique is the most reliable method that screens resistance under natural conditions. This technique mainly utilizes sporangia as infection propagules and deploys infector rows, test rows, and indicator rows for development of disease and identification of resistance (Williams et al., 1981). Disease incidence is represented as percent infected plants. Later Singh and Gopinath (1985) developed a technique which involved inoculation of potted seedlings at the coleoptile stage using a micro syringe. Greenhouse and laboratory screening techniques have been developed, for mass screening of breeding materials. Recently Shishupala et al. (1996) reported an Enzyme linked immuno sorbent assay (ELISA)-based technique to test downy mildew resistance in pearl millet cultivar. In this method a protein from a virulent pathotype of *Sclerospora graminicola* and a corresponding antibody were used in an ELISA to screen cells suspension of pearl millet cultivars for their resistance to the downy mildew pathogen.

Blast is becoming severe during the recent time and earlier screening programs mostly used natural field condition. Recently, Thakur et al. (2009) elaborated the field and greenhouse screening techniques for pearl millet blast. Like downy mildew, infector, indicator and tester row principles can be used for screening under field conditions using perfo-irrigation for maintaining high humidity (>90%). The greenhouse screening technique involved spray inoculation of 15-day-old potted seedlings with pathogen spore suspension and maintaining moderate temperature and high humidity through a misting system. Disease severity is recorded at the hard dough stage using a 1−9 progressive scale (1, no lesion; 9, all leaves dead). Similarly, screening for rust resistance can be performed following field or greenhouse screening techniques. Infector, indicator, and tester row principles, as in blast, can be followed but inoculation time and disease scoring scale vary. Water suspension of rust urediniospores are spray inoculated at 30−40 days after emergence and rust severity is scored at grain-filling stage using a modified Cobb's scale (Singh et al., 1997).

Ergot resistance can be screened under field conditions if high humidity can be assured using sprinkler-irrigation. Ergot infected panicle of previous year can be used for generating initial inoculum and inoculating early ergot susceptible lines. Honeydew developed in the early line can be used for spray inoculation of panicles of the test materials. Disease severity is scored by recording percent spikelet infection in a panicle. Similarly, smut resistance can be screened by using field-screening technique. Details about the technique is described by Thakur et al. (1992).

3.2.2.2 Resistant sources and utilization

Downy mildew is the most devastating disease of pearl millet. Much of the research effort in resistance breeding has been directed toward identifying resistance against this disease, while for other diseases efforts have been relatively less. International pearl millet downy mildew nursery was established during 1976 to coordinate multinational activities in Africa and India for screening for resistance and stability of resistance to downy mildew. Subsequently, a large number of germplasm and their selections from Africa and India were screened at various national and international downy mildew nurseries and many resistant sources were identified (Singh, 1990; Singh et al., 1994; Nutsugah et al., 2006) (Table 3.2). Singh et al. (1994) identified photoperiod insensitive germplasm ICML 22, resistant against three pathotypes (Mysore, Patancheru and Aurangabad) in India (4.1−10% disease). However, this line was susceptible when tested in Africa (58−100% disease). Later, emphasis has been given to identify downy mildew resistant parental lines for development of hybrid. Thakur et al. (2001) identified nine male sterile lines which were resistant to four pathotypes from representative hot spot locations (Patancheru, Mysore, Durgapur and Jalna) in India. Recently Manga and Kumar (2012), reported two maintainer lines (B-lines) ICMB 95444 and ICMB 95111 as possessing resistance to downy mildew. Apart from grain pearl millet, downy mildew resistance has been identified in forage lines. Ramesh et al. (2003) evaluated eleven forage lines against six pathotypes in multiple locations in India and found DRSB 6, DRSB 7, DRSB 10, Giant Bajra, IP 14188, and IP 14305 to be highly resistant to downy mildew (<5% disease).

Rust infection when it occurs at the early stage of the crop growth causes considerable damage. This warrants the development of rust resistant cultivars especially for Indian conditions where rust infection is quite common in the early stages. Search for rust resistance in pearl millet gathered steam during 1970s, when the international pearl millet rust nursery was established. Later many reports had come and rust resistances were identified in germplasm or their selections (Govindarajan et al., 1984; Singh, 1990), local land races (Wilson et al., 1989), advanced breeding lines (Jahagirdar et al., 2005) and hybrid parental lines (ICMB 96222, ICMR 06999, ICMP 451-P8, and ICMP 451-P6) (Sharma et al., 2009). Slow rusting or delaying rust development process is a genetically controlled phenomenon. A few sources for this trait have been identified (Sokhi and Singh, 1984; Pannu et al., 1996). Combined resistance of downy mildew and rust has been reported in the lines ICML Nos. 12, 13, 14, 15, and 16 (Singh, 1990).

Table 3.2 Downy Mildew Resistant Sources Reported in Pearl Millet

Screening Condition	Type of Resistance	Name of the Resistant Source	Reference
Natural field screening	Resistant to highly resistant against downy mildew & rust	African & Indian germplasm selections: D322/1/2-2, P1449-3, IP 147-4, P8695-1, P8899-3, P3281 (<5% DM); P310, P472, P1564 & 700516 (0% DM)	Singh (1990)
International Pearl Millet Downy Mildew Nursery in India, Burkina Faso, Senegal	Stable sources of resistance	Germplasm selection from Mali and Nigeria, ICML 12 (IP 6118, P7), ICML 13 (IP 8215, SDN 503), ICML 14 (IP 4984, 700251), ICML 15 (IP 5082, 700516), and ICML 16 (IP 8198, 700651). They have combined resistance against downy mildew & rust. P 310-17	Singh (1990), Thakur et al. (2004)
Natural field screening	Moderately resistant	Photoperiod insensitive germplasm ICML 22 resistant in India (4.1−10% DM) but susceptible in Africa (58−100% DM)	Singh et al. (1994)
Greenhouse & field condition	Resistant to highly resistant	Germplasm accession IP 8897, IP 14619, and IP 17311	Navi and Singh (1995)
Field screening	Resistant	Dwarf, restorer parental line ICMP 85410	Talukdar et al. (1998)
Greenhouse & field condition	Resistant in glasshouse conditions	Male sterile lines 863A, 841A and ICMA Nos. 88004, 94333, 98222, 98111, 92777, & 96666 were resistant against four pathotypes (Patancheru Mysore, Durgapur, Jalna)	Thakur et al. (2001)
Field nursery and greenhouse	Resistant	Forage lines DRSB 6, 7, 10, Giant Bajra, IP 14188, IP 14305 (<5% DM)	Ramesh et al. (2003)
Field nursery	Resistant to DM, Smut & Ergot	Germplasm lines from Africa Synth 16 C0, Synth 16 C1, SE 360, SE 2124, INMV 77, ICMV IS 90309, ICMV IS 90311	Nutsugah et al. (2006)
Field nursery	Resistant	Maintainer (B-line) ICMB 95444, ICMB 95111	Manga and Kumar (2012)

Smut is a common problem of pearl millet in Asia and Africa and lot of efforts have been invested in identifying resistance sources. Germplasm lines from different countries have been reported to possess good amount of resistance and many lines have been identified as stable in multiple locations. Six accessions (SSC FS 252-S-4, ICI 7517-S-1, ExB 132-2-S-5-2-DM-1, ExB 46-1-2-S-2, ExB 112-1-S-1-1, and P-489-S-13) out of 1500 tested showed stable resistance when evaluated in international pearl millet smut nursery (Thakur et al., 1986). Smut resistant inbred lines ICML Nos. 5, 6, 7, 8, 9, and 10 were registered (Thakur and King, 1988c). Now different types of smut resistant lines are available in pearl millet including elite lines (ICMPS 100-5-1, 900-9-3, 1600-2-4, and 2000-5-2); hybrid parents (ICMA/B 92444, ICMA/B 92777, and ICMA/B 92888) (Rao et al., 2006); and smut resistance combined with other disease (Synth 16 C0, Synth 16 C1, SE 360, SE 2124, INMV 77, ICMV IS 90309, ICMV IS 90311, ICMPES 28, 29, and 38) (Thakur and Sharma, 1990; Nutsugah et al., 2006).

Identification of resistant sources against blast has been emphasized only recently because of an increase in severity of the disease. Though in the past many sources have been identified to be resistant at natural field conditions, their authenticity needs to be verified by screening under epiphytotic conditions. Recently Thakur et al. (2011) evaluated 211 advanced breeding lines (designated B- & R-lines) for blast resistance and found three highly resistant (score 1.0 in a 1−9 scale) and 42 resistant (score 2.0−3.0) lines for foliar blast. Sharma et al. (2013) evaluated 238 mini-core accessions against five pathotypes of blast under greenhouse conditions and found 32 as resistant to at least one pathotype. Accessions IP 7846, IP 11036, and IP 21187 showed resistance to four pathotypes. These sources may be important for utilization in breeding program.

Ergot creates a hurdle in hybrid seed production and therefore resistance against this disease is specifically important for public and private seed agencies. Over the time many resistant sources have been identified and utilized in breeding programs for development of hybrid parents (Chahal et al., 1981; Thakur and King, 1988b; Thakur et al., 1993). However, there is a lack of recent research work on this topic, which needs to be looked at again.

3.2.2.3 Genetics of resistance

The knowledge on the magnitude of genetic variability and the nature of inheritance of disease resistance is important for deciding on the selection procedure and breeding methodology to be used for genetic improvement of a crop plant. Qualitative or major gene resistance is generally easy to transfer, but often face the threat of resistance break down. Polygenic or quantitative resistance, on the other hand, is durable in nature. Inheritance of downy mildew resistance in pearl millet is reported to be dominant, partially dominant, or complex (Kataria et al., 1994; Singh and Talukdar, 1998; Deswal et al., 1998). Pushpavathi et al. (2006a,b) observed that avirulence was dominant over virulence and suggested the presence of gene-for-gene interactions. Nature of inheritance of the blast and rust resistance in pearl millet is predominantly qualitative, though involvement of multiple gene actions is reported. Foliar blast resistance in governed by one or more dominant genes (Wilson and Hanna, 1992; Gupta et al., 2012). However, rust resistance might be governed by multiple dominant or recessive genes or their combinations depending on the host genotype (Andrews et al., 1985; Pannu et al., 1996; Godasara et al., 2010). Smut resistance in pearl millet is reported to be dominant or partially dominant with additive and additive × additive gene effects (Phookan, 1987; Chavan et al., 1988; Yadav et al., 2000). Regarding the genetics of ergot resistance, not much information is available. Thakur et al. (2011) suggested that ergot resistance was a recessive polygenic trait with significant cytoplasmic × nuclear interaction.

Quantitative trait loci (QTL) in response to diseases of pearl millet have been studied for downy mildew resistance (Jones et al., 1995; Gulia, 2004), while information is lacking for other diseases. QTLs for downy mildew resistance have been mapped to all seven pearl millet linkage groups. The position of the QTLs that are effective against particular downy mildew pathogen populations (pathotype) has been indicated in the linkage map (Hash and Witcombe, 2001). Jones et al. (2002) identified two same QTLs from the experiment conducted in two different environments along with two additional QTLs. One QTL had a major effect (explained up to 60% of the phenotypic variation), and the other had a minor effect (explained up to 16% of the variation). Number of loci contributed about 17.4% and 47.7% to total inheritance of resistance to downy mildew incidence and severity (Angarawai et al., 2009).

3.2.2.4 Mechanisms of resistance

Natural resistance of a plant to pathogen attack is exhibited by a combination of constitutive and induced defenses. Many morphological, physiological and biochemical mechanisms are known to play important role in disease resistance in crop plants. An understanding of the disease resistance mechanisms helps to develop cultivars with stable and durable resistance. The available information on the mechanism of downy mildew resistance in pearl millet suggests an important role of the cell wall structures like cuticle thickness, wax content (Celia Chalam, 1996), accumulation of hydroxyproline-rich glycoproteins (Shailasree et al., 2004; Deepak et al., 2007; Sujeeth et al., 2010), and silicon content (Deepak et al., 2008). Many living or inert substances have been used for induction of resistance against downy mildew in pearl millet. Among the chemical substances, L-methionine, vitamins, synthetic jasmonate analogon, nitric oxide donor, mannitol, thiamin, etc. have been reported to induce resistance against downy mildew (Deepak et al., 2007; Manjunatha et al., 2009). Substances or elicitors derived from living cells like Cellulysin, crude cellulose from *Trichoderma*, chitosan, Trichoshield, elicitors derived from yeast, *Datura* extract, have also been used for induction of resistance (Niranjan-Raj et al., 2005; Manjunatha et al., 2008; Devaiah et al., 2009; Hindumathy, 2012; Pushpalatha et al., 2013). Sharathchandra et al. (2006) observed that downy mildew resistance in pearl millet was governed by the specific recognition of pathogen-associated proteins (or elicitors) by the plant. They reported that the elicitor protein was able to elicit an array of defense responses in the form of higher activities of the enzymes phenylalanine ammonia lyase (PAL) and peroxidase in the resistant cell cultures.

Not much information is available on mechanisms of resistance to pearl millet blast, rust, ergot and smut. Phenolic compounds and plant defense enzymes are known to play an important role in blast resistance in other crops, but such information is lacking in pearl millet. Efficiency of urediniospore germination on leaf, speed of germ tube formation, penetration and establishment of early phase of the infection process are some of the important steps in rust infection and disease development (Siwecki and Werner, 1980). Therefore, histological study of these events reveals important clues for rust resistance. Studies of the infection court of rust infected pearl millet leaves of the resistant (Tift 85DB) and susceptible (Tift 23DB) cultivar suggested that a rapid cell necrosis and cell wall deposits might play an important role. In the resistant cultivar the fungal colonies had limited growth with no sporulation. Irrespective of resistance in the cultivars the mature leaves of each cultivar were more resistant to fungal colonization than seedling leaves (Taylor and Mims, 1991a,b). Induction of resistance to a virulent isolate of the rust fungus was observed by pretreatment of pearl millet with salicylic acid (Crampton et al., 2009). Girgi et al. (2006) reported that heterologous expression of antifungal protein gene of *Aspergillus giganteus* in pearl millet reduced downy mildew and rust diseases. In a hybrid seed production plot, early protruding stigmas of the seed parent have to wait in a receptive condition until the pollen is available from the pollinator. During this period the stigmas of the seed parent are exposed to spores of ergot and smut. Timely pollination and fertilization of pearl millet floret, therefore, reduces chance of ergot and smut infection. Thakur et al. (1986) observed that cultivars in which protogyny of floret of individual inflorescence lasted less than 49 h showed resistance to ergot infection. Disease escape was due to development of a localized stigmatic constriction that occurred 6 h after pollination. Little information is available on the mechanisms of smut resistance in pearl millet. The *tr* allele in pearl millet was reported to confer trichomeless plant structures such as trichomeless leaf and unbranched stigma, which made them less susceptible to smut (Wilson, 1995).

3.3 SMALL MILLETS

Although small millets are known to successfully cope up with losses due to biotic and abiotic stress, under vulnerable conditions some diseases take a heavy toll of these crops in various ecological zones. A list of common diseases of small millets is given in Table 3.3. Major diseases of small millets, their biology, epidemiology, and details about host-plant resistance vis-à-vis management are discussed in the following sections.

Table 3.3 Common Diseases of Small Millets

Crop	Disease	Pathogen
Finger millet	Blast	*Pyricularia grisea*
	Rust	*Puccinia substriata*
	Smut	*Melanopsichium eleusinis*
	Downy mildew	*Sclerophthora macrospora*
	Seedling & leaf blight	*Drechslera nodulosum*
	Cercospora leaf spot	*Cercospora eleusinis*
	Banded blight	*Rhizoctonia solani*
	Wilt or foot rot	*Sclerotium rolfsii*
	Bacterial leaf spot	*Xanthomonas eleusinae*
	Bacterial leaf blight	*Xanthomonas axonopodis* pv. *coracanae.*
	Bacterial leaf stripe	*Pseudomonas eleusinae*
	Ragi severe mosaic	Sugarcane mosaic virus
	Ragi mottle streak	Ragi mottle streak virus
	Ragi streak	Maize streak virus (Eleusine strain)
Foxtail millet	Blast	*Pyricularia setariae*
	Rust	*Uromyces setariae-italicae*
	Smut	*Ustilago crameri*
	Downy mildew	*Sclerospora graminicola*
	Udbatta	*Ephelis* sp.
	Bacterial leaf blight	*Pseudomonas avenae*
Kodo millet	Head smut	*Sorosporium paspali*
	Rust	*Puccinia substriata*
	Udbatta	*Ephelis* sp.
Barnyard millet	Head smut	*Ustilago crus-galli*
	Kernel smut	*Ustilago panici-frumentacei*
	Bacterial leaf blight	*Pseudomonas avenae*
Proso millet	Head smut	*Sphacelotheca destruens*
	Bacterial leaf blight	*Pseudomonas avenae*
Little millet	Rust	*Uromyces linearis*
Tef millet	Rust	*Uromyces eragrostidis*
	Damping off	*Helminthosporium poae*

3.3.1 DISEASE, BIOLOGY AND EPIDEMIOLOGY

This part provides a brief description of the major diseases of small millets. This is intended to give the readers an idea about the small millet diseases, that is, their importance, distribution, causal organism, and management practices, other than varietal resistance which has been discussed separately.

3.3.1.1 Fungal diseases

Fungal diseases are more important for small millets than the bacterial and viral diseases which are sporadic and negligible. These diseases may infect vegetative parts, reproductive parts, or even root and stalk. The smuts and rust are commonly observed on almost all the cereals of the small millet group including tef. Other fungal diseases like blast, brown spot, leaf blight, crazy top, udbatta, foot rot, sheath rot and banded sheath blight are somewhat specific to one or a few millets. Little information is available on diseases of tef and fonio. Bekele (1985) conducted a survey of tef diseases in Ethiopia along with diseases of other cereals and observed that tef was relatively free of plant diseases when compared to other cereals. Incidence of rust and head smut was noted in areas of high humidity. Ketema (1987) reported that early sown tef seedlings were affected by damping off caused by *Drechslera poae* and *Helminthosporium poae* (Baudys) Shoemaker.

3.3.1.1.1 Foliar diseases

Blast. Blast is reported in finger millet, foxtail millet, little millet, barnyard millet and proso millet. Of these, blast of finger millet caused by *Pyricularia grisea* (Cooke) Sacc. (Perfect stage: *Magnaporthe grisea* (Hebert) Barr.) is the most important production constraint. The disease was reported for the first time in India, from Tanjore delta of Tamil Nadu by Mc Rae (1920). Blast of foxtail millet (*Pyricularia setariae* Nisik) and proso millet (*Pyricularia grisea* (Cooke) Sacc.) were reported from Tamil Nadu and Bihar, respectively (Singh and Prasad, 1981) while that of little millet from mid-hills of Uttarakhand (Rawat et al., 2016). Finger millet blast is important in its growing regions in Eastern and Southern Africa and South Asia. The disease is now common in India in the states of Tamil Nadu, Karnataka, Andhra Pradesh, Maharashtra, Odisha, Bihar, Uttarakhand, Madhya Pradesh, Gujarat and Chattisgarh. The disease occurs almost every year during rainy season and yield loss varies depending on the time of onset of the disease, severity, plant variety and prevailing weather. The average yield loss varies from 28−36% and may go up to 90% in endemic areas (Rao, 1990; Nagaraja and Mantur, 2007).

The pathogen can infect the crop at all stages, from the seedling to the postflowering phase. Symptoms can develop on the leaf, peduncle and finger, depending on the stage of the crop. Elliptical or diamond-shaped lesions are formed on leaves with centers of lesions gray and water-soaked and lesions surrounded by chlorotic halo. Under highly congenial conditions such spots enlarge, coalesce, and leaf blades especially from the tip toward base give a blasted appearance. Seedlings may be killed under epidemic condition. The most damaging stage is neck blast, in which infection of the peduncle and development of elongated black color lesion mostly 1−2 inches below the ear takes place. The neck blast reduces grain number and grain weight and increases spikelet sterility. The pathogen also attacks ears, where few or several fingers appear brown resulting in shriveled and blackened seeds. Symptom starts at the tip and proceeds toward the base of the finger. Depending on the time and severity of infection the infected ears become completely chaffy or produce shriveled grains. Neck infection is the most serious phase of the disease that causes major loss in grain number, grain weight and increase in spikelet sterility.

Blast pathogen is very sensitive to minute changes of temperature and other environmental factors. The life cycle of the fungi in nature is very complicated. It can parasitize over 50 grasses. The pathogen is found in glumes, pericarp and endosperm. But embryo infection has not been observed. The initial inoculum most probably comes from weeds or some collateral hosts. The fungus spreads mainly by air-borne conidia. A temperature of 25–30°C, RH 90% and above, and cloudy days with intermittent rainfall are favorable for the rapid spread of the disease. Congenial climatic conditions for neck blast development are temperature of 15–25°C and relative humidity of more than 85% with intermittent rainfall. The disease incidence increases with increasing N levels (Kumar and Rashmi, 2012). Four different physiological races of fungus have been reported on the basis of pathogenicity, cultural, physiological and morphological characters of the fungus (Kulkarni and Patel, 1956). However, Yan et al. (1985) described seven races of the fungus from China. Gaikwad and D'Souza (1987) on the basis of size of conidia in culture medium and on host concluded that *Pyricularia setariae* that infects foxtail millet is different from the isolates that infect rice, pearl millet and finger millet.

Molecular markers like restriction fragment length polymorphism and simple sequence repeats (SSRs) have been used to study variability and diversity in blast pathogen *M. grisea* (Tanaka et al., 2009; Anjum, 2015). Amplified fragment length polymorphism analysis of *M. grisea* isolates causing leaf, neck, and panicle blast on finger millet suggested that the fungal isolates causing blast symptoms in different parts of the finger millet plant were genetically similar indicating that the same strains were capable of causing different expressions of blast (Tanaka et al., 2009). Kaye et al. (2003) designed 24 functional primer pairs by screening genomic library, public database, and one proprietary database and identified SSRs in *M. grisea*. The genome sequence data of this pathogen (*M. grisea* in rice) was released in 2005, and this led to the development of more SSRs and construction of genetic map (Dean et al., 2005). Molecular studies on blast pathogen from Japan suggested that *M. grisea* isolates from Japan belonged to a limited number of lineages (Sone et al., 1997; Suzuki et al., 2006, 2009). Similarly, studies on population structure by DNA fingerprinting of *Eleusine* isolates of *Magnaporthe oryzae* collected after an outbreak of finger millet blast in Japan during 1970s suggested that the outbreak was caused by seed transmission of a particular strain of *Eleusine* isolate (Tanaka et al., 2009). Kiran Babu et al. (2013a) studied genetic diversity and population structure of 72 *M. grisea* isolates collected from finger millet (56 isolates), foxtail millet, pearl millet and rice from major crop growing areas in India using 24 SSR markers. None of the SSRs detected polymorphism in the 7 isolates from pearl millet, while 17 SSR markers were polymorphic in the 65 non-pearl millet isolates. Analysis of molecular variance indicated that 52% of the total variation among the isolates was due to differences between the pathogen populations adapted to different hosts, 42% was due to differences in the isolates from the same host, and the remaining 6% due to heterozygosity within isolates.

Management practices like plant spacing and regulating the amount of nitrogenous fertilizer applied to the crop are important measures to minimize the occurrence of blast disease. In the absence of resistant varieties a vast number of chemicals are recommended. Spraying of fungicide like Carbendazim (@ 0.1% a.i.) or Tricyclazole (@ 0.05% a.i.) or combination of Mancozeb 63% + Carbendazim 12% is recommended. However, combination of seed treatment and spraying is known to give better management of the disease.

Cercospora leaf spot. Cercospora leaf spot caused by *Cercospora eleusinis* is one of the important foliar diseases of small millets that occur in the Himalayan foothills and mid-hills of

Nepal and India. The disease is known to occur at an altitude of 850–2000 m in Uttarakhand. It is important on finger millet although restricted to certain geographical regions. It was reported to assume destructive form in Zambia (Muyanga, 1995).

Symptoms are usually observed on the older leaves and then spread to the younger leaves. Initial symptoms appear as reddish-brown specks with a yellow halo. Later several such specks coalesce to form large lesions showing a burnt appearance. The disease is spread through rain splash. A high temperature and high relative humidity favors disease development. During rains the fungus sporulates and produces grayish white growth in the center of the spot and then it looks like brown spot. Infected crop residues or weeds might serve as a source of inoculum for the next season. Clean cultivation practices might reduce disease incidence.

Seedling blight and leaf blight. Seedling blight and leaf blight of finger millet caused by *Drechslera nodulosum* Berk and Curt. (Perfect stage: *Cochliobolus nodulosus*) is an important disease that causes considerable economic damage. The importance of the disease in finger millet is second only to the blast in terms of both severity and distribution. The disease was first noticed by Butler (1918) from different parts of India. Now it is reported from India, Japan, Philippines, countries in Africa (Uganda and other East African countries), and the United States. A severe infection causes adverse effects on grain yield.

The characteristic symptoms on the leaf lamina are the appearance of brown to dark brown spots (the disease is also known as brown spot). Symptoms are seen on leaf sheath, especially in older plants, wherein the woolly growth of the fungus can be seen at the center of the lesion (under high humidity). Often under humid conditions, the infection occurs on the neck and fingers. The neck may break and hang on to the plant. A severe infection causes chaffiness and discoloration of the seed.

The primary infection is caused by the seed-borne pathogen. The fungus remains viable in the stubbles and plant debris. Secondary spread is through air-borne conidia. The spores on the grain are reported to be viable for a year. According to Mc Rae (1932) the optimum temperature for infection is 30–32°C. High humidity caused by intermittent rains during ear emergence and before grain formation cause heavy ear infection and consequent yield reduction. The crop suffering from nutritional and moisture stress becomes vulnerable to attack by the pathogen.

Apart from finger millet, other small millets are also infected by many leaf-infecting fungi causing leaf spot or blotch or blight. Early infection may cause seedling blight. The fungi responsible for leaf spot/blotch/blight on different small millets are *Alternaria tenuissima* (kodo millet), *Cochliobolus setariae* (foxtail millet), *Drechslera nodulosa* (little millet, tef), *Exserohilum monoceras* (barnyard millet), and *Bipolaris panici-miliacei* (proso millet). *Curvularia lunata* is also reported to cause leaf spot on many small millets including finger millet (Shaw, 1921). However, all these are minor diseases and they are rarely responsible for yield loss.

Rust. Rust is a common disease on small millets particularly on finger millet (*Uromyces eragrostidis*), foxtail millet (*Uromyces setariae-italicae*), and also noted on kodo millet (*Puccinia substriata*), little millet (*Uromyces linearis*), and tef (*Uromyces eragrostidis*). The first report of finger millet rust was from Uttar Pradesh in India (Dublish and Singh, 1976). Rust on millet is not a significant disease and has little economic importance. In most cases incidences occur sporadically toward the end of the season causing little damage to the crop. But it is known to be prevalent in all the states in India, wherever this crop is grown. In recent years there are reports of the rust becoming serious and destroying the crop before the ears have formed.

The rust symptoms appear as minute to small, dark brown, broken pustules linearly arranged on the upper surface of the top leaves. The disease is more severe toward the top one-third portion of the upper leaf as compared to the lower and middle leaves. *Uromyces* rust produces small brown uredia that are arranged linearly on both surfaces of the leaves. Symptoms can be seen in all plant parts and infection can occur at any plant growth stage. Uredia produce brown color uredniospores that germinate readily to cause infection. The thick-walled teleutospores are produced later after urediniospores. *U. setariae-italicae* produce light yellow color single celled teleutospores whereas *U. linearis* produces black color teleutospores. *Puccinia substriata* develop small brown, oval and raised spots (uredia) on upper surfaces of leaf and leaf sheath. Uredia produce urediniospores that germinate readily in water drops and cause infection. Telia are usually produced on the lower surface of the leaf and leaf sheath. Red-brown pustules may be surrounded by a yellow halo. The pathogen can survive on grasses. Brinjal is known to act as an alternate host in which aecia is produced (de Carvalho et al., 2006). Tef rust is important in Ethiopia, widely distributed across the country and causes grain yield losses of 10−41% annually (Dawit and Andnew, 2005). Little information is available on systemic research on this disease on small millets. Varietal resistance needs to be worked out for its management.

3.3.1.1.2 Panicle diseases

Green ear or crazy top downy mildew. The green ear disease or crazy top downy mildew of finger millet is caused by *Sclerophthora macrospora* (Sacc.) Thirum., Shaw and Naras. The disease was first reported from Mysore, India, by Venkatarayan (1947). It could be highly destructive leading to a total crop failure owing to malformation of the affected ears. Green ear disease of finger millet has been reported from Japan, China, Russia, Manchuria and the southeastern countries of Europe, America and India (Ramakrishnan, 1971). It causes up to 50% yield loss in certain years. Takasugi and Akaisahi (1933) reported that the losses may go up to 20% in Manchuria.

The white cottony growth, characteristic of many downy mildews, is generally not seen in finger millet. The green ear manifests itself at the time of grain formation and completely converts the heads into green narrow leafy structures causing complete sterility. Partial or whole ears including lemma, palea and glumes are converted into narrow leafy structures. The proliferation takes place first in the basal spikelet and afterward others get affected. Finally, the whole ear gives a bush-like appearance displaying the typical "green ear" symptom. The pathogen has a wide host range including *Eleusine indica*, maize, wheat, oat, *Eragrostis pectinacea* and *Digitaria marginata*. Physiologic specialization and races of the pathogen are reported. The downy mildew or green ear disease in foxtail millet is caused by the fungus *Sclerospora graminicola*. Downy growth of the pathogen, shredding of infected leaves, and malformation of floral organs is common in foxtail millet. The disease is favored by heavy dew and low temperature during the period of crop development in rainy season. A temperature range of 20−25°C that occurs during night time and early morning favors spore germination. Seed treatment with fungicides is advisable for management of the disease.

Smut. All small millets are affected by various types of smuts in almost all growing regions. Grain smut is common on finger millet, foxtail, little and barnyard millet while head smuts are common on kodo, barnyard and proso millet, and tef. In general smuts are minor diseases but may cause significant damage especially on high yielding varieties. Most of the research works were undertaken a long time ago and recent literature is limited.

Grain smut of finger millet (*Melanopsichium eleusinis*) (Kulk) Mundk and Thirum., is gaining importance after the introduction of high yielding varieties. Air-borne spores cause infection of the flower. Smut sori develop in the grains, main rachis, or in the peduncle. The smutted grains can be seen generally a few days after flowering and the affected grains can be seen scattered at random in the ear. The affected ovaries are transformed into velvety greenish gall-like bodies which are several times bigger in size than the normal healthy grain. The greenish outer tunica of the sorus gradually turns pinkish green and finally to dirty black on drying. Although it was known to be a disease of summer crops, in the recent past it has occurred sparingly during the rainy season also. Smut disease in foxtail millet is caused by *Ustilago crameri* Korn. and occurs mainly in the southern states of India. The fungus is externally seed-borne and affects most of the grains in an ear. A certain amount of soil-borne infection has also been observed. Similar grain smut caused by *Mecalpinomyces sharmae* K. Vanky has also been reported in little millet and is severe in early maturing varieties.

Head smuts are common in kodo millet, barnyard millet, proso millet and tef. In kodo millet head smut (*Sorosporium paspali-thunbergii*) assumes significance as an important disease with considerable yield loss in Madhya Pradesh, Andhra Pradesh, Bihar, Tamil Nadu and Karnataka. The entire panicle is transformed into a long sorus. In some cases, it is enclosed in the flag leaf and may not emerge fully. The enclosing membrane bursts and exposes the black mass of spores. The disease is mainly seed-borne. The spores adhere to the surface of the grains and infect the seedlings. Two types of smuts are reported on barnyard millet, head smut (*Ustilago crus-galli* Tracy & Earle) and grain smut (*Ustilago panici-frumentacei* Brefeld). Head smut is recently reported from Uttarakhand in India (Kumar et al., 2008). Both of the diseases are seed-borne and can be managed through seed treatments.

Practice of clean cultivation and seed treatment with fungicides effectively controls the disease. Cultural practices are followed commonly and fungicidal seed treatment are used occasionally. Systematic breeding for the development of smut resistance variety has not been common.

Udbatta. Udbatta is a panicle disease commonly observed in India on foxtail millet, kodo millet and little millet. It is a minor disease and causes sporadic incidence. Infected plants are usually stunted and occasionally the white mycelia and conidia form narrow stripes on the flag leaves along the veins prior to panicle emergence. In this disease the affected panicle is transformed into a compact, silver colored, cylindrical spike that looks like an incense stick much resembling an *agarbatti* or udbatta and hence the name. An infected panicle fails to produce the normal grain and the panicle becomes sterile. The disease is caused by the fungus *Ephelis oryzae* Syd, Butler and Bisby and is prevalent in most of the small millet growing areas in India. It is a seed-borne disease and the fungus can also infect grasses like *Cynodon dactylon*, *Pennisetum* sp., and *Eragrostis tenella.*

3.3.1.1.3 Root and stalk diseases

Banded sheath blight. Banded sheath blight caused by *Rhizoctonia solani* Kuhn. (Basidial Stage: *Thanatephorus cucumeris* (Fr.) Donk) is a common disease among all the small millets. The disease was first recorded in finger millet in Kerala, India, during 1989. Since then it has been on the rise and is now an emerging problem on all other small millets.

The disease is characterized by oval to irregular light gray to dark brown lesions on the lower leaf and leaf sheath. The central portions of the lesions subsequently turn white with narrow reddish-brown margins. Later the spots get distributed irregularly on leaf lamina. Under favorable conditions, lesions enlarge rapidly and coalesce to cover large portions of the sheath and leaf lamina. At this stage the disease symptom is characterized by a series of copper or brown color bands across the leaves giving a very characteristic banded appearance. In severe cases, symptoms appear on peduncles, fingers and glumes as irregular to oval, dark brown to purplish to brown necrotic lesions. The mycelial growth along with sclerotia can be observed on and around the lesions. Seed treatment with propiconazole (@ 1 mL kg^{-1} seed) was found to reduce disease incidence in finger millet and increase yield (Patro and Madhuri, 2014).

Foot rot of finger millet. The foot rot of finger millet caused by *Sclerotium rolfsii* (Sacc.) Curzi. (Perfect Stage: *Pellicularia rolfsii*) is a sporadic disease that appears randomly. In India it is mostly observed in Maharashtra, Tamil Nadu, Karnataka, Odisha, and Gujarat. The infection occurs in and around the collar region and the infected area remains restricted to 2−3 inches above ground level. The basal portion of affected plant appears water-soaked and later on turns brown and subsequently dark brown with a concomitant shrinking of the stem in the affected region. Profuse white cottony mycelial growth occurs in this area with small roundish white velvety mustard seed like sclerotia bodies. Finally the leaves lose their luster, droop and dry and the plant dries up prematurely. *Sclerotium rolfsii* can infect more than 500 plant species, which increases the chance of heavy sclerotia build up in the soil. Sandy loam soils favor disease incidence and the pathogen survives better at low soil moisture levels. The disease incidence is greater in warm humid conditions.

Sheath rot of kodo millet. The disease is typically observed when kodo millet is grown after rainy season rice in India. It was first observed in Cuddalore district in Tamil Nadu. The disease is characterized typically by large brown discolored patches on the stem sheath. The fungus *Sarocladium oryzae* that infects rice also infects kodo millet. Other details about the disease are yet to be understood.

Kodo millet poisoning. Kodo millet grains produced during the rainy season when consumed sometimes cause giddiness, vomiting, unconsciousness, difficulty in swallowing, and rarely death of humans and cattle. The malady is supposed to be caused by a number of fungal species. *Phomopsis paspali* was isolated from kodo heads in the western Ghats of Maharashtra. *Aspergillus flavus* and *Aspergillus tamari* were also found associated with kodo grains. The affected plant parts, such as leaf, stem, ears and grains, show a whitish fungal growth during the reproductive stage of the crop. Initially, small patches of fungus are seen, which grow fast in a humid environment or just after rains.

3.3.1.2 Bacterial diseases

A small number of bacterial diseases are present on small millets grown in India and Africa. These diseases are of little significance and occur sporadically causing minor incidence. However, occasionally they may be potentially threatening under warm and humid conditions. In general, bacterial plant pathogens survive on plant residues in soil or seed and spread through irrigation water or rain splash. Seed treatment with antibiotics, crop rotation, and need-based spray can manage these diseases. A few diseases have been described in brief.

3.3.1.2.1 Bacterial leaf spot

The disease is caused by *Xanthomonas eleusinae* Rangaswami, Prasad, Eswaran. Linear spots are seen on both upper and lower surfaces of the leaf blade spreading along the veins. The spots measure 2 − 4 mm long, but often extend up to 25 mm or more. In the beginning, spots are light yellowish brown, but soon become dark brown. At the advanced stage, the leaf splits along the streak giving a shredded appearance. All the leaves, including the tender shoots, in a plant are affected. The bacterium mainly affects the leaves, but at times characteristic streaks may be found on the peduncle.

3.3.1.2.2 Bacterial blight

Bacterial leaf blight of finger millet is caused by *Xanthomonas axonopodis* pv. *coracanae*. It was first reported from Gujarat, India, by Desai et al. (1965). Later it was reported from Uganda, Africa (Adipala, 1980). Initial symptoms appear as water soaked, translucent, linear, pale yellow to dark greenish-brown streaks running parallel to the midrib of the lamina. The hyaline streak later develops into a broad yellowish lesion measuring 3−4 cm and turns brown. When the infection is heavy, particularly in the early stages, the entire leaf turns brown and withers away, the plants become yellow and show premature wilting. The plants are susceptible to infection at all stages of growth.

Another leaf blight of finger millet has been reported from Uganda (Mudingotto et al., 2002). They observed that seedlings of finger millet showed severe blight and necrotic stripe symptoms under controlled condition. The disease is caused by *Acidovorax avenae* ssp. *avenae*.

3.3.1.2.3 Bacterial leaf streak

Bacterial leaf streak of finger millet is caused by *Pseudomonas eleusinae*. The disease appeared in serious proportions in Karnataka, India (Billimoria and Hegde, 1971).

The common symptom of the disease is brown coloration of the leaf sheath especially from base upward. The affected portion of the lamina invariably involves the midrib and appears straw colored. This symptom spreads to about three-fourths of the lamina and then abruptly stops or in some cases reaches the leaf tip. Occasionally the strips of the infected areas are seen to proceed along the margin of the lamina, leaving the central portion, including the midrib healthy. The bacteria are readily detected in the phloem vessels. Infected plants can be recognized from a distance by their characteristic drooping of the leaves. Infected culms show a light brown discoloration along one side. In some cases, the discoloration begins from the base, but in most instances, it begins 5 − 7 cm above the base and extends to the leaf sheath proper. There is, however, no apparent reduction in girth or turgidity of the affected culms as compared to the healthy ones. Plants less than a month old are usually free from the disease. The bacterium is systemic and soil-borne.

3.3.1.3 Viral diseases

The small millets are infected by many viruses and the diseases are mostly expressed in sporadic form in favorable weather conditions. They are not of regular occurrence and are not economically significant. However, in certain environmental conditions that favor build-up of a virus-vector and disease development, they may be important.

3.3.1.3.1 Ragi severe mosaic

The finger millet crop in southern Karnataka and the border districts of Andhra Pradesh was affected by a severe mosaic during the rainy season of 1966. In certain pockets in Karnataka it was so severe that the farmers abandoned their crop, as heavily diseased plants failed to set seed (Joshi et al., 1966). The virus induces mosaic symptoms, which are more clear and pronounced on young leaves. Infected plants remain stunted and the ears of severely affected plants are malformed. Such plants produce few seeds of smaller size, which reduces the yield considerably. In addition, the affected plants appear pale yellow due to severe chlorosis and in severe cases become brownish-white. Thus, the entire field appears yellow and can be readily distinguished from noninfected stands from a distance. Stunted plants do not recover, develop roots at nodes, generally do not produce ears and if produced, are mostly sterile. Finger millet plants of all ages are susceptible, but the severity of infection decreases significantly with an increase in the age of the host.

The disease is caused by *Sugarcane mosaic virus*. Particles were flexuous rods with an average length of 667 ± 8 nm and an approximate diameter of 12–14 nm. The virus, thus, was identified as a strain of sugarcane mosaic virus (Subbayya and Raychaudhuri, 1970). The virus is neither seed-borne nor soil-borne. Continuous cropping of finger millet, coupled with abnormal weather factors, favors vector population and epidemic development (Keshavamurthy and Yaraguntaiah, 1977).

Paul Khurana et al. (1973) studied the virus-vector relationship employing *Longiunguis sacchari* as vector. The optimum acquisition feeding period was 5 min and optimum transmission feeding period was found to be 1 h. The aphid acquires the virus in 1 min. A preacquisition fasting increases the efficiency of the vector. The maximum transmission is obtained within 1½ to 2 h of fasting. Even a single aphid transmits the virus with an optimum of 10 aphids per plant. Post-acquisition fasting decreases the efficiency of the vector and the virus is found to be nonpersistent in *Rhopalosiphum maidis* since it was retained only for an hour after acquisition. The incubation period of the virus is found to be influenced by temperature but not by the age of the host.

3.3.1.3.2 Ragi mottle streak

Ragi mottle streak virus was first reported from Karnataka, India (Mariappan et al., 1973), and yield loss due to infection of this virus ranged from 50% to 100% in certain areas (Maramorosch et al., 1977). Symptoms are exhibited as regular dark-green areas all along the leaf veins when the plants are 4–6 weeks old. Other symptoms on the leaf include chlorosis and streak. In some cases occasional yellowing to almost albino symptoms are also observed. However, in the lower leaves, the symptoms are of mottle type in the form of white specks and the affected plants are generally stunted bearing small ears.

Ragi mottle streak virus has short rod-like, bacilliform particles which are abundant, measuring 80 nm in cross-section and 285 nm lengthwise. They were enveloped, bacilliform, and spiked corresponding to the morphology of rhabdoviruses. The virus is transmitted by two species of jassids, that is, *Cicadulina bipunctella* and *Cicadulina chinai*. *Cicadulina bipunctella* was able to transmit up to 82%. The minimum acquisition feeding period was 48 h and minimum inoculation feeding period was 24 h. The virus can persist in the insect for 8 days. Only a section of the population of the vector *C. bipunctella* transmitted the virus at a high percentage and the virus is carried in the leafhopper in a persistent manner.

3.3.1.3.3 Ragi streak

During the year 1974–75, a virus disease producing streaking and yellowing of leaves and stunting of finger millet plants in the fields around Bangalore was observed and the virus was found not to be transmitted through either mechanical sap inoculation or aphid species tested (Anon., 1975). The loss in grain yield depended on the age at which the virus infects the crop. Similarly, depending upon the stage of plant, the virus infection resulted in a drastic reduction in 1000-grain weight. The loss in 1000-grain weight was 84%, 63%, 27%, and 24% when the infection occurred at 30-, 40-, 50-, and 60-day-old seedlings, respectively. However, there was no significant change in number of tillers except where infection occurred on 10-day-old seedlings when there was a significant increase in seedling number (Nagaraju et al., 1982).

Symptoms appear on unfolding young leaves as pale specks or stripes of different size. The specks coalesce into larger areas resulting in chlorotic bands running almost the entire length of the leaf parallel to the midrib. These bands are occasionally interrupted by dark-green areas. The new emerging leaves of both the main shoot and the tiller show a number of well defined chlorotic streaks having almost uniform width running parallel to the midrib throughout the length of the leaf lamina. The infected plants in the field produce comparatively higher numbers of tillers and bear yellowish sickly ears, often bearing a few shriveled seeds. The plants infected very early in the crop growth stage die before they bloom.

The disease is caused by the eleusine strain of maize streak virus. Finger millet plants affected with streak, yellowing and stunting symptoms were noticed in a wider geographical area in the districts of Chitradurga, Mandya, Bangalore, Tumkur and Hassan in the subsequent surveys during 1977−78 and 1978−79. The incidence ranged from 5−45%. Subsequent studies revealed this virus to be different from all those reported earlier on finger millet and was found to be a strain of maize streak virus. Leaf hopper *Cicadulina chinai* transmitted the disease.

3.3.2 HOST-PLANT RESISTANCE

Since finger millet blast is the major disease affecting production and productivity of the crop, extensive studies on this topic have been made mostly on this disease, while systematic studies on other diseases are lacking.

3.3.2.1 Nature of resistance

The blast fungus *Pyricularia* can invade the host either through stomatal opening or by direct entry by piercing the epidermal cells. Substantial differences in the early events of infection were noticed among varieties (Madhukeshwara, 1990). The resistant varieties exhibit higher cytoplasmic granulation compared to susceptible varieties (Madhukeshwara et al., 1997). Fungus was present in the pericarp and endosperm but not in the embryo (Pande et al., 1994). In anatomical studies of infected and healthy plants, epidermis-cum-cuticle thickness was found to be significantly higher in leaves of highly resistant cultivars as compared to highly susceptible cultivars. However, cultivars with high resistance had significantly less stomatal frequencies per square mm and size in comparison to highly susceptible cultivars (Sanathkumar et al., 2002b).

3.3.2.2 Screening for resistance

An efficient screening technique is important for evaluation of germplasm and their use in breeding program. Though such technique is well established in other crops, little work has been done on small millet diseases. For identification of resistance against blast some works have been done in finger millet especially to identify blast resistance under natural infection (Mantur et al., 2001; Kumar et al., 2006; Kumar and Kumar, 2009; Nagaraja et al., 2010b). Recently field-screening technique involving growing of susceptible genotype at regular interval and artificial inoculation of plants at flowering has been developed. It was claimed to have developed a field-screening technique for neck and finger blast whereby finger millet germplasm can be screened in the field conditions. For leaf blast, a disease rating scale of 0−5 has been used, where a score of 0, immune or highly resistant; 0.1−2.0, highly resistant; 2.1−10, resistant; 10.1−25, moderately resistant; 25.1−50, susceptible; and >50, highly susceptible. Severity of neck and finger blast have been measured in percentage using formula for Neck blast (%) = (Number of ears showing infection on peduncle or neck/Total number of ears in all the plants in two rows) × 100, and for Finger blast (%) = (Number of fingers infected in randomly selected five plants/total number of fingers) × 100 (Nagaraja and Mantur, 2007; Kumar and Kumar, 2009).

3.3.2.3 Resistant sources and utilization

Among the small millets diseases most of the researches on resistant identification are concentrated on blast of finger millet, while in other diseases reports are scanty. Extensive researches have been invested for the identification of blast resistance sources. Almost all cultivars, both local and improved, presently under cultivation are susceptible to blast disease though the level of susceptibility reaction varies from cultivar to cultivar. Most of the local land races of finger millet in Nepal and India are susceptible to blast (Sherchan, 1989), while the African races are resistant. The repository of germplasm has been screened for the identification of resistance sources against blast of finger millet. Several reports of blast resistance in finger millet are available. The resistant sources available in various types of blast are given in Table 3.4. Most of the resistant sources have been identified under natural field or natural epiphytotic conditions, which sometimes may lead to the misidentification of true resistance. In spite of that, the variety GPU 28 developed by utilizing resistant sources at Project Coordinating Unit for Small Millets Improvement, Bengaluru, during the late 1990s remained highly resistant to neck and finger blast with only <2 incidence and occupied a vast area of almost 75% under finger millet in Karnataka (Nagaraja et al., 2008a). Disease reaction of a particular genotype to foliar blast may be different to that of neck blast (Madhukeshwara et al., 2004). However, there is significant positive relationship among these blasts in terms of disease rating (Kiran Babu et al., 2013b). Nagaraja et al. (2008b) found that blast incidence was low on white seeded entries in comparison with brown seeded ones.

SSR markers have been used extensively for identification of blast resistant genes in rice and a few reports are available in finger millet. Working on rice blast Selvaraj et al. (2011) identified 23 SSR markers putatively associated for the 6 traits studied. Among them, 3 SSR markers (RM 5757, RM 451, and RM 492 on chromosome 4 and 2) were linked to leaf blast resistance. Kalyan Babu et al. (2014a) used 58 genic SSRs for genetic diversity analysis for blast resistance in 190 genotypes in the world collection of finger millet. Three SSR markers (RM 23842, RM 5963, and RM 262) were found to be highly polymorphic loci for differentiating the global collections. Association mapping approach for blast resistance using 104 SSR markers (Kalyan Babu et al., 2014b) identified 3 QTLs for finger blast (RM262, FMBLEST32, and UGEP81 that explained phenotypic variance 7.5–10%). The QTLs for neck blast was associated with the genomic SSR marker UGEP18 and explained 11% of phenotypic variance. Forty finger millet genotypes were collected to study the genetic diversity using ten SSR and thirty Inter-simple sequence repeats (ISSR) markers. The genotypes were grouped into thirteen (SSR) and five clusters (ISSR), respectively, in Unweighted pairgroup method with arithmetic average (UPGMA) analysis (Prabhu and Ganesan, 2013).

A systematic resistance breeding program is lacking in the small millets. Only a few reports are available in literature. Three genotypes of barnyard millet, namely PRB 402, TNAU 92 and VL 216, showed complete resistance against grain and head smut as well as brown spot diseases, while seven genotypes of foxtail millet, that is, GPUS 27, SiA 3039, SiA 3059, SiA 3066, SiA 3088, TNAU 213 and TNAU 235, were found to be completely free from brown spot (Kumar and Kumar, 2009). Dawit and Andnew (2005) evaluated 2000 landrace accessions and 5000 mutant

Table 3.4 Different Types of Blast Resistant Sources Reported in Finger Millet

Type of Blast	Screening Condition	Level of Resistance	Name of the Resistant Source	Reference
Leaf blast		Resistant	GE 2400, 4913, 4914, 4915, 4929, 4966, 5102, 5126, 5148	Mantur et al. (2001)
	Field condition	Resistant	MR 1, GPU 56, GPU 53, GPU 58, VR 222, GPU 52, VL 317, VL 321, GPU 49, GPU 51	Ramappa et al. (2002)
		Partial resistance	Germplasm GE 632, 637, 659, 665, 669, 674, 676, 682, 696, 704, 705, 710, 728, 730	Sunil (2002)
		Low blast severity	ICRISAT lines KNE 620, 629, 688, 814, 1149; Farmers' variety 14, 29, 32, 44	Takan et al. (2004)
	Natural epiphytotic	Resistant	KFM 41, 120, 150, 174, 177, 181, 228, 230, 243, 253 (mid late group); KFM 183, 246 (late group)	Khot et al. (2008)
Neck blast		Resistant	MLC 29-5, 54-4, 63-4-1, 89-4	Nagaraja et al. (2010a,b)
		Resistant	18IE, GPU 28, GPU 28, VL 149, VL 253	Mantur et al. (2002)
	Natural epiphytotic	Resistant	Out of 2950 genotypes 630 were resistant	Madhukeshwara et al. (2004)
Finger blast	Field condition	Immune	GE 253, 357, 393 (long duration)	Kumar et al. (2006)
	Natural epiphytotic	Immune	GE 325	Kumar et al. (2007)
	Natural epiphytotic	Resistant	Out of 2950 genotypes 84 were resistant	Madhukeshwara et al. (2004)
Neck & finger blast	Natural epiphytotic	Moderately resistant	GE 326, GE 332	Kumar et al. (2007)
	Natural epiphytotic	Moderately resistant	GPU 26, GPU 28, AKE 1033, VL 149, MR 2	Rajanna et al. (2000)
	Natural epiphytotic	Resistant	IE 287, IE 976, IC 4335	Madhukeshwara et al. (2004)
	Field condition	Resistant	GE 326, GE 332	Kumar et al. (2007)
		Stable resistance	Germplasm GE 5183, 5203, 5205, 5209, 5212, 5215, 5218, 5227, 5230	Nagaraja and Mantur (2007)
	Natural epiphytotic	Resistant	WRC 1-12, GPUW 1, GE 4971, GE 5153	Nagaraja et al. (2008b)
	Natural epiphytotic	Resistant	VL 234, SANJI 1, PRM 9802, VL 328, VL 333, ED 201-5A, ICM 401, VR 708, VL 324	Kumar and Kumar (2009)
	Field condition	Highly resistant	Mini-core accessions VHC 3997, VHC 3996, VHC 3930 from north-western Himalayan region	Kiran Babu et al. (2013b)
		Incidence < check VR 708	Elite germplasm PCGF 18, PCGF 42, GPU 28, PCGF 26, PCGF 17, PCGF 13, PCGF 46	Patro et al. (2015)

lines of tef for resistance to rust and found that the majority of the lines were susceptible to rust and none was immune. Twenty-two landrace accessions, however, had low level of rust severity.

Finger millet is predominantly grown by poor and marginal farmers who have little means of controlling diseases. Development of varieties with inbuilt genetic resistance is the best means of disease control. A number of stabilized selections from a blast resistance program were tested for yield in different states (Gowda et al., 1986). After the high yielding Indaf varieties and others were found susceptible to neck and finger blast, the search for resistant varieties began in the early 1990s and as a result resistant varieties like GPU 28, VL 149, GPU 48, and L 5 were released in India. These varieties, however, had juvenile susceptibility to leaf blast (Nagaraja et al, 2012). GKVK center, Bangalore, utilized IE 1012 as the resistant parent and PES 176, HPB 7-6, PR 202, and Indaf 8 as high yielding adopted parents. Gautami, GPU 28, BM 9-1, GPU 26, GPU 45, Chilika, VL 315, and GPU 48 are some of the blast tolerant/resistant varieties which are presently cultivated in India. Blast fungi is a highly variable pathogen. Therefore, durable resistance is more desired for increasing the shelf life of a resistant cultivar. Other than conventional breeding, hardly any approach has been practiced for development of varieties resistant to blast of finger millet. Recently Ignacimuthu and Ceasar (2012) developed a leaf blast resistant finger millet transgenic using rice chitinase (*chi*11) gene.

3.3.2.4 Genetics of resistance

The line $\times$ tester analysis suggested the operation of an additive gene action for finger blast. The five generation mean analysis revealed nonadditive gene action for neck and finger blast. For neck blast both additive and dominance gene effects were significant with a predominance of dominance gene effects. Among epistatic gene effects additive $\times$ additive gene effects were important. All the three types, that is, additive, dominance, additive $\times$ additive and dominance $\times$ dominance, were found to be important for finger blast. In the cross GE 447 $\times$ GE 156 only additive and additive $\times$ additive gene effects were significant suggesting the possibility of early generation selection (Seetharam and Ravikumar, 1993). The estimates of narrow sense heritability to neck blast were low and medium for finger blast suggesting additive gene effects are not important in the expression of these characters. In further studies, Ravikumar and Seetharam (1994) observed moderate to high genetic variability for neck and finger blast in populations. The characters for neck and finger blast showed high heritability and high genetic advance, indicating the role of additive gene action. The phenotypic and genotypic coefficient of variation was large for finger and neck blast. Both additive and nonadditive gene effects played significant roles in the expression of blast disease. This disease manifested a high genetic advance suggesting the scope for improvement of this character with simple selection (Byregowda et al., 1997).

3.3.2.5 Mechanisms of resistance

3.3.2.5.1 Physical basis

Somappa (1999) studied the histology of resistant, moderately resistant, and susceptible varieties of healthy and diseased finger millet and found no significant differences among the varieties with regard to either the thickness of the leaf or that of the upper epidermis. However, there was a significant reduction in the length of germ tube, appressoria development, size of the lesion and spores per lesion in resistant genotypes compared to susceptible genotypes (Sanathkumar et al., 2002a).

3.3.2.5.2 Biochemical basis

Ever since Walker (1924) demonstrated the role of phenols in imparting resistance to *Colletotrichum circinans*, the cause of onion smudge, scientists all over the world started looking for similar phenomena in various host—pathogen combinations. It is imperative that plant species continuously fight against all odds, including biotic stresses, for their survival. In the course of such a fight and evolvements they get equipped with various defense mechanisms against invading pathogens. Several workers have investigated the basis of blast resistance in finger millet. Purushottaman and Marimuthu (1974) detected phytoalexins in infected finger millet plants. However, they did not identify them. Maheshkumar (2002) working with phytoalexin production in finger millet plants found four different phytoalexins but could identify only the most effective one, oryzaxelin. The infection resulted in an increase in protein content of the seed and a decrease in starch and ash. In all the cultivars, β-glucosidase activity was higher and glucose content was lower in the diseased part of the neck (Pall, 1992) and resulted in an imbalance of total carbohydrates (Pall, 1994). Byregowda et al. (1999) observed a significantly negative relationship between protein content and blast resistance. In contrast high phenol and tannin contents in the plant lowered disease incidence.

Ravikumar et al. (1991) observed a strong association of seed color with seed protein, phenol, and tannin with reaction to blast in finger millet. The brown colored seeds were more resistant than the white colored seeds. The total phenol and tannin content was higher in resistant lines. Total protein content in dry seeds was higher in susceptible lines. The occurrence of blast was not strongly influenced by the sugar content and peroxidase activity of the plant. The peroxidase isoenzyme description pattern was the same in resistant and susceptible genotypes. The amount of preformed total protein, total phenol and tannin content could serve as an indirect index for blast reaction in the field (Seetharam and Ravikumar, 1993; Ravikumar et al., 1995). Ramachandra et al. (1981) observed differences in the cathodic and anodic polyacrylamide disc gel electrophoresis pattern of isoperoxidase in seed of resistant and susceptible varieties and also showed a marked change during the cause of germination. According to Somappa (1999) the total phenol and tannin content was high in the resistant variety (GPU 28) whereas reducing sugar content was higher in dry seeds, 75-day-old seedlings, and the dough stage in the susceptible variety (K7). The infected susceptible genotype showed higher esterase activity compared with the resistant genotypes. Isoenzyme 12 was found in all the developmental stages. On infection, isoenzyme was suppressed in the susceptible genotype but isoenzymes 3 and 4 appeared in the resistant genotype. Based on the inhibitor studies, isoenzyme 12 can be classified as a general esterase and isoenzymes 3 and 4 as carboxyl and oryle esterases, respectively (Muralidharan et al., 1996). Ramaswamy (1995) recorded a reduction in the activity of proxidases, polyphenol oxidases and PAL and total phenol content in the blast infected finger millet plants. However, these enzymes were increased significantly when treated with plant extracts with the maximum occurring with *Prosophis juliflora* extract treatment.

3.4 CONCLUSIONS

Pearl millet diseases have drawn more research attention worldwide compared to the diseases of small millets. Downy mildew of pearl millet and blast of finger millet and pearl millet are the most important disease of millets, which have elicited world attention for a long time due to their devastating effect on these crops. Research information on them is available on all major aspects, which has contributed to their effective management. Considerable progress has been made in understanding host—pathogen interactions, refining disease screening methods, identifying and utilizing resistant sources, and breeding downy mildew resistant parental lines and hybrids. Other diseases, especially rust, smuts, and ergot, are also damaging but not as important as the former two. Exploration, identification, and utilization of host resistance against them have successfully checked their incidence and severity and increased yield. Apart from these, there are many other important diseases in millets, particularly small millets, including fungal and bacterial leaf spots, blights root and stalk diseases, and viral diseases, which have obviously received less focus because of their minor appearance so far. Basic information on the biology, epidemiology, and management strategies are lacking for the minor diseases, and efforts to identify resistance sources have not yet been systematized.

3.5 FUTURE PRIORITIES

Research priorities in pearl millet downy mildew include continuous monitoring of virulence, characterization of the pathogen populations, and continuous identification and utilization of new sources of resistance. A saturated genetic linkage map has been developed for pearl millet that might assist breeders to identify markers in selecting individual plants carrying specific resistance alleles of interest.

For finger millet blast, the development of varieties with durable resistance by utilizing the broad spectrum of resistance genes and pyramiding genes and QTL, identification of blast resistance in African germplasm and incorporation of the resistance to Indian cultivars through appropriate resistance breeding program, racial differentiation, and virulence mapping in blast fungi are a few research priorities. Information is lacking in the field of epidemiology, disease forecasting and greenhouse screening techniques. Studies in these areas should be strengthened for better understanding of the disease, identification of resistance, and efficient management of the blast disease in finger millet.

There is a lack of basic information on biology, epidemiology, and management including host-plant resistance on the minor diseases of small millets. Meager research has been done for understanding host—pathogen interaction, genetics of resistance and application of molecular tools in these crops. Any of the diseases may become severe under favorable conditions of disease development. There should be a minimum preparedness so that any sudden outbreak of any such disease can be managed. Efforts to locate resistance sources for diseases like sheath blight in all these crops have yielded very poor results, thus necessitating the investigation of suitable chemical alternatives for seed treatment of moderately resistant varieties. Furthermore, in the management of soil-borne diseases like foot rot and sheath rot, biocontrol agents, especially *Trichoderma* and *Pseudomonas* strains, have been useful and should be commercialized.

REFERENCES

Adipala, E., 1980. Diseases of Finger Millet (*Eleusine coracana* (L.) Gaertn) in Uganda (M.Sc thesis). Makerere University. 186 pp.

AICPMIP, 2013. Proceedings of the 48th Annual Pearl Millet Workshop All India Coordinated Pearl Millet Improvement Project, Junagadh Agricultural University, Junagadh, AICPMIP Mandor, Jodhpur, India, p. 51.

AICPMIP, 2014. Progress Reports 2011—14. All India Coordinated Pearl Millets Improvement Project, Indian Council of Agricultural Research, Mandore, Rajasthan, India.

Anaso, C.E., Anaso, A.B., 2010. Cost-effectiveness of seed dressing with a new formulation of metalaxyl (Apron Star 42 WS) for sustainable pearl millet production in northern Nigeria. Arch. Phytopath. Plant Prot. 43, 154—159.

Andrews, D.J., Rai, K.N., Singh, S.D., 1985. A single dominant gene for rust resistance in pearl millet. Crop Sci. 25, 565—566.

Angarawai, I.I., Kadams, A.M., Bello, D., Mohammed, S.G., 2009. Interaction of gene effect and number of loci on heritability and heterosis for downy mildew resistance in Nigerian elite pearl millet lines. American-Eurasian J. Agric. Environ. Sci. 5 (1), 106—114.

Anjum, S.,S., 2015. Variability in *Pyricularia grisea* Causing Blast of Finger Millet [*Eleusine coracana* (L.) Gaertn] (Ph. D. thesis). University of Agricultural Sciences, Bengaluru, p. 98.

Anonymous, 1975. Annual report of the virologist. In: AICRP on Small Millets 1974—75. University of Agricultural Sciences, Bangalore.

Bekele, E., 1985. A review of research on diseases of barley, tef, and wheat in Ethiopia. In: Abate, T. (Ed.), A Review of Crop Protection Research in Ethiopia. Proc. First Ethiopian Crop Prod. Symp. Dept. Crop Protection, Inst. Agr. Res., Addis Ababa, Ethiopia, pp. 79—108.

Bhide, V.P., Hegde, R.K., 1957. Ergot on bajra [*Pennisetum typhoides* (Burm) Stapf. & Hubbard] in Bombay State. Curr. Sci. 26, 116.

Billimoria, K.N., Hegde, R.K., 1971. A new bacterial disease of ragi, *Eleusine coracana* (Linn.) Gaertn in Mysore state. Curr. Sci. 40, 611—612.

Butler, E.J., 1918. Fungi and Diseases in Plants. Thacker Spinck and Co., Calcutta, p. 547.

Byregowda, M., Shankaregowda, B.T., Seetharam, A., 1997. Association of biochemical compounds with blast disease in finger millet. Extended Summary, National Seminar on Small Millets. ICAR and TNAU, Coimbatore, p. 54.

Byregowda, M., Seetharam, A., Shankaregowda, B.T., 1999. Selection for combining grain yield with high protein and blast resistance in finger millet (*Eleusine coracana*). Indian J. Genet. 59, 345—349.

Celia Chalam, V., 1996. Studies on the Mechanisms of Resistance in Pearl Millet Genotypes to Downy Mildew Disease (Ph.D. thesis). Acharya NG Ranga Agricultural University, Hyderabad, p. 132.

Chahal, S.S., Gill, K.S., Phul, P.S., Singh, N.B., 1981. Effectiveness of recurrent selection for generating ergot resistance in pearl millet. SABRAO J. 13, 184—186.

Chavan, S.B., Thakur, R.P., Rao, K.P., 1988. Inheritance of smut resistance in pearl millet. Plant Dis. Res. 3, 192—197.

Crampton, B.G., Hein, I., Berger, D.K., 2009. Salicylic acid confers resistance to a biotrophic rust pathogen, *Puccinia substriata*, in pearl millet (*Pennisetum glaucum*). Mol. Plant Pathol. 10 (2), 291—304.

Dawit, W., Andnew, Y., 2005. The study of fungicides application and sowing date, resistance, and maturity of *Eragrostis tef* for the management of teff rust (*Uromyces eragrostidis*). Can. J. Plant Pathol. 27 (4), 521—527.

Dean, R.A., Talbot, N.J., Ebbole, D.J., Farman, M.L., Mitchell, T.K., Orbach, M.J., et al., 2005. The genome sequence of the rice blast fungus *Magnaporthe grisea*. Nature 434, 980—986.

de Carvalho, A., Soares, D., Carmo, M., Costa, A., Pimentel, C., 2006. Description of the lifecycle of the pearl millet rust fungus—*Puccinia substriata* var. *penicillariae* with a proposal of reducing var. *indica* to a synonym. Mycopathologia 161, 331—336.

Deepak, S., Shailasree, S., Sujeeth, N., Kini, R.K., Shetty, H.S., Mithöfer, A., 2007. Purification and characterization of proline/hydroxyproline-rich glycoprotein from pearl millet coleoptiles infected with downy mildew pathogen *Sclerospora graminicola*. Phytochemistry 68, 298—305.

Deepak, S., Manjunath, G., Manjula, S., Niranjan-Raj, S., Geetha, N.P., Shetty, H.S., 2008. Involvement of silicon in pearl millet resistance to downy mildew disease and its interplay with cell wall proline/hydroxyproline-rich glycoproteins. Aus. Plant Path. 37, 498—504.

Desai, S.G., Thirumalachar, M.J., Patel, M.K., 1965. Bacterial blight disease of *Eleusine caracana* Gaertn. Indian Phytopath. 28, 384–386.

Deswal, D.P., Govila, O.P., Sheoran, R.K., 1998. Genetics analysis of downy mildew resistance in pearl millet (*Pennisetum glaucum*). Indian Phytopath. 51 (3), 261–264.

Devaiah, S.P., Mahadevappa, G.H., Shetty, H.S., 2009. Induction of systemic resistance in pearl millet against downy mildew (*Sclerospora graminicola*) by *Datura metel* extract. Crop Prot. 28, 783–791.

Dublish, P.K., Singh, P.N., 1976. Phytopathogenic fungi of Meerut, some new records of India. Curr. Sci. 45, 168.

Emchebe, A.M., 1975. Some aspects of crop disease in Uganda UAFRO, Seren (unpublished) International Rice Research Institute 1996. Standard Evaluation System for Rice, fourth ed. IRRI, Manila, Philippine.

Gaikwad, A.P., D'Souza, T.F., 1987. A comparative study on *Pyricularia* spp. J. Maharashtra Agril. Univ. 12, 134–135.

Girgi, M., Breese, W.A., Lörz, H., Oldach, K.H., 2006. Rust and downy mildew resistance in pearl millet (*Pennisetum glaucum*) mediated by heterologous expression of the afp gene from *Aspergillus giganteus*. Transgenic Res. 15 (3), 313–324.

Godasara, S.B., Dangaria, C.J., Savaliya, J.J., Pansuriya, A.G., Davada, B.K., 2010. Generation mean analysis in pearl millet [*Penisetum glaucum* (L.) R. BR.]. Agric. Sci. Digest 30, 50–53.

Govindarajan, K., Nagarajan, C., Raveendran, T.S., Prasad, M.M., Shanmugam, N., 1984. New source of resistance to pearl millet rust disease. Madras Agric. J. 71 (3), 210.

Gowda, B.T.S., Seetharam, A., Viswanath, S., Sannegowda, S., 1986. Incorporating blast resistance to Indian elite finger millet cultivars from African cv. IE 1012. SABRAO J. 18, 119–120.

Gulia, S.K., 2004. QTL Mapping for Improvement of Downy Mildew (*Sclerospora graminicola*) Resistance in Pearl Millet Hybrid Parental Line ICMB 89111 (Ph.D. thesis). Chaudhary Charan Singh Haryana Agricultural University, Hisar.

Gupta, S.K., Sharma, R., Rai, K.N., Thakur, R.P., 2012. Inheritance of foliar blast resistance in pearl millet (*Pennisetum glaucum*). Plant Breed. 13 (1), 217–219.

Gwary, D.M., Bdliya, B.S., Bdliya, J.A., 2009. Appraisal of yield losses in pearl millet due to downy mildew pathogen (*Sclerospora graminicola*) in Nigerian Sudan savanna. Arch. Phytopath. Plant Prot. 42, 1010–1019.

Hash, C.T., Witcombe, J.R., 2001. Pearl millet molecular marker research. Int. Sorghum Millet Newsl. 42, 8–15.

Hindumathy, C.K., 2012. The defense activator from yeast for rapid induction of resistance in susceptible pearl millet hybrid against downy mildew disease. Int. J. Agric. Sci. 4, 196–201.

Ignacimuthu, S., Ceasar, S.A., 2012. Development of transgenic finger millet (*Eleusine coracana* (L.) Gaertn.) resistant to leaf blast disease. J. Biosci. 37 (1), 135–147.

Jahagirdar, S., Lakshmana, D., Pawar, K.N., Guggari, A.K., 2005. Leaf rust resistance in some pearl millet advanced materials in northern Karnataka. J. Maha. Agric. Univ. 30 (1), 117.

Jones, E.S., Liu, C.J., Gale, M.D., Hash, C.T., Witcombe, J.R., 1995. Mapping quantitative trait loci for downy mildew resistance in pearl millet. Theor. Appl. Genet. 91, 448–456.

Jones, E.S., Breese, W.A., Liu, C.J., Singh, S.D., Shaw, D.S., Witcombe, J.R., 2002. Mapping quantitative trait loci for resistance to downy mildew in pearl millet: field and glasshouse screens detect the same QTL. Crop Sci. 42, 1316–1323.

Joshi, L.M., Raychaudhuri, S.P., Batra, S.K., Renfro, B.L., Ghosh, A., 1966. Preliminary investigations on a serious disease of *Eleusine coracana* in the states of Mysore and Andhra Pradesh. Indian Phytopath. 19, 324–325.

Kalyan Babu, B., Agrawal, P.K., Pandey, D., Sood, S., Chandrashekara, C., Kumar, A., 2014a. Molecular analysis of world collection of finger millet accessions for blast disease resistance using functional SSR markers. SABRAO J. Breed. Genet. 46 (2), 202–216.

Kalyan Babu, B., Dinesh, P., Agrawal, K.P., Sood, S., Chandrashekara, C., Bhatt, J.C., et al., 2014b. Comparative genomics and association mapping approaches for blast resistant genes in finger millet using SSRs. PLoS One 9 (6), e99182.

Kataria, R.P., Yadav, H.P., Beniwal, C.R., Narwal, M.S., 1994. Genetics of incidence of downy mildew (*Sclerospora graminicola*) in pearl millet. Indian J. Agric. Sci. 64 (9), 664−666.

Kaye, C., Milazzo, J., Rozenfeld, S., Marc-Henri Lebrun, Tharreau, D., 2003. The development of simple sequence repeat markers for *Magnaporthe grisea* and their integration into an established genetic linkage map. Fungal Genet. Biol. 40, 207−214.

Keshavamurthy, K.V., Yaraguntaiah, R.C., 1977. Virus Disease of Ragi in Symposium on Ragi (*Eleusine coracana*). Post Graduate College, University of Agricultural Sciences, Bangalore, pp. 100−105.

Ketema, S., 1987. Research recommendations for production and brief outline of strategy for the improvement of tef [*Eragrostis tef* (Zucc.) Trotter]. In: Proc. 19th Natl. Crop. Imp. Conf. IAR. Addis Ababa, Ethiopia.

Khairwal, I.S., Rai, K.N., Diwakar, B., Sharma, Y.K., Rajpurohit, B.S., Nirwan, B., et al., Pearl Millet: Crop Management and Seed Production Manual. International Crop Research Institute for Semi-Arid Tropics (ICRISAT), Patancheru, p. 104.

Khot, G.G., Kulkarni, S.R., Kulkarni, R.V., Patil, V.S., Tirmali, A.M., 2008. Response of promising finger millet genotypes against blast caused by Pyricularia setariae (Wallace) Ramkrishnen. J. Maha. Agric. Univ. 33, 93−95.

King, S.B., 1992. World review of pearl millet diseases: knowledge and future research needs. In: de Milliano, W.A.J., Frederiksen, R.A., Bengston, G.D. (Eds.), Sorghum and Millets Diseases: A Second World Review. ICRISAT, Patancheru, pp. 95−108.

Kiran Babu, T., Sharma, R., Upadhyaya, H.D., Reddy, P.N., Deshpande, S., Senthilvel, S., et al., 2013a. Evaluation of genetic diversity in *Magnaporthe grisea* populations adapted to finger millet using simple sequence repeats (SSRs) markers. Physiol. Mol. Plant Path. 84, 10−18.

Kiran Babu, T., Thakur, R.P., Upadhyaya, H.D., Reddy, P.N., Sharma, R., Girish, A.G., et al., 2013b. Resistance to blast (*Magnaporthe grisea*) in a mini-core collection of finger millet germplasm. Eur. J. Plant Path. 135 (2), 299−311.

Kulkarni, S., Patel, M.K., 1956. Study of the effect of nutrition and temperature on the size of spores in *Pyricularia setariae* Nishikado. Indian Phytopath. 9, 31−38.

Kumar, B., Kumar, J., Srinivas, P., 2008. First record of head smut of barnyard millet from mid-hills of Uttarakhand. J. Mycol. Pl. Pathol. 38, 142.

Kumar, B., Kumar, J., 2009. Evaluation of small millet genotypes against endemic diseases in mid-western Himalayas. Indian Phytopathol. 62, 518−521.

Kumar, B., Rashmi, Y., 2012. Influence of nitrogen fertilizer dose on blast disease of finger millet caused by *Pyricularia grisea*. Indian Phytopath. 65 (1), 52−55.

Kumar, V.B.S., Kumar, T.B.A., Bhat, S.A., Nagaraju, 2006. Screening of long duration finger millet (*Eleusine coracana* (L) Gaertn.) genotypes against neck and finger blast caused by *Pyricularia grisea* (Cke.) Sacc. J. Plant Prot. Environ. 3 (1), 136−139.

Kumar, V.B.S., Amruta Bhat, S., Nagaraju, 2007. Virulence analysis of *Pyricularia grisea* (Cke.) Sacc. on different finger millet genotypes and cultivars to determine race differential lines. Environ. Ecol. 25 (1), 190−192.

Lubadde, G., Tongoona, P., Derera, J., Sibiya, J., 2014. Major pearl millet diseases and their effects on-farm grain yield in Uganda. African J. Agric. Res. 9, 2911−2918.

Lukose, C.M., Kadvani, D.L., Dangaria, C.J., 2007. Efficacy of fungicides in controlling blast disease of pearl millet. Indian Phytopath. 60, 68−71.

Madhukeshwara, S.S., 1990. Studies on Variability in *Pyricularia grisea* (Cke) Sacc With Particular Reference to Virulence (M.Sc. thesis). University of Agricultural Sciences, Bangalore, p. 91.

Madhukeshwara, S.S., Viswanath, S., Mantur, S.G., 1997. Variability in different isolates of *Pyricularia grisea* (Cke) Sacc. with special reference to cytoplasmic granulation. In: Extended Summaries, National Seminar on Small Millets, Current Research Trends and Future Priorities as Food, Feed and in Processing for Value Addition, 23−24 April 1997. TNAU, Coimbatore, 73 pp.

Madhukeshwara, S.S., Mantur, S.G., Anil Kumar, T.B., Viswanath, S., Seetharam, A., 2004. Finger millet germplasm resistant to blast. Indian Phytopathol. 57 (1), 121.

Maheshkumar, 2002. Isolation, Purification and Characterization of Phytoalexin(s) From Blast Susceptible and Resistant Finger Millet (*Eleusine coracana*) Leaves (M.Sc. thesis). University of Agricultural Sciences, Bangalore.

Manga, V.K., Kumar, A., 2012. Research note development of downy mildew resistant maintainer inbreds (B lines) of pearl millet for arid environment. Electron. J. Plant Breed. 3 (2), 825−829.

Manjunatha, G., Roopa, K.S., Geetha, N.P., Shetty, H.S., 2008. Chitosan enhances disease resistance in pearl millet against downy mildew caused by *Sclerospora graminicola* and defence-related enzyme activation. Pest Manage. Sci. 64, 1250−1257.

Manjunatha, G., Niranjan-Raj, S., Geetha, N.P., Deepak, S., Amruthesh, K.N., Shetty, H.S., 2009. Nitric oxide is involved in chitosan induced systemic resistance in pearl millet against downy mildew disease. Pest Manage. Sci. 65, 737−743.

Mantur, S.G., Viswanath, S., Anil Kumar, T.B., 2001. Evaluation of finger millet genotypes for resistance to blast. Indian Phytopath. 54, 387.

Mantur, S.G., Madhukeshwara, S.S., Anil Kumar, T.B., Viswanath, S., 2002. Evaluation of pre-release and released varieties of finger millet for blast resistance and yield. Environment 12, 23−25.

Maramorosch, K., Govindu, H.C., Kondo, F., 1977. Rhabdo virus particles associated with mosaic disease of naturally infected *Eleusine coracana* (finger millet) in Karnataka state (Mysore) South India. Plant Dis. Rep. 61, 1029−1031.

Mariappan, V., Natarjan, C., Kandaswamy, T.K., 1973. Ragi streak disease in Tamil Nadu. Madras Agric. J. 60, 451−453.

Marley, P.S., Diourtk, M., Neya, A., Nutsugah, S.K., Skrkmk, P., Katilk, S.O., et al., 2002. Sorghum and pearl millet diseases in West and Central Africa. In: Leslie, J.F. (Ed.), Sorghum and Millets Diseases. Iowa State Press, Ames, IA, pp. 419−425.

McRae, W., 1920. Detailed administration report of the government mycologist for the year. Imperial Institute of Agriculture Research, Pusa, pp. 1919−20.

McRae, W., 1932. Report of the imperial mycologist. Scientist. Rep. Imper. Inst. Agric. Res. Pusa. 1930-31, 73−86.

Mehta, P.R., Singh, B., Mathur, S.C., 1953. A new leaf spot disease of bajra (*Pennisetum typhoides*) Staph and Hubbard, caused by a species of *Piricularia*. Indian Phytopath. 5, 140−143.

Michelmore, R.W., Pawar, M.N., Williams, R.J., 1982. Heterothallism in *Sclerospora graminicola*. Phytopathology 72, 1368−1372.

Miedaner, T., Geiger, H.H., 2015. Biology, genetics, and management of ergot (*Claviceps* spp.) in rye, sorghum, and pearl millet. Toxins 7, 659−678.

Mudingotto, P.J., Veena, M.S., Mortensen, C.N., 2002. First report of bacterial blight caused by *Acidovorax avenae* ssp. *avenae* associated with finger millet seeds from Uganda. Plant Path. 51, 396.

Muralidharan, J., John, E., Channamma, K.A.L., Theertha Prasad, D., 1996. Changes in esterases in response to blast infection in finger millet seedlings. Phytochemistry 43, 1151−1155.

Muyanga, S., 1995. Production and research review of small cereals in Zambia. In: Danial D.L. (Ed.), Breeding for Disease Resistance With Emphasis on Durability. Porc. Reg. Workshop for Eastern, Central and Southern Africa. Njaro, Kenya, Oct 2−6, 1994 Wageningen, Netherland.

Nagaraja, A., Mantur, S.G., 2007. Screening of *Eleusine coracana* germplasm for blast resistance. J. Mycopath. Res. 45 (1), 66−68.

Nagaraja, A., Gowda, J., Krishnappa, M., Krishne Gowda, K.T., 2008a. GPU 28: a finger millet variety with durable blast resistance. J. Mycopath. Res. 46 (1), 109−111.

Nagaraja, A., Mantur, S.G., Anil Kumar, T.B., 2008b. Evaluation of white seeded finger millet entries against blast disease and its influence on yield parameters. Seed Res. 36 (1), 102−104.

Nagaraja, A., Nanja Reddy, Y.A., Anjaneya Reddy, B., Patro, T.S.S.K., Kumar, B., Kumar, J., et al., 2010a. Reaction of finger millet recombinant inbred lines to blast. Crop Res. 39, 120−122.

Nagaraja, A., Nanja Reddy, Y.A., Gowda, J., Anjaneya Reddy, B., 2010b. Association of plant characters and weather parameters with finger millet blast. Crop Res. 39, 123−126.

Nagaraja, A., Kumar, B., Raguchander, T., Hota, A.K., Patro, T.S.S.K., Devaraje Gowda, P., et al., 2012. Impact of disease management practices on finger millet blast and grain yield. Indian Phytopath. 65 (4), 356−359.

Nagaraju, Viswanath, S., Reddy, H.R., Lucy Channamma, K.A., 1982. Ragi streak a leaf hopper transmitted virus disease in Karnataka. Mysore J. Agric. Sci. 16, 301−305.

Navi, S.S., Singh, S.D., 1995. New sources of resistance to pearl millet downy mildew. Indian J. Plant Prot. 23 (2), 142−145.

Niranjan-Raj, S., Shetty, N.P., Shetty, H.S., 2005. Synergistic effects of Trichoshield on enhancement of growth and resistance to downy mildew in pearl millet. Bio Control 50, 493−509.

Nutsugah, S.K., Atokple, I.D.K., Afribeh, D.A., 2006. Sources of resistance to downy mildew, smut and ergot of pearl millet and their stability in Ghana. Agric. Food Sci. J. Ghana 5, 381−394.

Pall, B.S., 1992. Biochemical studies on blast disease of finger millet (*Eleusine coracana*). Bioved 3, 153−154.

Pall, B.S., 1994. Biochemical studies of pathogenesis of finger millet blast. Res. Dev. Rep. 11, 43−47.

Pande, S., Mukuru, S.Z., Odhiambo, R.O., Karunakar, R.I., 1994. Seed-borne infection of *Eleusine coracana* by Bipolaris nodulosa and *Pyricularia grisea* in Uganda and Kenya. Plant Dis. 78, 60−63.

Pannu, P.P.S., Sokhi, S.S., Aulakh, K.S., 1996. Resistance in pearl millet against rust. Indian Phytopath. 49 (3), 243−246.

Patro, T.S.S.K., Madhuri, J., 2014. Management of banded blight of finger millet incited by *Rhizoctonia solani* (KUHN.). Indian J. Plant Sci. 3, 163−166.

Patro, T.S.S.K., Neeraja, B., Sandhya Rani, Y., Jyosthna, S., Keerthi, S., 2015. Evaluation of national elite germplasm genotypes of finger millet for their reaction to blast incited by *Pyricularia grisea* (Cke.) Sacc. In: National Symposium on Understanding Host-Pathogen Interaction Through Science of Omics, 16−17 March, p. 15.

Paul Khurana, S.M., Rayachaudhuri, S.P., Sundaram, N.V., 1973. Further studies on ragi mosaic in Delhi. Indian Phytopath. 26, 554−559.

Phookan, A.K., 1987. Studies on Pearl Millet Smut With Special Reference to Pathogenic Variability, Inheritance of Resistance, and Chemical Control (Ph.D. thesis). Haryana Agricultural University, Hisar, p. 116.

Prabhu, A.S., Filippi, M.C., Castro, N., 1992. Pathogenic variation among isolates of *Pyricularia grisea* infecting rice, wheat, and grasses in Brazil. Trop. Pest Manage. 38, 367−371.

Prabhu, R., Ganesan, N.M., 2013. Genetic diversity studies in ragi (*Eleusine coracana* (L.) Gaertn.) with SSR and ISSR markers. Mol. Plant Breed. 4 (17), 141−145.

Purushottaman, D., Marimuthu, T., 1974. Phytoalexin synthesis in ragi leaves infested with *Pyricularia setariae* as influenced by phenylalanine and glucose. Curr. Sci. 43, 47−49.

Pushpalatha, H.G., Sudisha, J., Shetty, H.S., 2013. Cellulysin induces downy mildew disease resistance in pearl millet driven through defense response. Eur. J. Plant Pathol. 137, 707−717.

Pushpavathi, B., Thakur, R.P., Chandrashekara Rao, K., 2006a. Fertility and mating type frequency in Indian isolates of *Sclerospora graminicola*, the downy mildew pathogen of pearl millet. Plant Dis. 90, 211−214.

Pushpavathi, B., Thakur, R.P., Chandrashekara Rao, K., 2006b. Inheritance of avirulence in *Sclerospora graminicola*, the pearl millet downy mildew pathogen. Plant Path. J. 5 (1), 54−59.

Rajanna, M.P., Rangaswamy, B.R., Basavaraju, M.K., Karegowda, C., Ramaswamy, G.R., 2000. Evaluation of finger millet genotypes for resistance to blast caused by *Pyricularia grisea* Sacc. Plant Dis. Res. 15, 199−201.

Ramachandra, G., Virupaksha, T.K., Shadakshara Swamy, M., 1981. Isoperoxidase changes in finger millet during development of blast disease. J. Sci. Food Agric. 32, 662−666.

Ramakrishnan, T.S., 1971. Diseases of Millets. Indian Council of Agricultural Research, New Delhi, pp. 83−100.

Ramappa, H.K., Ravishankar, C.R., Prakash, P., 2002. Estimation of yield loss and management of blast in finger millet (ragi). In: Proceedings of Asian Congress of Mycology and Plant Pathology, 1–4 October 2002, University of Mysore, Mysore, 195 p.

Ramaswamy, R., 1995. Studies on Blast Disease (*Pyricularia grisea* (Cooke) Sacc. of Finger Millet (*Eleusine coracana* (L) Gaertn) (Ph.D. thesis). Tamil Nadu Agricultural University, Coimbatore.

Ramesh, C.R., Sukanya, D.H., Thakur, R.P., Rao, V.P., 2003. Resistance to downy mildew (*Sclerospora graminicola*) in forage pearl millet (*Pennisetum glaucum*). Indian J. Agric. Sci. 73 (6), 327–330.

Ramteke, N.B., Mondhe, M.K., Gibhakate, P., Jadhav, B.R., 2011. Biological management of leaf blast (*Pyricularia grisea*) of pearl millet. J. Plant Dis. Sci. 6 (2), 202–203.

Rao, A.N.S., 1990. Estimates of losses in finger millet (*Eleusine coracana*) due to blast disease (*Pyricularia grisea*). J. Agric. Sci. 24, 57–60.

Rao, V.P., Thakur, R.P., Singh, A.K., Rai, K.N., 2006. Evaluation of pearl millet male-sterile lines and their maintainers for resistance to downy mildew and smut. Int. Sorghum Millet Newsl. 47, 131–133.

Ravikumar, R.L., Dinesh Kumar, S.P., Seetharam, A., 1991. Association of seed colour with seed protein, phenol, tannin and blast susceptibility in finger millet (*Eleusine coracana* Gaertn). SABRAO J. 23, 91–94.

Ravikumar, R.L., Seetharam, A., 1994. Genetic variation in yield and its components in relation to blast disease caused by *Pyricularia grisea* in finger millet (*Eleusine coracana*). Indian J. Agric. Sci. 64, 103–106.

Ravikumar, R.L., Seetharam, A., Shashidhar, V.R., 1995. Preformed internal defenses in finger millet (*Eleusine coracana* Gaertn) against blast fungus. Filopatol. Brasilania. 20, 13–19.

Rawat, L., Nagaraja, A., Bhatt, A., 2016. First record of leaf blast on little millet (Panicum sumatrense Roth ex Roemer and Schultes) from mid hills of Uttarakhand. J. Mycopathol. Res. 54, 145–147.

Safeeulla, K.M., 1976. Biology and Control of the Downy Mildews of Pearl Millet, Sorghum and Finger Millet. University of Mysore, Mysore, p. 304.

Sanathkumar, V.B., Anilkumar, T.B., Nagaraju, 2002a. Early events of infection by *Pyricularia grisea* in resistant and susceptible finger millet genotypes. In: Abstr. Proc. IPS Symp. Plant Disease Scenario in Southern India, 19–21 December, 15 p.

Sanathkumar, V.B., Anilkumr, T.B., Nagaraju, 2002b. Anatomical defence mechanism in finger millet leaves against blast caused by *Pyricularia grisea*. In: Abstr. Proc. IPS Symp. Plant Disease Scenario in Southern India, 19–21 December, 15 p.

Sastry, J.G., Ramakrishna, W., Sivaramakrishnan, S., Thakur, R.P., Gupta, V.S., Ranjekar, P.K., 1995. DNA fingerprinting detects genetic variability in the pearl millet downy mildew pathogen (*Sclerospora graminicola*). Theor. Appl. Genet. 91, 856–861.

Seetharam, A., Ravikumar, R.L., 1993. Blast resistances in finger millet—its inheritance and biochemical nature. In: Riley, K.W., Gupta, S.C., Seetaram, A., Mushonga, J.N. (Eds.), Advances in Small Millets. Oxford IBH Publishing Co. Ltd., New Delhi, pp. 449–465.

Selvaraj, C.I., Nagarajan, P., Thiyagarajan, K., Bharathi, M., Rabindran, R., 2011. Identification of microsatellite (SSR) and RAPD markers linked to rice blast disease resistance gene in rice (*Oryza sativa* L.). African J. Biotech. 10 (17), 3301–3321.

Shailasree, S., Kini, K.R., Deepak, S., Kumudini, B.S., Shetty, H.S., 2004. Accumulation of hydroxyproline-rich glycoproteins in pearl millet seedlings in response to *Sclerospora graminicola* infection. Plant Sci. 167, 1227–1234.

Shankara Rao, R., Rao, K.E.S., Singh, S.D., Reddy, M.S., Rao, M.V.B., Venkateswarlu, K., 1987. Fungicidal control of downy mildew of pearl millet. Indian J. Plant Prot. 15, 146–151.

Sharathchandra, R.G., Geetha, N.P., Amruthesh, K.N., Ramachandra Kini, K., Sarosh, B.R., Shetty, N.P., et al., 2006. Isolation and characterisation of a protein elicitor from *Sclerospora graminicola* and elicitor-mediated induction of defence responses in cultured cells of *Pennisetum glaucum*. Funct. Plant Biol. 33 (3), 267–278.

Sharma, R., Thakur, R.P., Rai, K.N., Gupta, S.K., Rao, V.P., Rao, A.S., et al., 2009. Identification of rust resistance in hybrid parents and advanced breeding lines of pearl millet. J. SAT Agric. Res. 7, 1−4.

Sharma, R., Rao, V.P., Varshney, R.K., Prasanth, V.P., Kannan, S., Thakur, R.P., 2010. Characterisation of pathogenic and molecular diversity in *Sclerospora graminicola*, the causal agent of pearl millet downy mildew. Arch. Phytopath. Plant Prot. 43, 538−551.

Sharma, R., Upadhyaya, H.D., Manjunatha, S.V., Rai, K.N., Gupta, S.K., Thakur, R.P., 2013. Pathogenic variation in the pearl millet blast pathogen *Magnaporthe grisea* and identification of resistance to diverse pathotypes. Plant Dis. 97 (2), 189−195.

Shaw, F.J.F., 1921. Report of the imperial mycologist scient. Rep. Agric. Res. Inst. Pusa 1920−21, 34−40.

Shetty, H.S., Mathur, S.B., Neergaard, P., 1980. *Sclerospora graminicola* in pearl millet seeds and its transmission. Trans. British Mycol. Soc. 74, 127−134.

Sherchan, K., 1989. Importance, genetic resources and varietal improvement of finger millet in Nepal. In: Seetharam, A., Riley, K.W., Harinarayana, G. (Eds.), Small Millets in Global Agriculture. Oxford and IBH Pub. Co. Pvt. Ltd., New Delhi, pp. 85−92.

Shishupala, S., Kumar, V.U., Shetty, H.S., Umeshkumar, S., 1996. Screening pearl millet cultivars by ELISA for resistance to downy mildew disease. Plant Pathol. 45 (5), 978−983.

Singh, R.S., Prasad, Y., 1981. Blast of proso millet in India. Plant Dis. 65, 442−443.

Singh, S.D., 1990. Sources of resistance to downy mildew and rust in pearl millet. Plant Dis. 74, 871−874.

Singh, S.D., Gopinath, R., 1985. A seedling inoculation technique for detecting downy mildew resistance in pearl millet. Plant Dis. 69, 582−584.

Singh, S.D., Singh, G., 1987. Resistance to downy mildew in pearl millet hybrid NHB 3. Indian Phytopath. 40 (2), 178−180.

Singh, S.D., Talukdar, B.S., 1998. Inheritance of complete resistance to pearl millet downy mildew. Plant Dis. 82, 791−793.

Singh, S.D., King, S.B., Weraer, J., 1993. Downy Mildew Disease of Pearl Millet. Information Bulletin No. 37. ICRISAT, Patancheru, p. 36.

Singh, S.D., Alagarswamy, G., Talukdar, B.S., Hash, C.T., 1994. Registration of ICML 22 photoperiod insensitive, downy mildew resistant pearl millet germplasm. Crop Sci. 34 (5), 1421.

Singh, S.D., Wilson, J.P., Navi, S.S., Talukdar, B.S., Hess, D.E., Reddy, K.N., 1997. Screening Techniques and Sources of Resistance to Downy Mildew and Rust in Pearl Millet. Information Bulletin No. 48. ICRISAT, Patancheru, p. 104.

Siwecki, R., Werner, A., 1980. Resistance mechanism involved in the penetration and colonization of poplar leaf tissues by *Melampsora* rust. Phytopathol. Mediterra. 27−29.

Sokhi, S.S., Singh, B.B., 1984. Components of slow rusting in pearl millet infected with *Puccinia penniseti*. Indian J. Mycol. Plant Pathol. 14, 190−192.

Somappa, K.M., 1999. Mechanism of Resistance and Biocontrol of Blast of Ragi Caused by *Pyricularia grisea* (Cke) Sacc. (M.Sc. thesis). University of Agricultural Sciences, Bangalore.

Sone, T., Abe, T., Yoshida, N., Suto, M., Tomita, F., 1997. DNA fingerprinting and electrophoretic karyotyping of Japanese isolates of rice blast fungus. Anal. Phytopath. Soc. Japan 63, 155−163.

Subbayya, J., Raychaudhuri, S.P., 1970. A note on a mosaic disease of ragi (*Eleusine coracana*) in Mysore, India. Indian Phytopath. 23, 144−148.

Sudisha, J., Ananda Kumar, S., Niranjana, S.R., Shetty, N.P., Shetty, H.S., 2009. Cloning and development of pathotype-specific SCAR marker associated with *Sclerospora graminicola* isolates from pearl millet. Aust. Plant Path. 38, 216−221.

Suh, J.P., Roh, J.H., Cho, Y.C., Han, S.S., Kim, Y.G., Jena, K.K., 2009. The Pi40 gene for durable resistance to rice blast and molecular analysis of Pi40-advanced backcross breeding lines. Phytopathology 99, 243−250.

Sujeeth, N., Deepak, S., Shailasree, S., Kini, R.K., Shetty, H.S., Hille, J., 2010. Hydroxyproline-rich glyco-proteins accumulate in pearl millet after seed treatment with elicitors of defense responses against *Sclerospora Graminicola*. Physiolo. Mol. Plant Pathol. 74, 230–237.

Sundaram, N.V., 1975. Ergot of Bajra. Advances in Mycology and Plant Pathology. Professor R.N. Tandon's Birthday Celebration Committee, New Delhi, pp. 155–160.

Sunil, 2002. Evaluation of Finger Millet Genotypes for Durable Resistance to Blast (M.Sc.(Ag.) thesis). University of Agricultural Sciences, Bangalore, p. 101.

Suzuki, F., Arai, M., Yamaguchi, J., 2006. DNA fingerprinting of *Pyricularia grisea* by rep-PCR using single primers designed from the terminal inverted repeat of each of the transposable elements Pot2 and MGR586. J. Gen. Plant Path. 72, 314–317.

Suzuki, F., Suga, H., Tomimura, K., Fuji, S., Arai, M., Koba, A., et al., 2009. Development of simple sequence repeats markers for Japanese isolates of *Magnaporthe grisea*. Mol. Ecol. Resour. 9, 588–590.

Takan, J.P., Akello, B., Esele, P., Manyasa, E.O., Obilana, A.B., Audi, P.O., et al., 2004. Finger millet blast pathogen diversity and management in East Africa: a summary of project activities and outputs. Int. Sorghum Millet Newsl. 45, 66–69.

Takan, J.P., Chipili, J., Muthumeenakshi, S., Talbot, N.J., Manyasa, E.O., Bandyopadhyay, R., et al., 2012. *Magnaporthe oryzae* populations adapted to finger millet and rice exhibit distinctive patterns of genetic diversity, sexuality and host interaction. Mol. Biotech. 50 (2), 145–158.

Takasugi, H., Akaisahi, Y., 1933. Studies on downy mildew of Italian millet in Manchuria. Res. Bull. S. Manchuria Rly. Co 11, 1–20.

Tanaka, M., Nakayashiki, H., Tosa, Y., 2009. Population structure of *Eleusine* isolates of *Pyricularia oryzae* and its evolutionary implications. J. Gen. Plant Path. 75, 173–180.

Taylor, J., Mims, C.W., 1991a. Fungal development and host cell responses to the rust fungus *Puccinia substriata* var. *indica* in seedling and mature leaves of susceptible and resistant pearl millet. Can. J. Bot. 69 (6), 1207–1219.

Taylor, J., Mims, C.W., 1991b. Infection of and colony development within leaves of susceptible and resistant pearl millet and two nonhosts by the rust fungus *Puccinia substriata* var. *indica*. Mycologia 565–577.

Thakur, R.P., King, S.B., 1988a. Ergot Disease of Pearl Millet. Information Bulletin No. 24. ICRISAT, Patancheru, p. 24.

Thakur, R.P., King, S.B., 1988b. Registration of four ergot resistant germplasms of pearl millet. Crop Sci. 28, 382.

Thakur, R.P., King, S.B., 1988c. Registration of six smut resistant germplasms of pearl millet. Crop Sci. 28, 382–383.

Thakur, R.P., Sharma, H.C., 1990. Identification of pearl millet lines with multiple resistance to ergot, smut and oriental armyworm. Indian J. Plant Prot. 18, 47–52.

Thakur, R.P., Rao, V.P., Williams, R.J., 1984. The morphology and disease cycle of ergot caused by *Claviceps fusiformis* in pearl millet. Phytopathology 74, 201–205.

Thakur, R.P., Subba Rao, K.V., Williams, R.J., Gupta, S.C., Thakur, D.P., Nafade, S.D., et al., 1986. Identification of stable resistance to smut in pearl millet. Plant Dis. 70, 38–41.

Thakur, R.P., Rao, V.P., King, S.B., 1989. Ergot susceptibility in relation to cytoplasmic male sterility in pearl millet. Plant Dis. 73, 676–678.

Thakur, R.P., King, S.B., Rai, K.N., Rao, V.P., 1992. Identification and Utilization of Smut Resistance in Pearl Millet. Research Bulletin No. 16. ICRISAT, Patancheru, p. 36.

Thakur, R.P., Rai, K.N., King, S.B., Rao, V.P., 1993. Identification and Utilization of Ergot Resistance in Pearl Millet. Research Bulletin No. 17. ICRISAT, Patancheru, p. 40.

Thakur, R.P., Rai, K.N., Rao, V.P., Rao, A.S., 2001. Genetic resistance of pearl millet male-sterile lines to diverse Indian pathotypes of *Sclerospora graminicola*. Plant Dis. 85, 621−626.

Thakur, R.P., Rao, V.P., Wu, B.M., Subbarao, K.V., Shetty, H.S., Singh, G., et al., 2004. Host resistance stability to downy mildew in pearl millet and pathogenic variability in *Sclerospora graminicola*. Crop Prot. 23 (10), 901−908.

Thakur, R.P., Sharma, R., Rai, K.N., Gupta, S.K., Rao, V.P., 2009. Screening techniques and resistance sources for foliar blast in pearl millet. J. SAT Agric. Res. 7, 1−5.

Thakur, R.P., Sharma, R., Rao, V.P., 2011. Screening Techniques for Pearl Millet Diseases. Information Bulletin No. 89. ICRISAT, Patancheru, p. 56.

Talukdar, B.S., Prakash Babu, P.P., Rao, A.M., Ramakrishna, C., Witcombe, J.R., King, S.B., et al., 1998. Registration of ICMP 85410: dwarf, downy mildew resistant, restorer parental line of pearl millet. Crop Sci. 38, 904−905.

Thomas, K.M., Ramakrishna, T.S., Srinivasan, K.V., 1945. The occurrence of ergot in south India. Proc. Indian Acad. Sci. B21, 93−100.

Upasana, R., Chaudhary, D.P., Srivastava, M., 2010. Downy mildew and its effect on quality of pearl millet hybrids. Crop Impro. 37 (2), 206.

Venkatarayan, S.V., 1947. Diseases of ragi (*Eleusine coracana*). Mysore Agric. J. 24, 50−57.

Walker, J.C., 1924. On the nature of disease resistance in plants. Trans. Wisc. Acad. Sci. Arts Lettr. 21, 225−247.

Williams, R.J., Singh, S.D., Pawar, M.N., 1981. An improved field screening technique for downy mildew resistance in pearl millet. Plant Dis. 65, 239−241.

Wilson, J.P., 1995. Mechanisms associated with the *tr* allele contributing to reduced smut susceptibility of pearl millet. Phytopathology 85, 966−969.

Wilson, J.P., Hanna, W.W., 1992. Effects of gene and cytoplasm substitutions in pearl millet on leaf blight epidemics and infection by *Pyricularia grisea*. Phytopathology 82, 839−842.

Wilson, J.P., Burton, G.W., Wells, H.D., Zongo, J.D., Dicko, I.O., 1989. Leaf spot, rust, and smut resistance in pearl millet landraces from central Burkina Faso. Plant Dis. 73, 345−349.

Wilson, J.P., Gates, R.N., Hanna, W.W., 1991. Effect of rust on yield and digestibility of pearl millet forage. Phytopathology 81, 233−236.

Yadav, H.P., Sagar, P., Sabharwal, P.S., 2000. Inheritance of smut resistance in pearl millet. Czech J. Genet. Plant Breed. 36 (2), 45−47.

Yadav, M.S., Duhan, J.C., Thakur, D.P., 1995. Evaluation of inoculation methods for screening pearl millet against smut disease. Indian Phytopath. 48 (2), 189−190.

Yan, W.Y., Xie, S.Y., Jin, L.X., Liu, H.J., Hu, J.C., 1985. A preliminary study on the physiological races of *Pyricularia setariae*. Scientia Agric. Sinica 3, 57−62.

INSECT PEST RESISTANCE IN SORGHUM

4

P.G. Padmaja

ICAR-Indian Institute of Millets Research, Hyderabad, India

4.1 INTRODUCTION

Sorghum [*Sorghum bicolor* (L.) Moench] is an important crop ranking fifth in the world cereal production with an annual production of 55.7 million tonnes (FAOSTAT, 2012). The productivity levels of sorghum under subsistence farming are quite low mainly because of biotic constraints. Nearly 150 insect species have been reported as pests on sorghum (Jotwani et al., 1980). About 32% of sorghum crop is lost due to insect pests during the rainy season (Borad and Mittal, 1983), and 26% during the postrainy season (Daware et al., 2012) in India. Sorghum shoot fly, *Atherigona soccata* (Rond.); spotted stemborer, *Chilo partellus* (Swin.); Oriental armyworm, *Mythimna separata* (Walk.); shoot bug, *Peregrinus maidis* (Ashm.); sugarcane aphid, *Melanaphis sacchari* (Zehnt.); sorghum midge, *Stenodiplosis sorghicola* (Coq.); head bugs, *Calocoris angustatus* (Leth.); and head caterpillars, *Helicoverpa armigera* and *Eublemma* are the major pests worldwide.

4.2 PEST BIOLOGY

4.2.1 SEEDLING PESTS

4.2.1.1 Shoot flies

Geographical distribution: Southern Europe, North and East Africa, India, and the Middle East.
Host plants: Sorghum, corn, various Gramineae (Poaceae).
Biology: Shoot fly, *Atherigona soccata* (Rond.) is one of the major pests that destabilize the sorghum production. The adult female measures about 3—3.5 mm in length. Body is gray-brown and abdomen and legs are yellow in color. Larva is about 6.5 mm long and has 2 posterior black spiracular lobes. The preoviposition period is 3—5 days. The females deposit a mean of 238 eggs with an average incubation period of 3 days. The larval development requires an average of 7.8 days, and the pupal stage is 7.1 days long. The females live an average of 30 days and the males about 20 days. The shoot fly females lay white, elongated, cigar-shaped eggs singly on the abaxial leaf surface of sorghum seedlings at 5—30 days after seedling emergence. Most of the eggs are laid between 08.00—12.00 h, and they hatch between 04.00—06.00 h. On emergence, the neonate larvae crawl to the plant whorl and move downward

Biotic Stress Resistance in Millets. DOI: http://dx.doi.org/10.1016/B978-0-12-804549-7.00004-4

between the folds of the young leaves. After reaching the growing point, it cuts the growing tip resulting in drying of the central leaf known as "deadheart". Deadheart formation leads to seedling mortality. The damage occurs 1−4 weeks after seedling emergence. The total life cycle from egg to adult is completed in 17−21 days (Kundu and Kishore, 1970; Zein el Abdin, 1981). The shoot fly females prefer the second leaf, followed by the third, first, and fourth leaves for egg laying under laboratory conditions, while the third leaf, followed by the second, fourth, fifth, sixth, first, and seventh leaf were preferred for oviposition under field conditions (Ogwaro, 1978; Davies and Reddy, 1981). Sorghum shoot fly is active throughout the year, and there may be 10−15 generations in a year (Jotwani, 1978). Bene (1986) recorded 5−6 generations in temperate areas in Italy. There is no diapause during the off-season (Rao and Rao, 1956). During the off-season, the insect survives on alternate hosts (*Sorghum* spp., *Echinochloa colonum*, *E. procera*, *Cymbopogon* spp., *Paspalum scrobiculatum*, and *Pennisetum glaucum*), tillers of ratoon crop, and volunteer or fodder sorghum (Reddy and Davies, 1979; Sharma and Nwanze, 1997).

4.2.2 LEAF FEEDERS

4.2.2.1 Grasshoppers

Many species of grasshoppers are present, but only a few have pest potential. Common pest species include: clearwinged grasshopper, *Camnula pellucida*; differential grasshopper, *Melanoplus differentialis*; migratory grasshopper, *Melanoplus sanguinipes*; two striped grasshopper, *Melanoplus bivitattus*; and red legged grasshopper, *Melanoplus femurrubrum*. Grasshoppers overwinter as egg laid in the soil in the form of an elongate shaped pod, which contains 20−120 egg. Eggs hatch into nymphs in late May and Jun., maturing in 2−3 months. By Aug., most of the nymphs become matured to the adult stage. They mostly feed on leaves during the day and rest during the afternoon and night on vegetation. Nymphs and adults of several species of grasshopper may chew holes in leaves causing a ragged appearance in sorghum. The grasshopper population builds up at the border of the field and move into it causing damage along the field margins. Their population usually is greater in dry than wet years.

4.2.2.2 Weevils

Ash weevils: *Myllocerus maculosus*, *M. viridanus*, *M. subfasciatus*, and *M. discolor*; (Curculionidae: Coleoptera).

> **Distribution:** Throughout India.
> **Host range:** Pearl millet, maize, sorghum.
> **Bionomics**
>> *Myllocerus viridanus:* Adult weevil with greenish white elytra.
>> *Myllocerus maculosus:* Adult weevil with greenish white elytra having dark lines.
>> *Myllocerus discolor:* Adult weevil is brown in color with white spot on the elytra. Grub is small, white apodous and found feeding on roots. Weevil appears during summer and lays ovoid, light yellow eggs in the soil. Female lays on an average 360 eggs over a period of 24 days. Eggs hatch in 3−5 days. Grub period is 1−2 months, pupation takes place in soil inside earthern cells and pupal period is 7−10 days. Life cycle is completed in 6−8 weeks, thereby completing 3−4 generations in a year. Adults live fairly long for 4−5 months in the winter.

Myllocerus subfasciatus: The adult weevil is light grayish to white with four black spots on the wing covers. The eggs are light yellow and laid deep in the soil. The grubs are fleshy, yellow-colored. Pupation occurs in earthen cells in the soil. The egg, larva, and pupal periods last for about 3−11, 3−42, and 5−7 days, respectively.

Damage symptoms: Leaf margins are notched resulting in wilting of plants in patches. Roots are eaten away by grubs and as a result plants come off easily when pulled. Adults feed on the leaves.

4.2.2.3 Hairy caterpillars
4.2.2.3.1 Red hairy caterpillar

The adult moth has white forewings and brownish markings and streaks. The white hind wings have black spots. There is a yellow band on the head and a yellow steak along the anterior margin of the forewings. Moth emergence occurs following rains. Cream or bright yellow eggs are laid in masses on the lower surfaces of leaves. The eggs may also be laid on the soil, stone, bits of wood, and on other vegetation. A moth can lay up to 2300 eggs in clusters of 50−100.

Damage symptoms: The dark hairy larvae feed gregariously on the lower surface of leaves, scraping them for 4−5 days. After about 10 days the larvae become ashy-brown and move slowly from plant to plant, and field to field, feeding voraciously. In about 40−50 days the larvae become fully grown. They are about 5 cm long and have white spots on the body, dense tufts of long hair and a red head. The larvae, which often groups in large numbers, are voracious leaf feeders.

4.2.3 BORERS

4.2.3.1 Spotted stemborer

Distribution: India, Pakistan, Sri Lanka, Indonesia, Iraq, Japan, Uganda, Taiwan, Sudan, Nepal, Bangladesh, and Thailand.

Host range: Maize, sorghum, sugarcane, pearl millet, rice, *Sorghum halepense*, finger millet, etc.

Bionomics: A female lays up to 500 eggs in batches of 10−80 near the midrib on the under surface of the leaves. Eggs hatch in 4−5 days. The larval development is completed in 19−27 days. Pupation takes place inside the stem and the adults emerge in 7−10 days. During the off-season, the larvae undergo diapause in plant stalks and stubbles. With the onset of rainy season, the larvae pupate and the adults emerge in 7 days.

Damage symptoms: It infests the crop a month after sowing and the damage continues up to the emergence of earheads. The first indication of stemborer infestation is the appearance of small-elongated windows in whorl leaves where the young larvae have eaten the upper surface of the leaves. Later, the plant presents a ragged appearance as the severity of damage increases. The third-instar larvae migrate to the base of the plant and bore into the shoot causing withering of the central shoot forming typical "deadheart" symptom. Normally, two leaves dry up as a result of stemborer damage. Larvae continue to feed inside the stem throughout the crop growth. Extensive tunneling of the stem and peduncle leads to drying up of the panicle, production of a partially chaffy panicle or peduncle breakage. Stemborer infestation starts about 20 days after seedling emergence, and deadhearts appear on 30−40 day-old crop.

4.2.3.2 *Pink stemborer:* Sesamia inferens *(Noctuidae: Lepidoptera)*

Geographical distribution: India, Pakistan, Malaysia, Taiwan, Burma, Bangladesh, Sri Lanka, South East Asia, China, Korea, Japan, and Indonesia.

Host range: Sorghum, maize, rice, wheat, sugarcane, pearl millet, ragi, barley, guinea grasses.

Bionomics: The adult moth is fawn-colored, with dark brown streaks on the forewings and white hind wings. The female lays about 150 creamy-white and hemispherical eggs that are arranged in two or three rows between the leaf sheath and the stem of the host plant. Egg period lasts for 7 days. The fully grown larvae measures about 25 mm and is pale yellow with a purple pink tinge and a reddish-brown head. Usually the larval period is 25 days but in cold months it may extend to 75 days. Pupation occurs in the larval tunnel in the stem and the adult emerges in 12 days. One generation may take 6−7 weeks. The life cycle is completed in 45−75 days. There are 4−6 generations per year.

Damage symptoms: The pink larva bores into the stem and damages the central shoot resulting in deadheart.

4.2.4 SUCKING PESTS

4.2.4.1 *Shoot bugs:* Peregrinus maidis *(Delphacidae, Hemiptera)*

Geographical distribution: Karnataka, Tamil Nadu, Andhra Pradesh, and Madhya Pradesh in India.

Host range: Sorghum, maize, rice, millets.

Bionomics: The adult is yellowish brown to dark brown with translucent wings. The brachypterous female is yellowish while the macropterous female is yellowish brown and male dark brown. It lays egg in group of 1−4 inside the leaf tissue, which remain covered with a white waxy substance. The fecundity of the bug is 97 eggs female^{-1} and the egg period lasts for 7 days. The nymphal stage undergoes 5 instars in 16 days. The total life cycle is completed in 18−31 days.

Damage symptoms: Shoot bugs pierce the vascular tissues and suck sap from the leaves, leaf sheaths, and stem during exploratory feeding. Adults and nymphs suck the plant sap, which causes reduced vigor, stunting, and yellowing of leaves. Under severe condition the leaf damage spreads downwards resulting in complete death of the plant. The infestation prior to preboot leaf stage usually causes girdling/twisting of top leaves, which bend downward and prevent panicle development and emergence. The damage is caused between 30−60 days (knee-high to preflowering stage). Excessive oviposition in the midrib causes eventual drying up of leaves and the tissue surrounding the ovipositing site becomes septic and turns reddish. The feeding punctures and wounds produced by ovipositor predispose sorghum plants to fungal infections. The copious excretion of honeydew by the insect, encourages sooty mold development. Further, it indirectly reduces the quality and quantity of plant biomass. Moreover, it is also a vector of several important viral diseases like maize mosaic, maize stripe, freckled yellow, and male sterile stunt virus.

Under favorable conditions, shoot bug completes several generations on sorghum within a season, and so can cause heavy damage. During the off-season, it survives on wild grasses like itch grass (*Rottboellia cochinchinensis*), goose grass (*Eleusine indica*), barnyard grass (*Echinochloa crusgalli* L.), bristle grass (*Setaria parviflora*), jungle rice (*Echinochloa colona*), Eastern gama grass

(*Tripsacum dactyloides*), and wild Sudan grass (*Sorghum bicolor* subsp. *drummondii*). In addition, maize, triticale, rye, sugarcane, and oats were also observed as alternate hosts for shoot bug. Initially, the macropterous adults emigrate to a healthy crop and colonize it for oviposition. The progeny emerging from these eggs may become either brachypterous (wingless) or macropterous (winged) adults under low and high populations. The adults discriminate and preferentially lay eggs on the abaxial surface of the midrib of mature leaves of the sorghum plants. Both adults and nymphs feed on sorghum plants by inserting their stylets into vascular tissue and sucking the plant sap. Repeated stylet insertions into the leaf and stem frequently result in a ring of damaged tissue known as "girdle". Girdled regions are characterized by necrotic epidermal tissue and disorganized vascular bundles. In addition, girdling also appears to interfere with translocation of photosynthates.

4.2.4.2 *Aphids:* Rhopalosiphum maidis, Melanaphis sacchari *(Aphididae: Hemiptera)*

Geographical distribution: All sorghum-growing areas of the world.
Host range: Sorghum, maize, finger millet.
Bionomics

> *Rhopalosiphum maidis:* The aphid is dark bluish-green in color. It is 2 mm long, with black legs, cornicles, and antennae. Winged and wingless forms occur. Females give birth to living young without mating and a generation requires only a week. The adult is yellow in color and has dark green legs.
> *Melanaphis sacchari:* The sugarcane aphid is yellow to buff in color. Their number increases rapidly during a dry spell or at the end of the rainy season. The female of the wingless form deposits 60−100 nymphs within its reproductive period of 13−20 days. The winged form produces slightly fewer nymphs. The life cycle is completed in about 5.5−7.0 days during the dry season. Colonies of aphids are seen in central leaf whorl, stems or in panicles. The young and adults suck the plant juice. This frequently causes yellowish mottling of the leaves and marginal leaf necrosis. The aphid produces an abundance of honeydew on which molds grow. In panicles, honeydew may hinder harvesting. The aphid transmits maize dwarf mosaic virus.

4.2.5 PANICLE PESTS

4.2.5.1 *Midges:* Contarinia sorghicola *(Cecidomyiidae: Diptera)*

Geographical distribution: India, Pakistan, Bangladesh, West Iran, Sri Lanka, Sudan, Java, Africa, South East Asia, South China, South America, West Indies, USA, and Italy.
Hosts: Sorghum and wild species.
Bionomics: The adult fly is small, fragile with a bright orange abdomen and a pair of transparent wings. It lays eggs singly in developing florets resulting in pollen shedding. A female lays about 30−35 eggs at the rate of 6−10 in each floret. The incubation period is 3−4 days. The maggot has 4 instars with duration of 8−10 days. Larvae are colorless, but when fully grown, they are dark orange in color. Larval period is about 9−11 days. The larval stage undergoes diapause in a cocoon during Dec.−Jan. within a spikelet. They pupate beneath the glume for about 3 days. When the adult emerges the white pupal skin remains at the tip of the spikelet. A generation is completed in 14−16 days. The insect's rapid developmental cycle permits 9−12 generations in a year.

Damage symptoms: A maggot feeds on the developing grains and pupates there. Damage symptoms appear as white pupal cases protruding out from the grains and formation of chaffy grains with holes.

4.2.5.2 Earhead bugs: Calocoris angustatus (Miridae: Hemiptera)

Geographical distribution: South India.
Host range: Pearl Millet, maize, tenai, sugarcane, and grasses.
Bionomics: Adult male is green in color and female is green with a brown margin. Blue cigar-shaped eggs are laid under the glumes or into the middle of the florets. Each insect lays between 150−200 eggs in about 7 days. Nymphs are slender, green in color. First instar larva is orange in color. The nymphal period is 10−14 days. The life cycle from egg to adult occupies less than 3 weeks. At least two generations of the bug can feed on the same crop when the panicles do not ripen at the same time.
Damage symptoms: The adults and nymphs damage the earheads by feeding on them. They suck the juice from the grains when they are in the milky stage. The sucked-out grains shrink and turn black in color and become ill-filled (or) chaffy. Older grain shows distinct feeding punctures that reduces grain quality.

4.2.5.3 Caterpillars

Earhead web worm: *Cryptoblabes gnidiella* (Pyraustidae: Lepidoptera).
   Host: Sorghum, Maize.
   Bionomics: The adult moth is small with brown forewings and light brown hind wings. Creamy white, round or conical eggs are laid singly on the spikelets and on grains in a panicle. The egg period is 3−4 days. The larva is light brown with a dark head and has dark lateral lines on the body. The larval duration is 9−10 days and the pupal period is 7 days. The larva constructs a silken cocoon and pupates within the silken webs. The life cycle is completed in 23−24 days.
   Damage symptoms: The larvae destroy the grain in the earhead. They produce webs of silken thread that remain on and inside the earhead. Heavily infested heads may be covered with webbing.
Gram caterpillar: *Helicoverpa armigera* (Noctuidae: Lepidoptera).
   Geographical distribution: Worldwide. It is a major pest on cotton, lablab, chillies, tomato, pulses, maize and a minor pest on sorghum.
   Host range: Cotton, sorghum, lablab, soybean, pea, safflower, chillies, tomato, groundnut, tobacco, gram, okra, maize, etc.
   Bionomics: Adult is a brown color moth with a "v" shaped speck on forewings and dull black border on the hind wing. The larva is green with dark broken gray lines and dark pale bands. It shows color variation of greenish to brown.
   Damage symptoms: Larvae hide within the earheads and feed on the grains. Earheads are partially eaten and appear chalky. Fecal pellets are visible within the earhead.

4.2.6 ROOT PESTS

4.2.6.1 White grubs

Geographical distribution: White grubs are common throughout North America, although species distribution varies greatly.

Host Plants: White grubs feed on the roots of corn, timothy, Kentucky bluegrass, sorghum, soybean, strawberry, potato, barley, oat, wheat, rye, bean, turnip, and to a lesser degree, other cultivated crops. They also infest various pasture grasses, lawns, and nursery plantings. The adults are strongly attracted to fragrant flowers and ripe fruits, and feed on the foliage of forest, shade, and fruit trees.

Bionomics: In spring, overwintering May beetles emerge from the ground at dusk, feed on the leaves of trees, and mate during the night. At dawn, they return to the ground, where the females lay 15−20 eggs in earthen cells several centimeters below the surface. Most of the May beetles lay eggs in grassy sod. Eggs hatch in about 3−4 weeks. The young grubs feed on plant roots throughout the summer; in the fall, they burrow below the frost line (to a depth of 1.5 m) and hibernate. The following spring, they return near the soil surface to feed and grow. In fall, the grubs again migrate downward to overwinter. The third spring, they move upward to feed on plant roots. By late spring, they are completely grown. These large grubs form earthen cells and pupate. In late summer, adults emerge from the pupal stage, but they do not leave the ground. These beetles overwinter, emerging the next spring to feed and mate. The usual length of time for one complete generation (adult to adult) is 2−4 years depending on latitude. Generations, however, are staggered so that the grubs and the beetles are present every year. Grubs are usually most numerous and damaging the second season following a large beetle flight.

Damage symptoms: Damage by white grubs is usually most severe when corn is planted following sod. In this case, root feeding can be so severe that plants may grow no taller than 30−60 cm. If the root system is badly damaged, injured plants will eventually die and can be easily pulled from the ground. Even light infestations usually result in increased lodging and reduce yield. White grubs are sensitive to differences in soil moisture and texture. Since these factors are not uniform throughout a given field, a white grub infestation, likewise, is not uniform. Therefore, within the same field, some areas may be completely destroyed while others remain undamaged.

4.3 HOST-PLANT RESISTANCE

Host-plant resistance is one of the most effective means of pest management in sorghum. It is an environmentally friendly method, which is compatible with other control strategies such as biological, cultural, and chemical control.

4.3.1 HOST FINDING AND ORIENTATION

Selection of a suitable site for oviposition by phytophagous insects is critical for successful development of the offspring. Host-plant recognition by insects is a cumulative effect of several stimuli. The leaf surface constitutes an interface between the external environment and the plant tissues. Most behavioral events that lead insects to lay eggs and feed on a host plant are associated with leaf surface contact sensory cues. The selection or avoidance of potential host plants by phytophagous insects is guided by a complex combination of physical and chemical stimuli. Color, shape, and olfactory cues may play a role in the initial orientation, whereas acceptance or rejection of a plant depends on texture as well as chemical stimulants or deterrents. Initiation of feeding is stimulated or deterred by the presence or absence of specific chemicals. The behavioral events leading to oviposition are mediated to a large extent by chemical cues associated with potential host plants. Orientation and landing are primarily guided by volatile constituents of a plant, whereas assessment of a leaf surface depends on contact stimuli. Perception of chemical cues that affects oviposition involves receptors on antennae, tarsi, mouthparts, or the ovipositor. Complex behavior such as tarsal "drumming" provides increased receptor contact with chemical stimuli. Abiotic and biotic environmental factors often influence the production or release of behavior-modifying chemicals by a plant, and therefore affect oviposition preferences. Insects have specialized sensory nervous systems that allow them to use a variety of cues to find and identify target organisms. Cues can be physical such as color, sound, shape, and size as well as chemical and these may be useful for long or short range attraction to prey.

4.3.1.1 Chemical cues

Females of sorghum shoot fly (*Atherigona soccata*) are attracted to the host plant for oviposition by the odors emitted from the host plant. Females are attracted both to the volatiles emitted by susceptible seedlings and to phototactic (optical) stimuli that facilitate orientation to the host for oviposition (Nwanze et al., 1998). Females elicited both electrophysiological and behavioral responses when exposed to the volatiles collected from the susceptible sorghum cultivar "Swarna". These compounds were identified as α-pinene, 6-methyl-5-hepten-2-one, octanal, (Z)-3-hexen-1-yl acetate, nonanal, methyl salicylate, decanal, (-)-(E)-caryophyllene (Padmaja et al., 2010a,b). The compounds, i.e., undecane 5-methyl, decane 4-methyl, hexane 2, 4-methyl, pentadecane 8-hexyl, and dodecane 2, 6, 11-trimethyl, present on the leaf surface of sorghum seedlings, were associated with susceptibility to shoot fly, while 4, 4-dimethyl cyclooctene was associated with resistance to shoot fly (Chamarthi et al., 2011).

Linalool, acetophenone, and 4-allylanisole were the electrophysiologically active components in volatiles released by sorghum that elicited large responses in *C. partellus* moth (Birkett et al., 2006). The ability of stemborer *Busseola fusca* (Fuller), an important pest of maize and sorghum in sub-Saharan Africa, to recognize and colonize a variety of plants is based on the interaction between its sensory systems and the physical and chemical characteristics of its immediate environment. Volatiles produced by the plant species did not appear to influence the general orientation of *B. fusca* toward the plant and the females were not able to recognize their preferred hosts from a distance. After landing, the female typically swept her ovipositor on the plant surface, simultaneously touching it with the tips of her antennae, and then oviposited. This behavior was more frequently observed on sorghum indicating that both antennal and ovipositor receptors are used by the female moths to evaluate the plant surface before deciding to oviposit. Females recognized their preferred hosts only after landing. Tactile and contact-chemoreception stimuli from the plants seemed to play a major role in oviposition decisions of *B. fusca* (Calatayud et al., 2008).

Female midges are attracted to sorghum crops via a combination of visual and chemical stimuli. Response of sorghum midge is greater to the host-plant odor when combined with red and yellow colors than to the host odor alone. Both visual and chemical stimuli possibly influence host selection (Sharma et al., 1990a,b, 2002; Sharma and Franzmann, 2001). Chemical stimuli from viable pollen and receptive stigmata were found to attract and direct sorghum midge oviposition.

4.3.1.2 Visual stimuli

Visual stimuli influence the orientation behavior of the sorghum midge, *Contarinia sorghicola*. The yellow, red, and white colors are attractive to the midge while blue and black are least attractive. Sorghum panicles covered with blue- or black-colored bags in a headcage showed maximum midge damage, while the reverse was true for panicles covered with yellow, red, and white colored bags (Sharma et al., 1990a,b). Variation in spectral reflectance between sorghum genotypes may influence host selection by the sorghum midge. Information on the color preference of sorghum midge females could be exploited for developing suitable traps to monitor its abundance in combination with kairomones or pheromones. This could play an important role in understanding the behavior and population dynamics of this insect.

4.3.2 SCREENING TECHNIQUES

Development of insect-resistant cultivar, use of marker-assisted selection, and development of transgenic plant with insect resistance depends largely on the precision of a resistance screening technique. Hot-spot locations, where the pest populations are known to occur naturally and regularly at levels that often result in severe damage, are ideal to test large numbers of germplasm accessions. Infester row, cage and leaf disk screening techniques have been standardized to evaluate sorghum germplasm, breeding material, and mapping populations for resistance to insect pests under field and greenhouse conditions (Table 4.1).

Table 4.1 Techniques to Screen for Insect Pest Resistance in Sorghum

Insect Pest	Screening Technique/Hot Spot
Shoot fly	Selection at hot spots
	Interlard fish-meal technique/infestor rows
Stemborer	Hot spots Hisar in Haryana and Warangal in Andhra Pradesh, India; Agfoi and Baidoa in Somalia; Panmure and Mezarbani in Zimbabwe; Kiboko in Kenya; and Golden Valley in Zambia.
	Artificial infestation
Sugarcane aphid	Hot spots
	Leaf cage technique/augmentation of aphids
Shoot bug	Hot spots
	Mylar cage technique (augmentation of shoot bugs)
Midge	Hot spots Dharwad, Bhavanisagar, and Pantnagar in India; Sotuba in Mali; Farako Bâ in Burkina Faso; Alupe in Kenya; Kano in Nigeria.
	Infester row technique
	Headcage technique (augmentation of midges)
Headbug	Hot spots ICRISAT-Patancheru, Bhavanisagar, Kovilpatti, Coimbatore, Palem, and Dharwad in India
	Headcage technique (augmentation of earhead bugs)

4.3.2.1 Shoot fly

The interlard-fish-meal technique (Soto, 1974) is used for screening the test material under field conditions. The moistened fishmeal in polyethylene bags are kept in interlards to attract shoot flies from the surrounding areas. The amines resulting from the biodegradation of fishmeal serve as the chemical cues for shoot fly attraction to the fishmeal, but the amine based chemical compounds have not been identified, which could be useful to understand the mode of attraction of sorghum shoot fly females to fishmeal and/or its host plant (Reddy et al., 1981). Shoot fly abundance in the field can also be monitored through fishmeal-baited traps to determine its peak period of activity, which helps in decision making for planting of the test material to expose the test material to optimum shoot fly density.

A cage-screening technique is developed to confirm the resistance to shoot fly under field conditions, and to study the resistance mechanisms. Shoot flies can also be collected from fishmeal-baited traps in the field, stored under greenhouse conditions, and used for screening sorghums for resistance to shoot fly under multi-, dual-, or no-choice tests (Sharma et al., 1992; Dhillon et al., 2005a,b,c). Rapid screening can also be carried out using a top-cage technique.

4.3.2.2 Stemborer

Screening under natural infestation: Hot-spot locations, where the pest populations are known to occur naturally and regularly at levels that often result in severe damage, are ideal to test large numbers of germplasm accessions. Sowing date is to be adjusted such that the crop is at a susceptible stage when the stemborer abundance is at its peak. In northern India, C. partellus is most abundant in Aug. to Sep., and the crop sown between the first and third week of Jul. suffers maximum stemborer damage.

Field infestation: Plants at 21 DAE are infested with five neonate larvae of C. partellus in the field. The freshly hatched neonate larvae are gently mixed with the carrier (*Papaver* sp. seed), and transferred into plastic bottles fixed to the bazooka applicator for field infestation. Plants are individually infested by placing the nozzle of the bazooka onto the leaf whorl. In each stroke, five larvae are released in the morning during 07.00−09.00 h into the whorls to cause an optimum level of leaf damage and deadheart formation. Generally 5−7 larvae per plant are sufficient to cause appreciable leaf feeding and deadhearts (>90% damage) in susceptible genotypes (Sharma et al., 1997).

4.3.2.3 Shoot bug

Screening for resistance to shoot bug can be carried out under natural infestation in field or in the greenhouse. For this purpose, the material should be planted during late rainy season in Jul. or early postrainy season in Oct. Under greenhouse conditions, shoot bug infestation can also be created by using leaf cages or by confining the shoot bug females to the whorl leaves.

4.3.2.4 Sugarcane aphid

The material should be planted during late rainy season in Jul. or early postrainy season in Oct. for screening for resistance to sugarcane aphid under natural infestation in the field. Aphid infestation can also be created under greenhouse conditions. Aphid multiplication and growth rates can be studied using leaf cages by confining aphid females with the leaves, and counting the numbers of aphids produced in 15 days. For this purpose, the leaf cages can be fixed on third or fourth leaf from the bottom.

4.3.2.5 Sorghum midge

For screening against midge resistance under natural conditions, it is necessary to determine the appropriate time for sowing at different locations. Periods of maximum midge density can be determined through fortnightly sowings of a susceptible cultivar. Sowing dates should be adjusted so that the flowering of the test material coincides with greatest insect density. Maximum midge damage has been observed in the crop planted during the third week of Jul. The peak in midge density occurs during Oct., and a second but smaller peak has been observed during Mar. in the postrainy season, for which planting is carried out during mid-Dec.

Infester row technique: Midge abundance can be increased through infester rows and spreading sorghum panicles containing diapausing midge larvae in the rows (Sharma et al., 1988). Infester rows of susceptible cultivars, such as CSH 1 and CSH 5 (1:1 mixture), are to be sown 20 days before the test material. Alternatively, early-flowering (40−45 days) lines (IS 802, IS 13249, and IS 24439) can be sown along with the test material. Four infester rows of the susceptible cultivar are to be sown after every 16 rows of the test material. Collect midge-infested chaffy panicles containing diapausing midge larvae at the end of the cropping season, store in gunny bags or in bins under dry conditions until the next season. Moisten the panicles for 10−15 days to stimulate the termination of larval diapause. Spread midge-infested sorghum panicles containing diapausing midge larvae at the flag leaf stage of the infester rows. Adults emerging from the diapausing larvae serve as a starter infestation in the infester rows to supplement the natural population. Midge population multiplies for 1−2 generations on the infester rows before infesting the test material. This technique increases the midge damage by 3−5 times. Infester rows alone are also effective in increasing midge infestation.
No-choice headcage technique: Caging midge flies with sorghum panicles permits screening for midge resistance under uniform insect pressure.

4.3.2.6 Headbug

Field screening: Screening for head-bug resistance can be carried out under field conditions during the periods of maximum bug density. Such screening, however, is influenced by variation in flowering, fluctuations in bug density and the effect of weather conditions on the bug population build-up and damage. Early- and late-flowering cultivars normally escape head-bug damage, while those flowering in mid-season are exposed to very high bug infestation. Head-bug density is very high during Sep.−Oct. and this period can be utilized to increase the screening efficiency under field conditions.

4.3.3 SOURCES OF PEST RESISTANCE

A large proportion of the world sorghum germplasm collection has been evaluated for resistance to insect pests, and a number of resistant lines have been identified (Table 4.2). Much progress has also been made in understanding the genetics and mechanisms of resistance to important pests of sorghum, that is, shoot fly, spotted stemborer, sorghum midge, and head bug, *Calocoris angustatus* (Lethiery). Identified resistance sources mostly come from maldandi (semicompact head type) or dagadi (compact head type) races grown in postrainy season (Rana et al., 1985). Around 36,700 sorghum germplasm accessions are in the ICRISAT genebank which serves as a global repository.

Table 4.2 Sources and Improved Cultivars of Sorghum with Resistance to Insect Pests

Insect	Sources of Resistance	Improved Varieties
Shoot fly	M35-1 (IS 1054), IS 1057, IS 2312, IS 2146, IS 4664, IS 2205, IS 5604, and IS 18551	Swati[a], CSV 8R[a], Phule Yashoda[a], ICSV 705, ICSV 700, ICSV 717
Stemborer	IS 1055 (BP 53), IS 1044, IS 2123, IS 2195, IS 2205, IS 2146, IS 5469, and IS 18551	CSV 8R[a], ISCV 700, ICSV 708, ICSV 714, and ICSV 93046
Shoot bug	Kafir Suma, Dwarf Hegari, I 753, H 109, GIB, 3677B, and BP 53(IS 1055), IS 19349, IS 18657, IS 18677, and PJ 8K(R)	
Midge	IS 2579C, IS 12666C, TAM 2566, AF 28, DJ 6514, IS 10712, IS 7005, IS 8891, and IS 8721	ICSV 197, ICSV 735, ICSV 745[a] (DSV 30, ICSV 88013, ICSV 758[a], and ICSV 88032
Headbug	IS 17610, IS 17643, IS 21443, IS 17618, and IS 14332	Malisor 84-7

[a]*Released for cultivation (Sharma et al., 1992, 2003; Singh and Rana, 1986, 1989).*

Resistant lines have been identified by undertaking extensive screening of the germplasm collections for key pests such as shoot fly, spotted stemborer, and midge and head bugs. Wild species of sorghum (*Sorghum purpureosericeum* and *S. versicolor*) possess very high levels of resistance to shoot fly (Mote, 1984). Wild accessions belonging to *Parasorghum* (*S. australience*, *S. purpureosericeum*, *S. brevicallosum*, *S. timorense*, *S. versicolor*, *S. matarankense*, and *S. nitidum*) and *Stiposorghum* (*S. angustum*, *S. ecarinatum*, *S. extans*, *S. intrans*, *S. interjectum*, and *S. stipoideum*) did not show any shoot fly damage under multichoice conditions in the field. *Heterosorghum* (*S. laxiflorum*) and *Chaetosorghum* (*S. macrospermum*) showed very low damage (Venkateswaran, 2003). IS 18226 (race *arundinaceum*) and IS 14212 (*S. halepense*) resulted in reduced survival and fecundity. The wild relatives also exhibited very high levels of antibiosis to shoot fly, while only low levels of antibiosis have been observed in the cultivated germplasm. Therefore, wild relatives with different mechanisms of resistance can be used as a source of alternate genes to increase the levels of resistance to shoot fly (Kamala et al., 2009).

IS 18584, IS 18577, and IS 2205 were identified as the most resistant to stemborer on the basis of deadhearts, leaf injury, stem tunneling, peduncle tunneling, and exit holes (Patel et al., 1996). The midge resistance present in Australian hybrids is largely drawn from North American sources (Henzell et al., 1994). The resistance genes within this material confer only one mechanism of resistance, ovipositional-antixenosis (Franzmann, 1993). DJ6514 and its derivatives from India are known to contain both antixenotic and antibiotic mechanisms of resistance (Sharma, 1985; Sharma et al., 2002). Sorghum lines IS 3461, IS 9807, IS 10712, IS 18563, IS 19476, IS 21873, IS 21881, IS 22806, PM 15936-2, and ICSV 197 have high levels of resistance to midge (Sharma et al., 2002). Plants with eggs, deadhearts, leaf glossiness, trichomes on the abaxial surface of the leaf, and leaf sheath pigmentation are the most reliable parameters, and these can be used as marker traits to screen and select for resistance to shoot fly. The genotypes showing resistance to stemborer leaf feeding, deadheart formation, stem tunneling, and/or compensation in grain yield can be used for sorghum improvement.

Table 4.3 Inheritance of Insect Pest Resistance in Sorghum		
Insect	**Genetics of Resistance**	**References**
Shoot fly	Complex and polygenically inherited with predominantly additive gene effects	Halalli et al. (1982), Dhillon et al. (2006a,b), Gibson and Maiti (1983)
	Recessive trait conditioned by a single gene	
Stemborer	Additive and partially dominant over susceptibility	Pathak (1985), Pathak and Olela (1983), Rana et al. (1984), Sharma et al. (2007), Nour and Ali (1998)
	Foliar damage, deadheart, stem tunneling, and number of exit holes are governed by additive gene action	
	The general combining ability (GCA) and specific combining ability (SCA) estimates suggested that leaf feeding score, number of nodes, overall resistance score, panicle initiation, recovery score, and stalk length are governed by additive type of gene action	
Midge	Recessive trait and is controlled by two or more loci	Boozaya-Angoon et al. (1984), Rossetto and Igue (1983)
Greenbug	Resistance is controlled largely by additive gene action	Widstrom et al. (1984), Sharma et al. (1996, 2000)
Headbug	Resistance to biotypes C, E, F, and I is inherited as an incompletely dominant trait controlled by a few major genes; partially dominant trait controlled by both additive and nonadditive gene action	Weibel et al. (1972), Puterka and Peters (1995), Tuinstra et al. (2001), Sharma et al. (2000)

4.3.4 GENETICS AND INHERITANCE OF RESISTANCE

The inheritance of resistance to insects in sorghum shows that some resistance traits are controlled qualitatively by dominant genes while others by recessive genes and quantitative inheritance (Table 4.3). Expression of resistance in the F_1 hybrids is influenced by cytoplasmic male sterility and resistance is needed in both parents to produce sorghum hybrids with resistance to shoot fly, sugarcane aphid, midge, and headbug (Sharma et al., 1996, 2005; Dhillon et al., 2006a,b).

4.3.5 MECHANISM OF RESISTANCE

4.3.5.1 Basic mechanisms

The three major mechanisms of resistance, that is, oviposition nonpreference (antixenosis), antibiosis, and recovery resistance contribute to host-plant resistance to sorghum insect pests (Table 4.4). Resistant plants may contain one or a combination of the three mechanisms that collectively contribute to the level of insect resistance. Each mechanism of resistance acts at some stage of the insect/plant relationship and is contributed by physical or chemical plant characters that may be referred to as components of resistance. Nonpreference or antixenosis mechanisms of resistance

Table 4.4 Resistance Mechanism to Insect Pest in Sorghum

Insect	Resistance Mechanism	Reference
Shoot fly	Antixenosis for oviposition; antibiosis; recovery resistance in terms of productive tillers	Dogget et al. (1970), Soto (1974), Raina et al. (1981), Taneja and Leuschner (1985), Sharma and Nwanze (1997), Dhillon et al. (2005a,b,c, 2006a,b)
Stemborer	Antibiosis tolerance	Jotwani et al. (1978), Verma et al. (1992), Padmaja et al. (2012)
Shoot bug	Antixenosis for adult colonization and oviposition; antibiosis for nymphal population buildup; tolerance for plant damage/transmission of viruses	Chandra Shekar et al. (1992, 1993)
Sugarcane aphid	Antixenosis for adult colonization; antibiosis for population buildup; tolerance to plant damage	
Midge	Ovipositional antixenosis; antibiosis to feeding larvae	Franzmann (1988), Rossetto et al. (1984), Sharma and Vidyasagar (1994)
Earheadbug	Nonpreference	Sharma and Lopez (1990)

may occur during the colonization phase as insects approach, make contact with, arrest, and oviposit on host plants. Antibiosis resistance occurs at the utilization phase of insect $\times$ plant interaction where larval growth, survival, and adult fecundity may be affected as the insect ingests, assimilates, and converts food. Finally, as emerging adults disperse from their host plant they may either reinfest the crop, or emigrate to another distant host.

4.3.5.1.1 Ovipositional antixenosis/nonpreference

Nonpreference by insects is the property of the plant to render it unattractive for oviposition, feeding, or shelter. Absence of physicochemical stimuli that are involved in selection of host plant or presence of repellents, deterrants, and antifeedants contribute to the antixenosis mechanism of resistance. Oviposition nonpreference is considered to be a primary mechanism of resistance to major insect pests of sorghum, that is, shoot fly, head bug, and midge (Blum, 1967; Sharma and Lopez, 1990; Sharma and Vidyasagar, 1994).

Nonpreference is an important mechanism of resistance to shoot fly. Oviposition nonpreference has been identified to be one of the components of resistance to spotted stemborer in sorghum (Singh and Rana, 1984; Alghali, 1985; Saxena, 1990; van den Berg and van der Westhuizena, 1997; Padmaja et al., 2012). Antixenosis for colonization, oviposition, and/or feeding is one of the predominant mechanisms of resistance to planthopper in sorghum (Chandra Shekar, 1991; Chandra Shekar et al., 1992, 1993). Resistant genotypes of sorghum receive 5−10 times fewer eggs, and those deposited are arranged in a disorderly manner, compared to the susceptible hybrid, even in no-choice conditions (Chandra Shekar, 1991; Chandra Shekar et al., 1993). There is also a marked preference for oviposition on mature leaves of older plants compared to very few eggs on young leaves (Napompeth, 1973) even in no-choice conditions (Fisk, 1978; Singh and Rana, 1992).

Compact and tightly wrapped whorl leaves around the stem of some sorghum genotypes impart resistance (Agarwal et al., 1978). High levels of nitrogen, sugar, and total chlorophyll contents have been shown to be strongly associated with susceptibility and genotypes with high phosphorus, potash, and polyphenol content are less preferred by planthopper (Mote and Shahane, 1994).

Ovipositional antixenosis has been recognized as the major mechanism of resistance in midge-resistant hybrids (Franzmann, 1988; Rossetto et al., 1984; Sharma and Vidyasagar, 1994; Waquil et al., 1986). Antixenosis mechanism of resistance may be closely linked with the structural morphology of spikelets (Henzell et al., 1994); small glume size, the extent of glume closure (Bergquist et al., 1974; Jadhav and Jadhav, 1978; Rossetto et al., 1975). The midge susceptibility of a number of sorghum lines was positively and significantly correlated with glume, palea, lemma, anther, and style length (Sharma, 1985; Sharma et al., 1990a,b, 2002), while glumes of spikelets of resistant varieties were more tightly closed than those in susceptible varieties (Waquil et al., 1986). Searching time was shorter (5.9 s) for midge females on a resistant hybrid compared to a susceptible hybrid (7.2 s), probing time and successful oviposition took four times as long in resistant hybrids (Waquil et al., 1986). Midge females did not lay eggs in the spikelets of *Sorghum amplum*, *S. bulbosum*, and *S. angustum* compared to 30% spikelets with eggs in *S. halepense* when infested with five midge females per panicle under no-choice conditions. However, one egg was laid in *S. amplum* when infested with 50 midges per panicle. A larger number of midges were attracted to the odors from the panicles of *S. halepense* than to the panicles of *S. stipoideum*, *S. brachypodum*, *S. angustum*, *S. macrospermum*, *S. nitidium*, *S. laxiflorum*, and *S. amplum* in dual-choice olfactometer tests. The differences in midge response to the odors from *S. halepense* and *S. intrans* were not significant. Under multichoice conditions, when the females were also allowed a contact with the host, more females were attracted to the panicles of *S. bicolor* compared with *S. amplum*, *S. angustum*, and *S. halepense*. More midges responded to the panicles of IS 10712 compared with *S. halepense*, whereas the differences in midge response to the panicles of ICSV 197 (*S. bicolor*) and *S. halepense* were not apparent, indicating that *S. halepense* is as attractive to sorghum midge females as *S. bicolor*. The wild relatives of sorghum (except *S. halepense*) were not preferred for oviposition, and they were also less attractive to the sorghum midge females. Thus, wild relatives of sorghum can prove to be an alternative source of genes for resistance to sorghum midge (Sharma and Franzmann, 2001).

Scizaphis graminum is deterred from feeding by the phenolics procyanidin, P-hydroxybenzaldehyde, and dhurrin in resistant sorghum genotypes (Dreyer et al., 1981). Cultivar preference or nonpreference for feeding is one of the components of resistance to the head bug. Reduced oviposition is an important component of resistance to head bugs in IS 17645, IS 17610, and IS 17618 (Sharma and Lopez, 1990).

4.3.5.1.2 Antibiosis

Antibiosis includes the adverse effect of the host-plant on the biology of the insects and their progeny (survival, development, and reproduction). Both chemical and morphological plant defenses mediate antibiosis. The death of early instars, reduced size or low weight, prolonged periods of development of the immature stages, reduced adult longevity and fecundity, and death in the prepupal or pupal stage are the effects of antibiosis.

Retardation of growth and development, prolonged larval and pupal periods, and poor emergence of adults on resistant genotypes provides an evidence of antibiosis to sorghum shoot fly (Sharma et al., 1997; Singh and Jotwani, 1980a,b,c; Raina et al., 1981). The larvae on the resistant genotypes are generally smaller, and the mortality of the first-instars was higher than on the susceptible genotypes. The mortality was the highest (90%) during the first 24 h (Zein el Abdin, 1981). Highest larval survival has been observed on 2-week old plants, followed by very young seedlings, and lowest in >50-day-old plants (Ogwaro and Kokwaro, 1981). Antibiosis of shoot fly offers exciting possibilities of exerting biotic pressure against insect feeding and development, resulting in low-larval survival on resistant varieties (Dahms, 1969; Soto, 1974). Resistance to shoot fly is a cumulative effect of nonpreference and antibiosis (Raina et al., 1981). Survival and development were adversely affected when shoot flies were reared on resistant varieties (Jotwani and Srivastava, 1970; Narayana, 1975; Raina et al., 1981; Unnithan and Reddy, 1985) compared with susceptible genotypes (Singh and Narayana, 1978). Retardation of growth and development prolonged the larval and pupal periods and caused poor emergence of adults on resistant varieties indicating direct evidence of antibiosis (Narayana, 1975; Raina et al., 1981). Survival and fecundity were also better on highly susceptible varieties (Singh and Narayana, 1978) but adversely affected on resistant varieties (Taneja and Leuschner, 1985). The larval and total growth indices were significantly lower in resistant than susceptible varieties. The percentage pupation on the resistant lines was significantly lower than on susceptible lines (Dhawan et al., 1993; Dhillon et al., 2005a,b,c).

Biochemical deficiencies or the presence of chemical factors in resistant cultivars might adversely affect the development and survival of shoot fly larvae (Raina, 1985). Patil et al. (2006) observed high enzyme activity (peroxidase and polyphenol oxidase) in resistant lines as well as resistant × resistant and resistant × susceptible crosses. The higher enzyme activity might be inducing and activating the antibiosis mechanism, leading to reduction in damage caused by the shoot fly. Trichomeless cultivars accumulate more dew and stay wet longer facilitating the movement of freshly hatched larvae to the base of the central shoot. On the other hand, trichomed cultivars would tend to dry faster, making the downward journey of the larvae more difficult (Raina et al., 1981). Antibiosis was also attributed to early deposition of irregular shaped silica bodies in the abaxial epidermis of the leaf sheaths and distinct lignification and thickening of walls of cells enclosing the vascular bundle sheaths within the central whorl of young leaves (Ponnaiya, 1951a,b; Blum, 1968). The resistant varieties, as compared to the susceptible one, possessed a much greater density of silica bodies (dumbbell-shaped, intercostal, and silicified prickle hairs) in the abaxial epidermis at the base of the first, second, and third leaf sheaths. The density increased from the first to the third leaf sheath (Blum, 1968).

Tolerance and antibiosis mechanisms are operative in stemborer resistant cultivars (Jotwani, 1976). Resistant lines exhibited adverse effects on survival and development of the borer under field conditions (Lal and Pant, 1980; Singh and Verma, 1988; Woodhead and Taneja, 1987). The larval, pupal, and the total development period is also prolonged (Jotwani et al., 1978; Saxena, 1992; Verma et al., 1992; Padmaja et al., 2012). Antibiosis is also expressed in terms of reduced pupal weight (Lal and Sukhani, 1982; Verma, et al., 1992) and low pupation and adult emergence (Singh and Verma, 1988). The antibiotic effects of the resistant genotypes on the development of the borer may be because of secondary plant substances in the leaves and/or poor nutritional quality of the food. Low sugar content (Swarup and Chaugale, 1962) and greater amounts of amino acids, tannins, total phenols, neutral detergent fiber (NDF), acid detergent fiber (ADF), and lignins (Khurana and Verma, 1982, 1983) and silica content (Narwal, 1973) are associated with resistance to stemborer in sorghum. Larval mortality is greater in diet impregnated with petroleum ether extract of the borer-resistant lines. Methanolic extracts from the susceptible line IS 18363 showed greater feeding stimulation than the extracts from the less susceptible cultivar, IS 2205. IS 18363 had greater phenolic and sugar contents than the less susceptible cultivar, IS 2205 (Torto et al., 1990). These biochemical constituents might influence the insect survival and development adversely.

Antibiosis is expressed as an increased mortality rate, prolonging of nymphal development, and reduction of fecundity of the shoot bug in sorghum (Chandra Shekar, 1991; Chandra Shekar et al., 1993). Antibiosis to larvae feeding within the spikelet is a second less common mechanism of midge resistance observed in sorghum, and results in the death of larvae before they cause the seed to abort (Sharma, 1985). There is evidence for antibiosis to midge in sorghum germplasm, leading to decreased rates of postembryonic growth, survival, and adult fecundity. Larvae reared on a number of resistant varieties were smaller in size and weight compared to larvae reared on susceptible varieties (Sharma et al., 1993; Wuensche, 1980). A greater proportion of larvae developed through late instars on a susceptible hybrid compared to resistant hybrids and a greater proportion of larva positioned against the caryopsis in the susceptible hybrid (Waquil et al., 1986). In addition to delayed emergence there was decreased fecundity and lower rates of progeny production in females reared on midge-resistant lines (Sharma et al., 1993). Postembryonic stages of life cycle were smaller, lighter, and took longer to complete development when reared on a number of midge-resistant lines (Natarajan and Chelliah, 1985). There is also evidence for higher larval mortality in midge-resistant lines from a number of studies undertaken world-wide (Sharma, 1985; Sharma et al., 1993; Teetes and Johnson, 1978). A positive correlation exists between growth rates of grain (caryopsis), and midge resistance (Sharma, 1985, 1993) and between tannin contents and midge resistance, with the exception of the highly antibiotic line DJ6514. Antibiosis, or mortality of the immature stages of midge development, has been identified in a number of lines, including line DJ6514, and the related line ICSV745, both of which produced over 60% larval mortality (Sharma et al., 1993).

Postembryonic development of head bug was extended by 1−2 days when nymphs were fed on the genotypes, IS 17645, IS 17610, and IS 17618. Fifth-instar nymphs and adults had lower weights when reared on IS 17610 and IS 9692. Nymphal survival was relatively lower on IS 9692, IS 17610, and IS 17645. Fourth-instar nymphs were relatively less efficient in food utilization when fed on grain of IS 2761 and IS 6984 as compared with the susceptible cultivars, CSH 5 and Swarna. Stage of grain development influenced the indices of consumption and utilization of food by head bugs. Consumption index and growth rate were lower on 20-day-old grain of IS 2761 as compared with 12-day-old grain (Sharma and Lopez, 1990). Reduced survival and establishment will reduce the insect population and the resultant crop damage. Prolongation of development period will also result in reduction of number of generations in a season/year.

4.3.5.1.3 Tolerance/recovery resistance

Tolerance or recovery resistance is where the plant is able to withstand or recover from damage caused by insect. Synchronized tillering after the main shoot is killed is potentially a form of recovery resistance because the tillers in some genotypes express higher levels of resistance than the main shoots (Dogget, 1972). This form of resistance has been referred to as tiller survival (Blum, 1969a,b) or recovery resistance (Dogget et al., 1970). Tall seedlings and high-plant recovery were reported as the characteristics of resistant varieties by Sharma et al. (1977), which may not have definite relation with the height of the plant, as some of the tolerant germplasm lines were dwarf, medium tall, or very tall (Shivankar et al., 1989; Dhillon, 2004). Resistant cultivars of sorghum had a very high rate of tiller survival compared with susceptible cultivars. It was also suggested that the frequency of tiller survival was related to the rate of tiller growth, so that the faster a tiller grew, the greater were its chances of avoiding infestation (Blum, 1972). The shoot fly-resistant genotypes had significantly less tiller deadhearts than the susceptible ones. Tiller development consequent to deadheart formation in the main shoot, and its survival depend on the level of primary resistance and shoot fly abundance (Dogget et al., 1970; Dhillon, 2004). Varieties with high recovery resistance yield more under shoot fly infestation (Rana et al., 1985).

Faster tiller growth leads to the minimizing shoot fly infestation in tillers (Blum, 1972). Varieties with high recovery resistance compensate for yield loss under shoot fly infestation (Rana et al., 1985). The *Serena* and *Namatrare* varieties recovered well even when more than 90% of the main plants were killed by shoot fly attack (Dogget and Mjisu, 1965; Dogget et al., 1970). Seedling vigor and high rate of recovery are important characteristics of resistant cultivars (Sharma et al., 1977), which may not be related with seedling height, because some of the tolerant germplasm lines are dwarf, medium tall, or very tall (Shivankar et al., 1989; Dhillon et al., 2005a,b,c). Recovery resistance does not appear to be a useful mechanism of resistance particularly when shoot fly population increases progressively as the rainy season continues (Singh and Rana, 1986). The damaged plants produce axial tillers, which serve as a mechanism of recovery resistance. However, the axial tillers often mature later than the main plants and often suffer greater damage by sorghum midge, head bugs, and birds or may not be able to produce grain under drought stress (Dhillon, 2004). In Africa, it was reported that farmers actually preferred an initial infestation of their sorghum by shoot fly that led to profuse tillering and a good harvest (Dogget, 1972). However, tolerance can be greatly influenced by the growth conditions of the plant and thus may not always be predictable at various locations, particularly those with irregular patterns of rainfall (Raina, 1985). Various studies have shown that yield compensation occurs within panicles of all sorghum lines when spikelets or developing kernels are physically removed (Fisher and Wilson, 1975; Hamilton et al., 1982; Henzell and Gillieron, 1973). Compensation was generally positive in both midge resistant and susceptible hybrids and greatest between 30−50% seed set. While Sharma et al. (2002) also recorded variable but generally positive levels of yield compensation in a wide range of midge-resistant lines compared to negative compensation in a susceptible line.

4.3.5.2 *Factors associated with resistance*
4.3.5.2.1 Climatic and edaphic factors
The most desirable form of insect resistance is the one that is stable across locations and seasons. Several climatic and edaphic factors influence the level and nature of resistance to insect pests. Sorghum plants with water stress suffer greater damage by the spotted stemborer and sugarcane aphid (Sharma et al., 2005). Nutrition plays an important role in plant resistance to insects. Application of nitrogenous fertilizers decreases the damage by shoot fly and stemborer (Reddy and Rao, 1975; Chand et al., 1979). A decrease in shoot fly damage has also been observed after application of phosphatic fertilizers (Sharma, et al., 1997). Changes in nutrient supply also affect the resistance to greenbug (Schweissing and Wilde, 1979). Expression of resistance to midge is influenced by temperature and the relative humidity (Sharma et al., 2003). Lower temperatures resulting in reduced rate of growth of the developing grain increase susceptibility to midge (Sharma et al., 1999). Differences in genotypic susceptibility to greenbug increase with an increase in temperature (Schweissing and Wilde, 1978).

4.3.5.2.2 Morpho-physiological traits

Glossiness: Genotypes with glossy leaf trait are resistant to shoot fly (Maiti, 1994; Dhillon et al., 2005a,b,c; Sharma et al., 2006; Padmaja et al., 2010a,b). The leaf glossiness at seeding stage has a strong influence on the orientation of shoot fly females due to reflection of light in sorghum (Blum, 1972; Maiti and Bidinger, 1979). The lower amount of chlorophyll in the leaves renders them less attractive to the shoot fly females for oviposition (Patil et al., 2006). Differences between glossy and nonglossy genotypes can be detected by the adherence of water sprayed on leaf blades (Nwanze et al., 1990). There is a negative correlation between leaf glossiness, oviposition, and deadhearts (Jadhav et al., 1986; Vijayalakshmi, 1993; Dhillon et al., 2005a,b,c). Maiti (1980) suggested that presence of trichomes and glossy traits are independent, and apparently have an additive effect in reducing the incidence of shoot fly.

Trichomes: Trichomes are common anatomical features on leaves, stem, and/or reproductive structures in higher plants. Trichomes on the abaxial and adaxial leaf surfaces of sorghum may inhibit the movement of young larvae in the whorl, which may prolong the time to reach the growing point or result in mortality of the neonate shoot fly larvae (Maiti et al., 1980; Gibson and Maiti, 1983; Raina, 1985). There is negative association between trichome density and insect feeding, oviposition responses, and nutrition of larvae (Levin, 1973). Trichome density has a positive correlation with resistance to shoot fly in sorghum (Moholkar, 1981; Omori et al., 1983; Dhillon et al., 2005a,b,c). The percentage of plants with eggs and the number of eggs per plant were negatively correlated with trichome density at 14 days after emergence (Dhillon et al., 2005a,b,c; Patil et al., 2006). Although there is highly significant and negative correlation between the trichome density and shoot fly infestation, it seems that trichomes do not have a direct role in reducing the deadhearts, but are associated with reduced oviposition (Karanjkar et al., 1992). Trichomes on both leaf surfaces can be used as a reliable selection criterion to select for resistance to shoot fly (Maiti, 1994). Level of resistance to shoot fly is higher when both glossy and trichome traits occurred together (Agrawal and House, 1982; Dhillon et al., 2005a,b,c; Sharma et al., 2006).

Seedling vigor: Faster seedling growth and toughness of the leaf sheath are associated with resistance to shoot fly (Singh and Jotwani, 1980a,b,c; Kamatar and Salimath, 2003). Blum (1972) observed that shoot fly-resistant sorghum lines grew faster than susceptible ones. Seedling vigor was significantly and negatively associated with deadhearts and oviposition (Taneja and Leuschner, 1985). Faster seedling growth and longer shoot length causes the larvae to take more time to reach the base of the shoot. Singh (1998) concluded that rapid seedling growth and long and thin leaves during the seedling stages makes plants less susceptible to shoot fly. Karanjkar et al. (1992) suggested that seedling vigor can be used to select for resistance to shoot fly. Jayanthi et al. (2002) showed that shoot fly-resistant parental lines and their hybrids showed significantly higher seedling vigor compared to susceptible parental lines and their hybrid groups. The negative association of seedling vigor and plant height with shoot fly resistance seems to be influenced by shoot fly damage in resistance screening trials, rather than the direct effect of seedling vigor on shoot fly damage. The seedling vigor scores in shoot fly screening trials are affected by shoot fly damage. Under shoot fly damage the shoot fly susceptible lines apparently appear to be less vigorous as a result of deadheart formation.

Leaf surface wetness: The role of leaf surface wetness (LSW) in plant resistance to insects was first studied by Rivnay (1960), who observed the role of morning dew in the movement of freshly hatched shoot fly larvae through the leaf sheath to the growing point. LSW originates from the plant, and it is not due to condensation of atmospheric moisture (Sree et al., 1992). This was further confirmed by radioactive labeling using tritinium and C^{14} (Shivaramakrishnan et al., 1994). Tritiated water applied to the soil of potted seedlings was translocated to the surface of the whorl leaf. There were significant differences in the amount of tritiated water collected from susceptible (CSH 5) and resistant (IS 18551) genotypes. LSW is associated with shoot fly resistance (Nwanze et al., 1990). Cultivars with a high transpiration rate are preferred for oviposition (Mate et al., 1988). The dew or moisture accumulation in the central whorl leaf, through which the larvae move downward from the site of oviposition to the growing point has an important role in shoot fly resistance (Blum, 1963; Raina, 1981). The shoot fly larvae take less than 30 min from egg hatch to arrival at the funnel, and >3 h from the funnel to the growing point. Larval survival is affected by the wetness of the central shoot rather than the central expanded leaves on which eggs are laid. Initial contact with moisture enhances larval movement and survival. A waxy surface will permit an even spread of water on leaf surface, but may not retain water in large droplets as a nonwaxy surface does. A smooth amorphous wax layer and sparse wax crystals characterize shoot fly resistant and moderately resistant genotypes, while susceptible genotypes possess a dense mesh of crystalline epicuticular wax (Nwanze et al., 1992). Hence, a highly waxy leaf retains more water as droplets than a nonwaxy leaf and vice versa (Nwanze et al., 1990). LSW could be the result of some form of cuticular movement of water to the leaf surface (Sree et al., 1994). LSW trends are also positively associated with shoot fly abundance, crop infestation, rainfall, temperature, and relative humidity (Nwanze et al., 1992). There was no consistent variation in the relationship between plant water potential and soil metric potential of resistant and susceptible sorghum genotypes (Soman et al., 1994). However, soil metric potential affects the water status of the shoot fly susceptible plant, which is associated with the appearance of water droplets in the central leaf whorl of the susceptible cultivar, CSH 1. No water droplets were observed on the central whorl leaf of the resistant genotypes indicating that the production of water droplets is not solely the result of internal water status of the plant.

Other plant traits associated with resistance: The shoot fly incidence has been found to be positively correlated with days to flowering and days to maturity, but negatively correlated with number of leaves per plant and plant height (Rao et al., 2000). The taller varieties with more leaves are desirable for minimizing the shoot fly incidence. Leaf width and stem thickness were positively associated, and number of leaves per plant and leaf length were negatively associated with shoot fly deadhearts in maize, while there was no significant influence of these plant characters on the oviposition (Rao and Panwar, 2001). Tall, late-maturing genotypes with high glossy intensity were the most resistant to shoot fly (Maiti et al., 1994). The percentage of plants with eggs and the number of eggs per plant were negatively correlated with seedling height (Patil et al., 2006). Shoot fly oviposition was negatively correlated with seedling height, leaf length, and stem length, but positively correlated with number of leaves per plant, leaf width, stem girth, and panicle initiation, while shoot fly deadhearts were negatively correlated with seedling height, leaves per plant, leaf length, leaf width, and stem length but positively correlated with stem girth and panicle initiation (Verma and Singh, 2000). The plumule and leaf sheath pigmentation in sorghum were found to be associated with resistance to shoot fly (Dhillon, 2004; Dhillon et al., 2005a,b,c, 2006a,b).

4.3.5.2.3 Biochemical factors of resistance

Cyanogenic glycosides: Sorghum cultivars with low shoot fly infestation had low HCN in leaves (Chavan et al., 1990). The cyanide content in both seedlings and older plants depends on growth conditions and genetic background (Gillingham et al., 1969; Gorz et al., 1987). The occurrence of *p*-hydroxybenzaldehyde, produced by enzymatic degradation of dhurrin in sorghum seedlings of CSH 1, on the leaf surface was suspected to act as oviposition stimulant for adults and/or feeding activator for the maggots of shoot fly (Alborn et al., 1992). There is a negative correlation between HCN content and shoot fly deadhearts, and its antibiotic effects against sorghum shoot fly (Kumar and Singh, 1996).

Sugars: Resistance to shoot fly is associated with low levels of reducing and total sugars in sorghum seedlings (Singh et al., 2004). Reducing sugars increased slightly between 17 and 20 days after seedling emergence in shoot fly-resistant sorghum genotypes, but decreased in susceptible varieties (Bhise et al., 1997). Concentrations of reducing and total sugars influenced the resistance of little millet genotypes to *Atherigona pulla* (Wiedemann). Low sugar content in sorghum was associated with susceptibility to shoot fly (Swarup and Chaugale, 1962). Development of sugarcane aphid, *Melanaphis sacchari* (Zhent.), and delphacid, *Peregrinus maidis* (Ashm.), populations were more pronounced in varieties with higher sugar content in leaves (Mote and Shahane, 1994). Total sugars, reducing and nonreducing sugars, and amino acids are two times higher in midge susceptible than in the resistant genotypes (Naik et al., 1996).

Nutritional elements: Several micronutrients play an important role in the host-plant resistance to shoot fly. Low levels of N (Singh and Narayana, 1978; Singh and Jotwani, 1980a,b,c; Khurana and Verma, 1983; Chavan et al., 1990), P, and K (Bhise et al., 1997) and high levels of Ca (Chavan et al., 1990) were associated with the shoot fly resistance in sorghum. Higher amounts of Mg and Zn, and lower amounts of Fe were associated with the expression of resistance to shoot fly. Sorghum cultivars with low shoot fly infestation have low N and Mg contents, and high Si and Ca contents (Chavan et al., 1990). Concentrations of Si and K also influence the resistance of little millets to *A. pulla* (Kadire et al., 1996). High amounts of P, K, Fe, and Si contribute to stemborer resistance in maize (Arabjafari and Jalali, 2007). However, varieties with high content of P and K were less preferred by delphacids and aphids in sorghum (Mote and Shahane, 1994).

Amino acids, sugars, tannins, phenols, NDF, ADF, lignins, and silica content are associated with resistance to the stemborer (Sharma and Nwanze, 1997). Larval mortality is higher when larvae are fed on a diet impregnated with a petroleum ether extract of borer-resistant lines. Methanolic extracts from the susceptible line IS 18363 caused greater feeding stimulation than did extracts from a less susceptible cultivar, IS 2205. IS 18363 has greater phenolic and sugar contents than IS 2205 (Torto et al., 1990).

4.3.5.2.4 Plant defense traits

Plants respond to herbivore attack through a defense system which includes structural barriers and secondary metabolites that have toxic, repellent, and/or antinutritional effects on the herbivores. Plant structural traits such as leaf surface wax, thorns or trichomes, and cell wall thickness and lignification form the first physical barrier to feeding by the herbivores, and the secondary metabolites, such as terpenoids, alkaloids, anthocyanins, phenols, and quinones, that either kill or retard the development of the herbivores form the next barriers that defend the plant from subsequent attack.

Morphological structures: Plant structures are the first line of defense against herbivory, and play an important role in host-plant resistance to insects. Structural defense includes morphological and anatomical traits of the plant which deter the herbivores from feeding. The first line of plant defense against insect pests is the erection of a physical barrier through the formation of either a waxy cuticle and/or trichomes.

Epicuticular lipids: Plant epicuticular lipids play an important role in insect–plant interactions. Surface lipids are the potential plant surface cues on the leaf that contribute to plant defense by adversely affecting insects by interfering with their movement. Epicuticular lipids may influence herbivores directly, or indirectly, through their chemical composition, their morphology, or both. Glossy and normal genotypes of sorghum cultivars differ in susceptibility to key pests. Resistant genotypes have a smooth, amorphous wax layer with few wax crystals, while susceptible genotypes have significantly more wax in the epicuticle (Padmaja et al., 2010a,b). Leaf glossiness, plumule, and leaf sheath pigmentation were responsible for shoot fly resistance in sorghum (Chamarthi et al., 2010).

Resistance was related to the chemical composition of glossy leaf lipids rather than physical structure alone. These lipids appeared to deter neonate larvae of the sorghum stemborer from establishing feeding sites on glossy *Sorghum bicolor*. Chemical factors in the epicuticular lipids of glossy resistant sorghum also contribute to resistance by influencing insect behavior. In their climb up the culm of sorghum plants, stemborer larvae sometimes stray out onto leaves (Bernays et al., 1983, 1985; Chapman et al., 1983). They must then reorient downward to regain the culm, before turning upward again to the whorl where they begin feeding. The larvae reorient successfully more frequently on models coated with sorghum leaf surface lipid extract than on controls treated only with the solvent. However, reorientation is much less frequent when surface lipids from a glossy resistant sorghum are used instead of those from a susceptible sorghum (Bernays et al., 1985). The chemical differences between the epicuticular lipids of the two genotypes have been proposed as the basis of the behavioral response (Taneja and Woodhead, 1989). The aphid-resistant sorghum had higher levels of triterpenols in the surface wax than did the susceptible plant. Free and esterified triterpenols increased aphid resistance in sorghum when present at high levels. Concentrations of amyrins (major wax components in waxes of some plant species) in surface waxes are correlated with aphid resistance in sorghum (Heupel, 1985). Mutagenesis of genes affecting cuticular lipids provides a means for identifying genes involved in biosynthesis of cuticular lipids. Mutagenesis has localized 24 loci in sorghum.

Trichomes: Trichomes play an important role in plant defense against insect pests. There is great variation in the structure of trichomes including length of trichome, tip shape and size, gland shape and size, and number of cells in base. Trichome density negatively affects the ovipositional behavior, feeding and larval nutrition of insect pests (Handley et al., 2005). Dense trichomes and trichome morphology affect the herbivory mechanically, and interfere with the movement of insects on the plant surface. Density of nonglandular pointed trichomes on leaf surfaces offers barrier to the movement of shoot fly maggots. The presence of bicellular trichomes in nonglossy lines increases the shoot fly attack (Padmaja et al., 2010a,b). Sorghum genotypes having nonglandular trichomes are more tolerant to shoot fly than the genotypes with glandular trichomes (Maiti and Gibson, 1983). Glandular trichomes secrete secondary metabolites including flavonoids, terpenoids, and alkaloids that can be poisonous, repellent, or trap insects, thus forming a combination of structural and chemical defense. Trichomes on the abaxial surface of the sorghum leaves have been reported to be associated with resistance to shoot fly (Blum, 1968; Maiti et al., 1980). Egg laying property of shoot fly was significantly and negatively associated with trichomes and leaf glossiness (Omori et al., 1983) and there was higher level of resistance to shoot fly when leaf glossiness and trichomes occurred together in a genotype.

Secondary metabolites: Secondary metabolites play an important role in host-plant resistance to insects. Plants are known to produce certain chemical compounds in different quantities and proportions, which affect the behavior and biology of phytophagous insects (Painter, 1958). An important group of defense chemicals in sorghum is the polyphenols, particularly flavonoids and their oligomers, and the condensed tannins.

Polyphenols: Polyphenols are widely distributed in plants, but they are not directly involved in any metabolic process, and therefore, are considered to be secondary metabolites. Phenolic compounds in sorghum caryopsis are associated with resistance to insects and fungal pathogens (Dreyer et al., 1981). The presence of phenolic compounds in young sorghum seedlings and their decline at later stages of crop growth plays a significant role in the physiological relationships between shoot fly larvae and seedlings (Woodhead and Bernays, 1978; Woodhead and Cooper, 1979; Woodhead et al., 1980). Shoot fly resistance is associated with high amounts of phenolic compounds in sorghum seedlings (Khurana and Verma, 1983; Kumar and Singh, 1998). Resistance to shoot fly is associated with low levels of polyphenol oxidase and peroxidase (Bhise et al., 1996). Amounts of protocatechuic acid, syringic acid, and p-coumaric acid were negatively correlated, while p-hydroxybenzoic acid, vanillic acid, and ferulic acid contents were positively correlated with shoot fly deadheart incidence (Pandey et al., 2005). Important phenolic acids in sorghum are listed in the Table 4.5.

Flavonoids: Flavonoids are derivatives of the monomeric polyphenol flavan-4-ol, and are known as anthocyanidins. Flavonoids and isoflavonoids are known to confer resistance against insect attack in several plant species (Hedin and Waage, 1986; Grayer et al., 1992). The two flavonoids identified to be abundant in sorghum grains are luteoforol (Bate Smith, 1969) and apiforol (Watterson and Butler, 1983). The latter compound was also found in sorghum leaves.

Table 4.5 Phenolic Acids in Sorghum Grains

Phenolic Acid	Reference
Hydroxybenzoic acid	
Gallic	Hahn et al. (1983), Subba Rao and Muralikrishna (2002)
Gentisic	McDonough et al. (1986), Waniska et al. (1989)
p-Hydroxybenzoic	Hahn et al. (1983), McDonough et al. (1986)
Salicylic	Waniska et al. (1989)
Syringic	Waniska et al. (1989), McDonough et al. (1986)
Hydroxycinnamic acids	
Cinnamic	Hahn et al. (1983), McDonough et al. (1986)
Ferulic, caffeic, and p-Coumaric	Hahn et al. (1983), McDonough et al. (1986), Subba Rao and Muralikrishna (2002)
Sinapic	Waniska et al. (1989), McDonough et al. (1986)

Table 4.6 Flavonoids and Proanthocyanidins in Sorghum Grains

Compound	Reference
Anthocyanins	
Apigeninidin and luteolinidin	Nip and Burns (1971), Gous (1989)
Apigeninidin 5-glucoside	Nip and Burns (1969, 1971), Wu and Prior (2005)
Luteolinidin 5-glucoside	Nip and Burns (1971), Wu and Prior (2005)
5-Methoxyluteolinidin 7-glucoside and 7-methoxyapigeninidin 5-glucoside	Wu and Prior (2005)
7-Methoxyapigeninidin	Pale et al. (1997), Seitz (2004), Wu and Prior (2005)
5-Methoxyapigeninidin and 7-methoxyluteolinidin	Seitz (2004)
Flavan-4-ols	
Luteoforol	Bate Smith (1969)
Apiforol	Watterson and Butler (1983)
Flavones	
Apigenin	Gujer et al. (1986), Seitz (2004)
Luteolin	Seitz (2004)
Flavanones	
Eriodictyol	Kambal and Bate-Smith (1976)
Eriodictyol 5-glucoside and naringenin	Gujer et al. (1986)
Flavonols	
Kaempferol 3-rutinoside-7-glucuronide	Nip and Burns (1969)
Dihydroflavonols	
Taxifolin and taxifolin 7-glucoside	Gujer et al. (1986)
Proanthocyanidin monomers/dimers	
Catechin, procyanidin B-1, and epicatechin- (epicatechin)–catechin	Gupta and Haslam (1980), Gujer et al. (1986)
Prodelphinidin	Brandon et al. (1982), Krueger et al. (2003)
Proapigeninidin and proluteolinidin	Krueger et al. (2003)

Flavonoids play a vital role in insect feeding and oviposition behavior. Insect can discriminate among flavonoids, and these modulate the feeding and oviposition behavior of insects (Simmonds, 2001). Flavonoids and proanthocyanidins available in sorghum grains are presented in Table 4.6. **Tannins:** Tannins are polymers resulting from condensation of flavan-3-ols. Tannin content in sorghum decreases after germination (Osuntogun et al., 1989). Inheritance of tannin content as a component of resistance to shoot fly was studied (Kumar and Singh, 1998). Exploitation of heterosis to increase tannin content is needed to confer resistance (Kamatar et al., 2003). Sorghum genotypes IS 1056C, IS 2177C, IS 2246C, IS 4023C, IS 7399C, and IS 12680C had a significantly higher antibiotic resistance and high amounts of acid detergent, and NDF or tannin content in the leaves (Diawara et al., 1992). Short floral parts, faster rate of grain development and high tannin content of grain were apparently associated with resistance to sorghum midge (Sharma et al., 1990a,b). Tannin content was generally double the normal in sorghum midge-resistant genotypes as compared to the susceptible ones.

Oxidative enzymes: Modifications in plant protein profiles and alterations in plant oxidative enzyme levels are among a plant's first response to insect herbivory (Green and Ryan, 1972; Hildebrand et al., 1986; Felton et al., 1994; Ni et al., 2001). These enzymes, because of their potential roles in plant signaling, synthesis of defense compounds and/or oxidative stress tolerance, have been implicated in plant resistance to insect herbivores. Resistant sorghum plants to shoot fly exhibited higher levels of peroxidase and polyphenol oxidase activities compared with susceptible plants. Resistant genotypes may be able to tolerate shoot fly feeding by increasing their peroxidase and polyphenol oxidase activities (Padmaja et al., 2014).

Proteins: Resistance to sorghum shoot fly (Mote et al., 1979; Kamatar et al., 2002), and stemborer is associated with low levels of proteins (Rao and Panwar, 2002, 2001). Maiti et al. (1994) isolated three polypeptides (106 kDa, 82 kDa, 54 kDa) from protein extracts of six glossy sorghum and one nonglossy sorghum leaves. The 54 kDa polypeptide was present in several glossy lines, while the nonglossy lines contained polypeptides of a higher molecular weight (106 kDa). Presence of 54 kDa band in the glossy lines may be related to shoot fly resistance in sorghum.

Volatiles: Green leaf volatiles, generally occurring in C_6 alcohols, aldehydes, and acetates from plants, play an important role in plant−plant communication. These compounds induce plants to produce jasmonic acid (JA) and defense-related gene expression, and the release of volatile compounds. Composition of young sorghum seedlings analyzed for volatile compounds by GC-MS indicated that the sorghum headspace samples had α-pinene, 6-methyl-5-hepten-2-one, octanal, (Z)-3-hexen-1-yl acetate, nonanal, methyl salicylate, decanal, and (-)-(E)-caryophyllene (Padmaja et al., 2010a,b). Females of *A. soccata* are attracted to the volatiles emitted by the susceptible sorghum seedlings (Nwanze et al., 1998; Padmaja et al., 2010a,b). A large number of sorghum midge females were attracted to the odors from the panicles of *S. halepense*.

Gene Expression: When attacked by a phloem-feeding greenbug aphid, *Schizaphis graminum*, sorghum activates JA- and salicylic acid (SA)-regulated genes, as well as genes outside known wounding and SA signaling pathways (Zhu-Salzman et al., 2004). Activation of certain transcripts regulated exclusively by greenbug infestation was observed and the expression patterns may represent unique signal transduction events independent of MJ- and SA-regulated pathways. Transcriptional changes in a parallel system, greenbug resistant, and susceptible genotypes of sorghum, led to detection of the abundance of the transcripts corresponding to 2304 sorghum genes during the infestation by virulent greenbug biotype I (Park et al., 2006). The experiments showed comprehensive gene activation resulting from upregulating, or activating existing defense pathways in sorghum seedlings in response to greenbug feeding. Among the induced genes identified in this study, the expression level of 38 genes was threefold or more, while that of 26 genes was significantly less. This can enhance the understanding of plant defense mechanisms against insect pests, and also accelerate the identification of resistance genes or specific targets for improvement of plant resistance. More studies are needed to fully unravel interactions between the phloem-feeding insect and the host plant.

4.3.6 DEVELOPMENT AND USE OF PEST-RESISTANT CULTIVAR

Exploitation of host-plant resistance through genetic enhancement has always been the first approach in addressing the insect problem.

4.3.6.1 Conventional breeding

Breeding for crop resistance to insect pests is a safe and inexpensive control method. Sorghum improvement programs have intensified their efforts to breed insect-resistant and stable yielding varieties and hybrids. Incorporation of insect resistance into elite parental lines and hybrids has been the goal of many sorghum breeding programs. Insect–plant interactions, life cycle of the insect, population dynamics of the insect species are the aspects that are needed to be considered in a breeding program for insect resistance. A large proportion of the world sorghum germplasm has been evaluated for insect resistance and this has resulted in the identification of several lines with reasonable levels of resistance to shoot fly, stemborer, midge, and head bugs. Many of the resistant sources identified in the germplasm are poor in productivity due to a physiologically inefficient plant type, and are poor combiners for shoot fly resistance and the traits associated with resistance (Aruna and Padmaja, 2009). Since these resistant sources are poor in productivity, the plant breeding programs were aimed to transfer their resistance to an improved plant background. Using these donors, several maintainer and restorer lines were developed. For the development of shoot fly-resistant hybrids, resistance is required in both male and female parents (Dhillon et al., 2005a,b,c). Intensive breeding efforts were initiated to incorporate resistance into sorghum parental lines, by crossing the elite parental lines with shoot fly-resistant germplasm. A number of improved resistant sources were developed and tested in multilocation trials. The lines NRCSFR 09-1, NRCSFR 09-2, NRCSFR 09-3, and NRCSR 09-4 with shoot fly resistance and improved agronomy can be used as sources in the shoot fly resistance breeding program (Aruna et al., 2014). Some important lines for shoot fly resistance include IS 2123, ICSV 705, ICSV 708, SPSFR 94019, SPSFR 94006, SPSFR 94007, SPSFR 94011, SPSFR 94034, ICSV 93127, SPSFR 96069, SPSFR 86065, PS 23585, ICSR 89058 and that for stemborer include IS 5448, IS 5470, IS 2205, IS 18573, ICSV 700 (sweet stalk), and ICSV 93046 (sweet stalk). These lines have been used in crop improvement program in Asia, Africa, USA, and Australia.

4.3.6.2 Marker-assisted selection

Through conventional breeding, it takes 5–6 generations to transfer a trait within a species into high-yielding cultivars. Plant biotechnology including molecular genetics, genomics, and plant transformation has provided a powerful means to supplement traditional breeding approaches. Marker-assisted selection will allow rapid introgression of the resistance genes, and ultimately gene pyramiding, into the high-yielding varieties and hybrids. With the availability of dense linkage maps of the sorghum genome, progress in the identification of genes or Quantitative trait loci (QTLs) linked to plant resistance to insects has been made. Polymorphic simple sequence repeat (SSR) loci associated with resistance to shoot fly have been identified (Folkertsma et al.,

2003; Dhillon et al., 2006a,b; Satish et al., 2009; Aruna et al., 2011). These QTLs are now being transferred into the locally adapted hybrid parental lines via SSR-based marker-assisted selection. QTLs associated with resistance to sorghum head bug (*Eurystylus oldi* Poppius) have also been identified (Deu et al., 2005). In silico analysis of the regions/QTLs associated with stemborer resistance on chromosome SBI 07, SBI 04, and SBI 02 showed homology with maize chromosome 1 genomic regions containing spotted stemborer resistance. Resistance to greenbug is contributed by multiple genomic regions depending on the resistance source. Some of the alleles are biotype specific, while others are biotype nonspecific (Table 4.7).

Table 4.7 Insect Resistant QTLs Identified in Sorghum

Insect	QTLs	Associated With	Reference
Shoot fly	Twenty-nine QTLs	Four each for leaf glossiness and seedling vigor, seven for oviposition, six for deadhearts, two for adaxial trichome density, and six for abaxial trichome density	Satish et al. (2009)
Stemborer	Twenty-nine QTLs	Deadhearts, stem tunneling, leaf feeding, recovery resistance, and overall resistance	Vinayan (2010)
		The putative QTL on SBI 07 strongly associated with stem tunneling	
		QTLs for seedling basal pigmentation, plant color, testa pigmentation, mesocarp thickness, and leaf angle are identified on SBI 06 and SBI 04	
		Antixenosis explaining 12−15% of the total variation in egg numbers/spikelet	
Midge	Two QTLs on different linkage groups (SBI-03 and SBI-09)	Antibiosis and explained 34.5% of the variation of the difference of egg and pupal counts	Tao et al. (2003)
	One region on SBI-07		
Greenbug	Three loci present on SBI-05, SBI-06, and SBI-07 conferring resistance to green bug biotype I		Katsar et al. (2002)
	Nine QTL affecting both resistance and tolerance to biotype I and K with individual QTL accounting for 5.6−38.4% of phenotypic variance		Agrama et al. (2002)
	Three QTL on SBI-01 and SBI-04 for biotype I resistance and tolerance using chlorophyll loss as an indicator to greenbug damage		Nagaraj et al. (2005)
	Major QTL on SBI-09 for resistance to biotype I		Wu and Huang (2008)

4.3.6.3 Transgenics

Bacillus thuringiensis (Bt) is a naturally occurring, gram positive, spore forming soil bacterium. During sporulation, many Bt strains produce crystal proteins called δ-endotoxins, that have insecticidal action. The use of genes that encode insecticidal proteins in transgenic crops has the potential to benefit agricultural crop production. The increasing pressure to use nonhazardous, environmentally compatible pest control measures have spurred interest in the use of natural insecticides such as *Bacillus thuringiensis* insecticidal crystal proteins in a number of countries. Advances in gene identification and gene transfer techniques allow the incorporation of beneficial genes into crop plants. These new tools enable plant breeders to design new varieties by installing desired foreign genes, such as insect resistance genes, into elite breeding lines in a considerably short period of time.

Development of insect-resistant sorghum by transferring Bt genes is one of the best ways available today to overcome insect attack and to improve the yield of sorghum. Sorghum has been widely considered as a recalcitrant major crop in terms of tissue culture and genetic transformation (Grootbroom et al., 2010). Microprojectile and *Agrobacterium*-mediated transformation methods are two main approaches that have been developed and applied for sorghum transformation. The first report of successful transgenic sorghum by using particle bombardment was published in 1993 (Casas et al., 1993). Seven years later, the first *Agrobacterium*-mediated transgenic sorghum was reported (Zhao et al., 2000). There are many reports on successful transformation of sorghum utilizing particle bombardment (Able et al., 2001; Grootbroom et al., 2010) or *Agrobacterium*-mediated transformation (Zhao et al., 2000; Gurel et al., 2009).

Sorghum plants having the cry1Ac gene have been developed under the control of a wound-inducible promoter from a maize protease inhibitor gene (mpi) for resistance to spotted stemborer (Seetharama et al., 2001; Harshavardhan et al., 2002; Girijashankar et al., 2005). The expression and inheritance of the Bt genes were confirmed in T_1 plants by partial tolerance against first instar larvae of stemborer (Girijashankar et al., 2005). Zhang et al. (2009) utilized *Agrobacterium*-mediated transformation to transfer the Cry1Ab gene into three sorghum cultivars. Transgenic plants with a high content of Bt protein displayed a tolerance to pink rice borer (*Sesamia inferens*). Transgenic sorghum plants produced through particle bombardment and *Agrobacterium* methods in two elite, but recalcitrant, genotypes of sorghum carrying Bt toxin genes, Cry1Aa, and Cry1B showed 20−30% of damage in comprehensive insect bioassays for tolerance to stemborer through leaf disk and whole plant assays (Visarada et al., 2014). An *Agrobacterium*-mediated gene transformation developed using shoot apices with fully modified synthetic *Cry1C* coding sequences along with *hpt* and *gus* genes were highly resistant to stemborer as revealed by insect bioassay with 100% insect mortality rate (Ignacimuthu and Premkumar, 2014). While the modem biotechnology has been recognized to have a great potential, adoption of biosafety protocol is necessary to protect human health and environment from the possible adverse effects of the products of genetic engineering.

4.4 CONCLUSIONS

Host-plant resistance has been used for successful management of several insect pests in sorghum with certain limitations. Although several sources of resistance have been identified against important insect pests of sorghum, only a few of them are being deployed in the development of insect-resistant varieties as it takes several years and involves many resources and expertise.

A good beginning has been made in developing genetic linkage maps of many crops, but the accuracy and precision of phenotyping for resistance to insect pests remains a critical constraint. Improved phenotyping systems will have a substantial impact on both conventional and biotechnological approaches to breeding for resistance to insect pests, in addition to the more strategic research that feeds into these endeavors.

4.5 FUTURE PRIORITIES

The progress in improving the resistance levels of sorghum cultivars for the last 40 years using the identified resistance sources and other wild resistant genotypes through conventional selection methods has been slow, largely due to complex inheritance of resistance, genotype-environment interaction, and difficulty in crossing with wild genotypes. All the efforts to breed an insect resistant line have not been fruitful to the required extent due to nonavailability of stable, consistent resistant sources. Searching for new sources of resistance is essential to diversify the genetic background. Identification of resistant sources in elite backgrounds as well as in parental lines in the development of hybrids, and knowledge on the mechanisms would greatly assist in developing cultivars with stable sources of resistance to the insect pests of sorghum, under high yield background. Molecular markers tagged to resistance genes may thus be useful to plant breeders to support the introgression of the resistance alleles into elite high-yielding inbred lines. The vast pool of genes existing in the wild relatives of cultivated sorghum will provide a new resource for genetic improvement for insect resistance. The transfer of resistance genes from wild relatives of sorghum is of particular relevance to shoot fly resistance. Future improvement of sorghum production also depends on the development and availability of new technologies. There is a need to understand the herbivore-specific signal molecules, their identification, mode of action, and further signal transduction. An understanding of induced resistance in sorghum plant can be utilized for interpreting the ecological interactions between plants and herbivores and for exploitation in pest management in sorghum. The future challenge is to exploit the elicitors of induced defense in sorghum for pest management and identify the genes encoding proteins that are up- and/or downregulated during plant response to the herbivore attack, which can be deployed for conferring resistance to the herbivores through genetic transformation. The work in the development of transgenic sorghums for shoot fly resistance has not been attempted primarily due to nonavailability of suitable Bt toxins for dipterans. Hence, future research efforts should be directed toward the development of transgenic Bt sorghum against shoot fly. Future research must focus on environmentally sound pest management strategies that are compatible with the needs and limitations of sorghum farmers.

REFERENCES

Able, J.A., Rathus, C., Godwin, I.D., 2001. The investigation of optimal bombardment parameters for transient and stable transgene expression in sorghum. In Vitro Cell Dev. Biol. 37, 341–348.
Agarwal, R.K., Verma, R.S., Bharaj, G.S., 1978. Screening of sorghum lines for resistance against shoot bug, *Peregrinus maidis* Ashmead (Homoptera: Delphacidae). JNKVV Res. J. 12, 116.

Agrama, H., Widle, G., Reese, J., Campbell, L., Tuinstra, M., 2002. Genetic mapping of QTLs associated with greenbug resistance and tolerance in *Sorghum bicolor*. Theor. Appl. Genet. 104, 1373−1378.

Agrawal, B.L., House, L.R., 1982. Breeding for resistance in sorghum. In: House, L.R., Mughogho, L.K., Peacock, J.M. (Eds.), Sorghum in Eighties. Proceeding, International Symposium on Sorghum, 2−7 November 1981. ICRISAT, Patancheru 502324, India, pp. 435−446.

Alborn, H., Stenhagen, G., Leuschner, K., 1992. Biochemical selection of sorghum genotypes resistant to sorghum shoot fly (*Atherigona soccata*) and stem borer (*Chilo partellus*): role of allelochemicals. In: Rizvi, S.H., Rizvi, V. (Eds.), Allelopathy: Basic and Applied Aspects. Chapman & Hall, London, pp. 101−117.

Alghali, A.M., 1985. Insect-host plant relationships: the spotted stalk-borer, *Chilo partellus* (Swinhoe) (Lepidoptera: Pyralidae) and its principal host, sorghum. Insect Sci. Appl. 6, 315−322.

Arabjafari, K.H., Jalali, S.K., 2007. Identification and analysis of host plant resistance in leading maize genotypes against spotted stem borer, *Chilo partellus* (Swinhoe) (Lepidoptera: Pyralidae). Pak. J. Biol. Sci. 10, 1885−1895.

Aruna, C., Padmaja, P.G., 2009. Evaluation of genetic potential of shoot fly resistant sources in sorghum (*Sorghum bicolor* (L.) Moench). J. Agric. Sci. 147 (1), 71−80.

Aruna, C., Bhagwat, V.R., Madhusudhana, R., Sharma, V., Hussain, T., Ghorade, R.B., et al., 2011. Identification and validation of genomic regions that affect shoot fly resistance in sorghum (*Sorghum bicolor* (L.) Moench). Theor. Appl. Genet. 122, 1617−1630.

Aruna, C., Padmaja, P.G., Bhagwat, V.R., Subbarayudu, B., Patil, J.V., 2014. Improved sources of resistance to shoot fly. Sorghum Times 10 (2), 5−7.

Bate Smith, E.C., 1969. Luteoforol (3′,4,4′,5,7-pentahydroxyflavan) in *Sorghum vulgare* L. Photochemistry 8, 1803−1810.

Bene, G.D., 1986. Note sulla biologia di *Atherigona soccata* Rondani (Diptera: Muscidae) in Toscanae Lazio. Redia 69, 47−63.

Bergquist, R.R., Rotar, P., Mitchell, W.C., 1974. Midge and anthracnose head blight resistance in sorghum. Tropical Agric. (Trinidad) 51, 431−435.

Bernays, E.A., Chapman, R.F., Woodhead, S., 1983. Behaviour of newly hatched larvae of *Chilo partellus* (Swinhoe) (Lepidoptera: Pyralidae) associated with their establishment in the host-plant sorghum. Bull. Entomol. Res. 73, 75−83.

Bernays, E.A., Woodhead, S., Haines, L., 1985. Climbing by newly hatched larvae of the spotted stalk borer *Chilo partellus* to the top of sorghum plants. Entomol. Exp. Appl. 39, 73−79.

Bhise, H.T., Desai, B.B., Chavan, H.D., 1996. Effect of chemical constituents on resistance of shoot fly in sorghum. J. Maha. Agric. Univ. 21, 293−294.

Bhise, H.T., Desai, B.B., Chavan, U.D., 1997. Assessments of some biochemical parameters responsible for shoot fly resistance in sorghum. J. Maha. Agric. Univ. 21, 127−129.

Birkett, M.A., Chamberlain, K., Khan, Z.R., Pickett, J.A., Toshova, T., Wadhams, L.J., 2006. Electrophysiological responses of the lepidopterous stemborers *Chilo partellus* and *Busseola fusca* to volatiles from wild and cultivated host plants. J. Chem. Ecol. 32, 2475−2487.

Blum, A., 1963. The penetration and development of the sorghum shoot fly in susceptible sorghum plants. Hassadesh 44, 23−25.

Blum, A., 1967. Varietal resistances of sorghum to the sorghum shoot fly (*Atherigona varia soccata*). Crop Sci. 7, 461−462.

Blum, A., 1968. Anatomical phenomena in seedlings of sorghum varieties resistant to the sorghum shoot fly, (*Atherigona varia soccata*). Crop Sci. 8, 388−391.

Blum, A., 1969a. Factors associated with tiller survival in sorghum varieties resistant to the sorghum shoot fly (*Atherigona varia soccata*). Crop Sci. 9, 508−510.

Blum, A., 1969b. Ovipositional preference by the sorghum shoot fly (*Atherigona varia soccata*) in progenies of susceptible x resistant sorghum crosses. Crop Sci. 9, 695−696.

Blum, A., 1972. Sorghum breeding for shoot fly resistance in Israel. In: Jotwani, M.G., Young, W.R. (Eds.), Control of Sorghum Shoot Fly. Oxford and IBH Publishing Co., New Delhi, pp. 180−191.

Boozaya-Angoon, D., Starks, K.J., Weibel, D.E., Teetes, G.L., 1984. Inheritance of resistance in Sorghum, *Sorghum bicolor*, to the sorghum midge, *Contarinia sorghicola* (Diptera: Cecidomyiidae). Environ. Entomol. 13, 1531−1539.

Borad, P.K., Mittal, V.P., 1983. Assessment of losses caused by pest complex to sorghum hybrid, CSH 5. In: Krishnamurthy Rao, B.H., Murthy, K.S.R.K. (Eds.), Crop Losses Due to Insect Pests. Indian J. Entomol. (special issue), pp. 271−278.

Brandon, M.J., Foo, L.Y., Porter, L., Me redith, P., 1982. Proanthocyanidins of barley and sorghum composition as a funation of maturity of barley ears. Phytochemistry 12, 2953−2957.

Calatayud, P.A., Guénégo, H., Ahuya, P., Wanjoya, A., Le Rü, B., Silvain, J.F., et al., 2008. Flight and oviposition behaviour of the African stemborer, *Busseola fusca*, on various host plant species. Entomol. Exp. Appl. 129, 348−355.

Casas, A.M., Kononowicz, A.K., Zehr, U.B., Tomes, D.T., Axtell, J.D., Butler, L.G., et al., 1993. Transgenic sorghum plants via microprojectile bombardment. Proc. Natl. Acad. Sci. USA 90, 11212−11216.

Chamarthi, S.K., Sharma, H.C., Sahrawat, K.L., Narasu, L.M., Dhillon, M.K., 2010. Physico-chemical mechanisms of resistance to shoot fly, *Atherigona soccata* in sorghum (*Sorghum bicolor*). J. Appl. Entomol. 135, 446−455.

Chamarthi, S.K., Sharma, H.C., Vijay, P.M., Narasu, L.M., 2011. Leaf surface chemistry of sorghum seedlings influencing expression of resistance to sorghum shoot fly, *Atherigona soccata*. J. Plant Biochem. Biotechnol. 20, 211−216.

Chand, P., Sinha, M.P., Kumar, A., 1979. Nitrogen fertilizer reduces shoot fly incidence on sorghum. Sci. Cult. 45, 61−62.

Chandra Shekar, B.M., 1991. Mechanisms of Resistance in Sorghum to Shoot Bug, *Peregrinus maidis* (Ashmead) (Homoptera: Delphacidae) (MSc. thesis). Andhra Pradesh Agricultural University, Hyderabad, India, p. 106.

Chandra Shekar, B.M., Dharma Reddy, K., Singh, B.U., Reddy, D.D.R., 1992. Components of resistance to corn planthopper, *Peregrinus maidis* (Ashmead), in sorghum. Resist. Pest Manage. Newsl. 4, 25.

Chandra Shekar, B.M., Singh, B.U., Reddy, K.D., Reddy, D.D.R., 1993. Antibiosis component of resistance in sorghum to corn planthopper, *Peregrinus maidis* (Ashmead) (Homoptera: Delphacidae). Int. J. Trop. Insect Sci. 14, 559−569.

Chapman, R.F., Woodhead, S., Bernays, E.A., 1983. Survival and dispersal of young larvae of *Chilo partellus* (Swinhoe) (Lepidoptera: Pyralidae) in two cultivars of sorghum. Bull. Entomol. Res. 73, 65−74.

Chavan, M.H., Phadnawis, B.N., Hudge, V.S., Salunkhe, M.R., 1990. Biochemical basis of shoot fly tolerant sorghum genotypes. Ann. Plant Physiol. 4, 215−220.

Dahms, R.G., 1969. Theoretical Effects of Antibiosis on Insect Population Dynamics. United States Department of Agriculture, ERO, Beltsville, p. 5.

Davies, J.C., Reddy, K.V.S., 1981. Observations on oviposition of sorghum shoot fly, *Atherigona soccata* Rond. (Diptera: Muscidae). Sorghum Entomology, Progress Report 4. ICRISAT, Patancheru, India.

Daware, D.G., Bhagwat, V.R., Ambilwade, P.P., Kamble, R.J., 2012. Evaluation of integrated pest management components for the control of sorghum shoot pests in rabi season. Indian J. Entomol. 74, 58−61.

Deu, M., Ratnadass, M.A., Hamada, M.A., Noyer, J.L., Diabate, M., Chantereau, J., 2005. Quantitative trait loci for head-bug resistance in sorghum. Afr. J. Biotechnol. 4, 247−250.

Dhawan, P.K., Singh, S.P., Verma, A.N., Arya, D.R., 1993. Antibiosis mechanism of resistance to shoot fly, *Atherigona soccata* (Rondani) in sorghum. Crop Res. 6, 306−310.

Dhillon, M.K., 2004. Effects of Cytoplasmic Male−Sterility on Expression of Resistance to Sorghum Shoot Fly, *Atherigona soccata* (Rondani) (Ph.D. thesis). Department of Entomology, Chaudhary Charan Singh Haryana Agricultural Univeristy, Hisar, Haryana, India, p. 382.

Dhillon, M.K., Sharma, H.C., Reddy, B.V.S., 2005a. Agronomic characteristics of different cytoplasmic male-sterility systems and their reaction to sorghum shoot fly, *Atherigona soccata*. ISMN 46, 52−55.

Dhillon, M.K., Sharma, H.C., Ram Singh, Naresh, J.S., 2005b. Mechanisms of resistance to shoot fly, *Atherigona soccata* in sorghum. Euphytica 144 (3), 301−312.

Dhillon, M.K., Sharma, H.C., Reddy, B.V.S., Ram-Singh, Naresh, J.S., Kai-Zhu, 2005c. Relative susceptibility of different male-sterile cytoplasms in sorghum to shoot fly, *Atherigona soccata*. Euphytica 144 (3), 275−283.

Dhillon, M.K., Sharma, H.C., Folkertsma, R.T., Chandra, S., 2006a. Genetic divergence and molecular characterization of sorghum hybrids and their parents for reaction to *Atherigona soccata* (Rondani). Euphytica 149 (1/2), 199−210.

Dhillon, M.K., Sharma, H.C., Reddy, B.V.S., Ram-Singh, Naresh, J.S., 2006b. Inheritance of resistance to sorghum shoot fly, *Atherigona soccata*. Crop Sci. 46 (3), 1377−1383.

Diawara, M.M., Wiseman, B.R., Isenhour, D.J., Hill, N.S., 1992. Sorghum resistance to whorl feeding by larvae of the fall armyworm (Lepidoptera: Noctuidae). J. Agric. Entomol. 9, 41−53.

Dogget, H., 1972. Breeding for resistance to sorghum shoot fly in Uganda. In: Jotwani, M.G., Young, W.R. (Eds.), Control of Sorghum Shoot Fly. Oxford and IBH Publishing Co., New Delhi, pp. 192−201.

Dogget, H., Starks, K.J., Eberhart, S.A., 1970. Breeding for resistance to the sorghum shoot fly. Crop Sci. 10, 528−531.

Doggett, H., Mjisu, B.N., 1965. Sorghum breeding research. Ann. Rep. East Afr. Agric. Forestry Res. Org. 70−79.

Dreyer, D.L., Reese, J.C., Jones, K.C., 1981. Aphid feeding deterrents in sorghum: bioassay isolation and characterization. J. Chem. Ecol. 7, 273−284.

FAOSTAT, 2012. Available at: <http://faostat.fao.org/default.aspx?lang1/4en>.

Felton, G.W., Summers, C.B., Mueller, A.J., 1994. Oxidative responses in soybean foliage to herbivory by bean leaf beetle and three−corned alfalfa leaf hopper. J. Chem. Ecol. 20, 639−650.

Fisher, K.S., Wilson, G.L., 1975. Studies of grain production in *sorghum bicolor* L. Moench: III. Aust. J. Agric. Res. 26, 31−41.

Fisk, J., 1978. Resistance of *Sorghum bicolor* to *Rhopalosiphum maidis* and *Peregrinus maidis* as affected by differences in the growth stage of the host. Entomol. Exp. Appl. 23, 227−236.

Folkertsma, R.T., Sajjanar, G.M., Reddy, B.V.S., Sharma, H.C., Hash, C.T., 2003. Genetic mapping of QTL associated with sorghum shoot fly (Atherigona soccata) resistance in sorghum (Sorghum bicolor). In: Final Abstracts Guide, Plant & Animal Genome XI, 11−15 January 2003, San Diego, CA, USA, p. 42. <http://www.intl-pag.org/11/abstracts/P5d_P462_XI.html>.

Franzmann, B.A., 1988. Components of resistance to sorghum midge in grain sorghum. Proc. Ninth Aust. Plant Breed. Conf. 277−278.

Franzmann, B.A., 1993. Ovipositional antixenosis to *Contarinia sorghicola* Coq., (Diptera: Cecidomyiidae) in grain sorghum. J. Aust. Entomol. Soc. 32, 59−64.

Gibson, P.T., Maiti, R.K., 1983. Trichomes in segregating generations of sorghum matings. I. Inheritance of presence and density. Crop Sci. 23, 73−75.

Gillingham, J.T., Shirer, M.M., Starnes, J.J., Page, N.R., McClain, E.F., 1969. Relative occurrence of toxic concentrations of cyanide and nitrate in varieties of sudangrass and sorghum-sudangrass hybrids. Agron. J. 61, 727−730.

Girijashankar, V., Sharma, H.C., Sharma, K.K., Swathisree, V., Prasad, L.S., Bhat, B.V., et al., 2005. Development of transgenic sorghum for insect resistance against the spotted stem borer (*Chilo partellus*). Plant Cell Rep. 24, 513−522.

Gorz, H.J., Haskins, F.A., Morris, R., Johnson, B.E., 1987. Identification of chromosomes that condition dhurrin content in sorghum seedlings. Crop Sci. 27, 201–203.

Gous, F., 1989. Tannins and Phenols in Black Sorghum. Ph.D. Dissertation. Texas A&M University, College Station, TX, USA.

Grayer, R.J., Kimmins, F.M., Padgham, D.E., Harborne, J.B., Rangarao, D.V., 1992. Condensed tannin levels and resistance of groundnuts (*Arachis hypogea*) against *Aphis craccivora*. Phytochemistry 31, 3795–3800.

Green, T.R., Ryan, C.A., 1972. Wound-induced proteinase inhibitor in plant leaves: a possible defense mechanism against insects. Science 175 (4023), 776–777.

Grootbroom, A.W., Mkhonza, N.I., O'Kennedy, M.M., Chakauya, E., Kunert, K., Chikwamba, R.K., 2010. Biolistic-mediated sorghum (*Sorghum bicolor* L. Moench) transformation via Mannose and Bialaphos based selection systems. Int. J. Bot. 6, 89–94.

Gujer, R., Magnolato, D., Self, R., 1986. Lucosylated flavonoids and other phenolic compounds from sorghum. Phytochemistry 25, 1431–1436.

Gupta, R.K., Haslam, E. 1980. Vegetative tannins: structure and biosynthesis. In: Hulse, J.H., (Ed.), Polyphenols in cereals and legumes, pp. 15–24.

Gurel, S., Gurel, E., Kaur, R., Wong, J., Meng, L., Tan, H.Q., et al., 2009. Efficient, reproducible *Agrobacterium*-mediated transformation of sorghum using heat treatment of immature embryos. Plant Cell Rep. 28, 429–444.

Hahn, D.H., Faubion, J.M., Rooney, L.W., 1983. Sorghum phenolic acids, their performance liquid chromatography separation and their relation to fungal resistance. Cereal Chem. 60, 255–259.

Halalli, M.S., Gowda, B.T.S., Kulkarni, K.A., Goud, J.V., 1982. Inheritance of resistance to shoot fly (*Atherigona soccata* Rond.) in sorghum [*Sorghum bicolor* (L.) Moench]. SABRAO J. 14, 165–170.

Hamilton, R.I., Subramanian, B., Reddy, M.N., Rao, C.H., 1982. Compensation in grain yield components in a panicle of rainfed sorghum. Ann. Appl. Biol. 101, 119–125.

Handley, R., Ekbom, B., Agren, J., 2005. Variation in trichome density and resistance against a specialist insect herbivore in natural populations of *Arabidopsis thaliana*. Ecol. Entomol. 30, 284–292.

Harshavardhan, D., Rani, T.S., Sharma, H.C., Arora, R., Seetharama, N., 2002. Development and testing of Bt transgenic sorghum. International Symposium on Molecular Approaches to Improve Crop Productivity and Quality, 22–24 May 2002. Tamil Nadu Agricultural University, Coimbatore, India.

Hedin, P.A., Waage, S.K., 1986. Role of flavonoids in plants resistance to insects. In: Cody, V., Middleton, E., Harborne, J.B. (Eds.), Plant Flavonoids in Biology and Medicine. Vol I Biochemical, Pharmacological and Structure Activity Relationships. Alan R. liss, New York, NY, pp. 87–100.

Henzell, R.G., Gillieron, W., 1973. Effect of partial and complete panicle removal on the rate of death of some *Sorghum bicolor* genotypes under moisture stress. Queensland J. Agric. Anim. Sci. 30, 291–299.

Henzell, R.G., Franzmann, B.A., Brengman, R.L., 1994. Sorghum midge resistance research in Australia. ISMN 35, 41–47.

Heupel, R.C., 1985. Varietal similarities and differences in the polycyclic isopentenoid composition of sorghum. Phytochemistry 24, 2929–2937.

Hildebrand, D.F., Rodriguez, J.G., Brown, G.C., Luu, K.T., Volden, C.S., 1986. Peroxidative responses of leaves in two soybean genotypes injured by twospotted spider mites (Acari: Tetranychidae). J. Econ. Entomol. 79, 1459–1465.

Ignacimuthu, S., Premkumar, A., 2014. Development of transgenic *Sorghum bicolor* (L.) Moench resistant to the *Chilo partellus* (Swinhoe) through *Agrobacterium*-mediated transformation. Mol. Biol. Genet. Eng. 2 (1). Available from: http://dx.doi.org/10.7243/2053-5767-2-1.

Jadhav, R., Jadhav, L.D., 1978. Studies on preliminary screening of some sorghum hybrids and varieties against earhead midge (*Contarinia sorghicola* Coq.,). J. Maharashtra Agric. Univ. 3, 187–188.

Jadhav, S.S., Mote, U.N., Bapat, D.R., 1986. Biophysical plant characters contributing to shoot fly resistance. Sorghum Newsl. 29, 70.

Jayanthi, P.D.K., Reddy, B.V.S., Gour, T.B., Reddy, D.D.R., 2002. Early seedling vigour in sorghum and its relationship with resistance to shoot fly, *Atherigona soccata* Rondani. J. Entomol. Res. 26, 93−100.

Jotwani, M.G., 1976. Host plant resistance with special reference to sorghum. Proc. Natl. Acad. Sci. (India) 468, 42−48.

Jotwani, M.G., 1978. Investigations on Insect Pests of Sorghum and Millets with Special Reference to Host Plant Resistance. Final Technical Report, 1972−77. Indian Agricultural Research Institute, New Delhi, p. 114.

Jotwani, M.G., Srivastava, K.P., 1970. Studies on sorghum lines resistant against shoot fly *Atherigona soccata* Rond. Indian J. Entomol. 32, 1−3.

Jotwani, M.G., Chaudhari, S., Singh, S.P., 1978. Mechanism of resistance to *Chilo partellus* (Swinhoe) in sorghum. Indian J. Entomol. 40, 273−276.

Jotwani, M.G., Young, W.R., Teetes, G.L., 1980. Elements of integrated control of sorghum pests. Plant Production and Protection Paper. FAO, Rome, Italy, p. 159.

Kadire, G., Jagadish, P.S., Gowda, K.N.M., Ramesh, S., Gowda, K., 1996. Biochemical basis of resistance to shoot fly (*Atherigona pulla Wiede*) in little millet. J. Agric. Sci. 9, 56−62.

Kamala, V., Sharma, H.C., Manahor Rao, D., Varaprasad, K.S., Bramel, P.J., 2009. Wild relatives of sorghum as sources of resistance to sorghum shoot fly, *Atherigona soccata*. Plant Breed 128 (2), 137−142.

Kamatar, M.Y., Salimath, P.M., 2003. Morphological traits of sorghum associated with resistance to shoot fly, *Atherigona soccata* Rondani. Indian J. Plant Prot. 31, 73−77.

Kamatar, M.Y., Salimath, P.M., Nayakar, N.Y., Deshpande, V.P., 2002. Biochemical parameters responsible for resistance to sorghum shoot fly (Atherigona soccata (R.). In: Proceedings, National Seminar on Resource Management in Plant Protection in 21st Century, 14−15 November 2002, National Plant Protection Association of India, Hyderabad, India. pp. 49.

Kamatar, M.Y., Salimath, P.M., Ravi Kumar, R.L., Swamy Rao, T., 2003. Heterosis for biochemical traits governing resistance to shoot fly in sorghum [*Sorghum bicolor* (L.) Moench]. Indian J. Genet. 63, 124−127.

Kambal, A.E., Bate Smith, E.C., 1976. Genetic and biochemical study on pericarp pigments in a cross between two cultivars of grain sorghum. *Sorghum bicolor*. Heredity 37, 413−416.

Karanjkar, R.R., Chundurwar, R.D., Borikar, S.T., 1992. Correlations and path analysis of shoot fly resistance in sorghum. J. Maharashtra Agric. Univ. 17, 389−391.

Katsar, C.S., Paterson, A.H., Teetes, G.L., Peterson, G.C., 2002. Molecular analysis of sorghum resistance to the greenbug (Homoptera: Aphididae). J. Econ. Entomol. 95, 448−457.

Khurana, A.D., Verma, A.N., 1982. Amino acid contents in sorghum plants, resistance/susceptible to stemborer and shoot fly. Indian J. Entomol. 44, 184−188.

Khurana, A.D., Verma, A.N., 1983. Some biochemical plant characters in relation to susceptibility of sorghum to stem borer and shoot fly. Indian J. Entomol. 45, 29−37.

Krueger, C.G., Vestling, M.M., Reed, J.D., 2003. Matrix-assisted laser desorption/ionization time of flight mass spectrometry of heteropoly-flavan-3-ols and glucosylated heteropolyflavans in sorghum (*Sorghum bicolor* (L.) Moench). J. Agric. Food. Chem. 51, 538−543.

Kumar, S., Singh, R., 1996. Combing ability for shoot fly resistance in sorghum. Crop Improv. 23, 217−220.

Kumar, S., Singh, R., 1998. Inheritance of tannin in relation to shoot fly resistance in sorghum. Cereal Res. Commun. 26, 271−273.

Kundu, G.G., Kishore, P., 1970. Biology of the sorghum shoot fly, *Atherigona varia soccata* Rondani (Anthomyiidae: Diptera). Indian J. Entomol. 32, 215−217.

Lal, G., Pant, J.C., 1980. Laboratory and field testing for resistance in maize and sorghum varieties to *Chilo partellus* (Swinhoe). Indian J. Entomol. 42, 606−610.

Lal, G., Sukhani, T.R., 1982. Antibiotic effects of some resistant lines of sorghum on post-larval development of *Chilo partellus* Swinhoe. Indian J. Agri. Sci. 52, 127−129.

Levin, D.A., 1973. The role of plant trichomes in plant defense. Quart. Rev. Biol. 48, 3−15.

Maiti, R.K., 1980. Role of glossy trichome trait in sorghum crop improvement. In: Annual Meeting, All India Sorghum Improvement Workshop, 12−14 May, 1980, Coimbatore, Tamil Nadu, India, pp. 1−14.

Maiti, R.K., 1994. The roles of morpho-physiological traits in shoot fly resistance in sorghum. ISMN 35, 107−108.

Maiti, R.K., Bidinger, F.R., 1979. A simple approach to identification of shoot fly tolerance in sorghum. Indian J. Plant Prot. 7, 135−140.

Maiti, R.K., Gibson, R.W., 1983. Trichomes in segregating generations of sorghum matings, II. Associations with shoot fly resistance. Crop Sci. 23, 76−79.

Maiti, R.K., Bidinger, F.R., Seshu Reddy, K.V., Gibson, P., Davies, J.C., 1980. Nature and occurrence of trichomes in sorghum lines with resistance to the sorghum shoot fly, Joint Progress Report of Sorghum Physiology/Sorghum Entomology, vol. 3. ICRISAT, Patancheru, India.

Maiti, R.K., Rao, K.V., Swaminathan, G., 1994. Correlation studies between plant height and days to flowering to shoot fly (*Atherigona soccata* Rond.) resistance in glossy and nonglossy sorghum lines. ISMN 35, 109−111.

Mate, S.N., Phadanwis, B.A., Mehetre, S.S., 1988. Studies on growth and physiological factors in relation to shoot fly attack on sorghum. Indian J. Agric. Res. 22, 81−84.

McDonough, C.M., Rooney, L.W., Earp, C.F., 1986. Structural characteristics of *Eleusine coracana* (fingermillet) using scanning electron and fluorescence microscopy. Food Microstruct. 5, 247−256.

Moholkar, P.R., 1981. Investigations on the Resistance to Shoot Fly, *Atherigona soccata* (Rondani) and Stem Borer *Chilo partellus* (Swinhoe) in Sorghum (Ph.D. thesis). Indian Agricultural Research Institute, New Delhi, India.

Mote, U.N., 1984. Sorghum species resistant to shoot fly. Indian J. Entomol. 46, 241−243.

Mote, U.N., Shahane, A.K., 1994. Biophysical and biochemical characters of sorghum variety contributing resistance to delphacid, aphid, and leaf sugary exudation. Indian J. Entomol. 56, 113−122.

Mote, U.N., Phadnis, B.N., Taley, Y.M., 1979. Studies on some physiological factors of shoo fly resistance in sorghum. Sorghum Newsl. 22, 66−67.

Nagaraj, N., Reese, J.C., Tuinstra, M.R., Smith, M.C., Amand, P.S., Kirkham, M.B., et al., 2005. Molecular mapping of sorghum genes expressing tolerance to damage by greenbug (Homoptera: Aphididae). J. Econ. Entomol. 98 (2), 595−602.

Naik, M.R., Hiremath, I.G., Salimath, P.M., Patil, S.U., 1996. Relationship of biochemical compounds with midge, *Contarinia sorghicola* (Coq.) resistance in sorghum. Mysore J. Agric. Sci. 30, 142−148.

Napompeth, B., 1973. Ecology and Population Dynamics of the Corn Plant Hopper, *Perigrinus maidis* (Ashmead) (Homoptera: Delphacidae) in Hawaii (Ph.D. thesis). University of Hawaii, Honolulu, Hawaii, p. 257.

Narayana, N.D., 1975. Characters contributing to sorghum shoot fly resistance. Sorghum Newsl. 18, 21.

Narwal, R.P., 1973. Silica bodies and resistance to infection in jowar (*Sorghum vulgare* Pers.). Agra university. J. Res. (Sci.) 22, 17−20.

Natarajan, K., Chelliah, S., 1985. Studies on the sorghum grain midge, *Contarinia sorghicola* Coquillet, in relation to environmental influence. Trop. Pest Manage. 31, 276−285.

Ni, X., Quisenberry, S., Heng-Moss, T., Markwell, J., Sarath, G., Klucas, R., et al., 2001. Oxidative responses of resistant and susceptible cereal leaves to symptomatic and non symptomatic cereal aphid (Hemiptera: Aphididae) feeding. J. Econ. Entomol. 94, 743−754.

Nip, W.K., Burns, E.E., 1969. Pigment characterization in grain sorghum. I. Red varieties. Cereal Chem. 46, 490−495.

Nip, W.K., Burns, E.E., 1971. Pigment characterization in grain sorghum. II. White varieties. Cereal Chem. 48, 74–80.

Nour, A.M., Ali, A.E., 1998. Genetic variation and gene action on resistance to spotted stem borer, *Chilo partellus* (Swinhoe) in three sorghum crosses. Sudan J. Agric. Res. 1, 61–63.

Nwanze, K.F., Reddy, Y.V.R., Soman, P., 1990. The role of leaf surface wetness in larval behaviour of the sorghum shoot fly, *Atherigona soccata*. Entomol. Exp. Appl. 56, 187–195.

Nwanze, K.F., Pring, R.J., Sree, P.S., Butler, D.R., Reddy, Y.V.R., Soman, P., 1992. Resistance in sorghum to the shoot fly, *Atherigona soccata*: epicuticular wax and wetness of the central whorl leaf of young seedlings. Annal. Appl. Biol. 120, 373–382.

Nwanze, K.F., Nwilene, F.E., Reddy, Y.V.R., 1998. Evidence of shoot fly, *Atherigone soccata* Rondani (Dipt., muscidae) oviposition response to sorghum seedlings volatiles. J. Appl. Entomol. 122, 591–594.

Ogwaro, K., 1978. Oviposition behaviour and host plant preference of the sorghum shoot fly *Atherigona socaata* (Diptera: Anthomyiidae). Entomol. Exp. Appl. 23, 189–199.

Ogwaro, K., Kokwaro, E.D., 1981. Development and morphology of the immature stages of the sorghum shoot fly, *Atherigona soccata* Rondani. Insect Sci. Appl. 1, 365–372.

Omori, T., Agrawal, B.L., House, L.R., 1983. Componental analysis of the factors influencing shoot fly resistance in sorghum (*Sorghum bicolor* (L.) Moench). Jap. Agric. Res. Quart. 17, 215–218.

Osuntogun, B.O., Adewusi, S.R.A., Ogundiwin, J.O., Nwasike, C.C., 1989. Effect of cultivar, steeping, and malting on tannin, total polyphenol, and cyanide content of Nigerian sorghum. Cereal Chem. 66, 87–89.

Padmaja, P.G., Christine, M., Woodcock, Toby, J.A., Bruce, 2010a. Electrophysiological and behavioral responses of sorghum shoot fly (*Atherigona soccata*), to sorghum volatiles. J. Chem. Ecol. 36 (12), 1346–1353.

Padmaja, P.G., Madhusudhana, R., Seetharama, N., 2010b. Epicuticular wax and morphological traits associated with resistance to shoot fly (*Atherigona soccata* Rondani) in sorghum, (*Sorghum bicolor*). Entomon. 34, 137–146.

Padmaja, P.G., Aruna, C., Patil, J.V., 2012. Evidence of genetic transmission of antibiosis and antixenosis resistance of sorghum to the spotted stemborer, *Chilo partellus* (Lepidoptera: Pyralidae). Crop Prot. 31, 21–26.

Padmaja, P.G., Shwetha, B.L., Swetha, G., Patil, J.V., 2014. Oxidative enzyme changes in sorghum infested by the sorghum shoot fly, *Atherigona soccata*. J. Insect Sci. 14 (1), 193.

Painter, R.H., 1958. Resistance of plants to insects. Ann. Rev. Entomol. 3, 267–290.

Pale, E., Kouda Bonafos, M., Mouhoussine, N., Vanhaelen, M., Vanhaelen Fastre, R., Ottinger, R., 1997. 7-O-methylapigeninidin, ananthocyanidin from *Sorghum Caudatum*. Phytochemistry 45, 1091–1092.

Pandey, P.K., Singh, R., Shrotria, P.K., 2005. Reversed-phase HPLC separation of phenolic acids in sorghum and their relation to shoot fly (*Atherigona soccata* Rond.) resistance. Indian J. Entomol. 67, 170–174.

Park, S.J., Huang, Y., Ayoubi, P., 2006. Identification of expression profiles of sorghum genes in response to greenbug phloem-feeding using cDNA subtraction and microarray analysis. Planta 223, 932–947.

Patel, G.M., Sukhani, T.R., Patel, M.B., Singh, S.P., 1996. Relative susceptibility of promising sorghum genotypes to stemborer in Delhi and Hisar conditions. Indian J. Entomol. 57, 279–284.

Pathak, R.S., 1985. Genetic variation of stemborer resistance and tolerance in three sorghum crosses. Insect Sci. Appl. 6, 359–364.

Pathak, R.S., Olela, J.C., 1983. Genetics of host plant resistance in food crops with special reference to sorghum stemborers. Insect Sci. Appl. 4, 127–134.

Patil, S.S., Narkhede, B.N., Barbate, K.K., 2006. Effects of biochemical constituents with shoot fly resistance in sorghum. Agric. Sci. Digest 26, 79–82.

Ponnaiya, B.W.X., 1951a. Studies in the genus *Sorghum:* I. Field observations on sorghum resistance to the insect pest *Atherigona indica* M. Madras Univ. J. 21, 96–117.

Ponnaiya, B.W.X., 1951b. Studies in the genus *Sorghum:* II. The cause of resistance in sorghum to the insect pest *Atherigona indica* M. Madras Univ. J. 21, 203–217.

Puterka, G.J., Peters, D.C., 1995. Genetics of greenbug (Homoptera: Aphididae) virulence to resistance in sorghum. J. Econ. Entomol. 88, 421–429.

Raina, A.K., 1981. Movement, feeding behavior and growth of larvae of the sorghum shoot fly, *Atherigona soccata*. Insect Sci. Appl. 2, 77–81.

Raina, A.K., 1985. Mechanisms of resistance to shoot fly in sorghum: a review. In: Kumble, V. (Ed.), Proceedings of the International Sorghum Entomol. Workshop, 15–21 July 1984. Texas A&M University, ICRISAT, College Station, Texas, U.S.A., Patancheru, India, pp. 131–136.

Raina, A.K., Thindwa, H.K., Othieno, S.M., Cork-Hill, R.T., 1981. Resistance in sorghum to the sorghum shoot fly: larval development and adult longevity and fecundity on selected cultivars. Insect Sci. Appl. 2, 99–103.

Rana, B.S., Singh, B.U., Rao, V.J.M., Reddy, B.B., Rao, N.G.P., 1984. Inheritance of stemborer resistance in sorghum. Indian J. Genet. 44, 7–14.

Rana, B.S., Singh, B.U., Rao, N.G.P., 1985. Breeding for shoot fly and stemborer resistance in sorghum. Proceedings of the International Sorghum Entomology. Workshop, 15–21 July 1984. Texas A&M University, College Station, TX, USA, pp. 347–360.

Rao, C.N., Panwar, V.P.S., 2001. Morphological plant factors affecting resistance to *Atherigona* spp. in maize. Indian J. Genet. 61, 314–317.

Rao, C.N., Panwar, V.P.S., 2002. Biochemical plant factors affecting resistance to *Chilo partellus* (Swinhoe) in maize. Ann. Plant Prot. Sci. 10, 28–30.

Rao, P., Rao, D.V.N., 1956. Studies on the sorghum shoot-borer fly *Atherigona indica* malloch (Anthomyiidae: Diptera) at Siruguppa. Mysore Agric. J. 31, 158–174.

Rao, S.S., Basheeruddin, M., Sahib, K.H., 2000. Correlation studies between the plant characters and shoot fly resistance in sorghum. Crop Res. 19, 366–367.

Reddy, K.V.S., Davies, J.C., 1979. Pests of Sorghum and Pearl Millet, and their Parasites and Predators, Recorded at ICRISAT Center, India up to August 1979. ICRISAT, Patancheru, India.

Reddy, S., Rao, N., 1975. Effect of nitrogen application on the shoot fly incidence and grain maturity in sorghum. Sorghum Newsl. 18, 23–24.

Reddy, K.V.S., Skinner IIJ.D., Davies, J.C., 1981. Attractants for *Atherigona* spp. including the sorghum shoot fly, *Atherigona soccata* Rond. (Muscidae: Diptera). Insect Sci. Appl. 2, 83–86.

Rivnay, E., 1960. Field crop pests in the near East (Pests of graminaceous and leguminous crops and stored products). Mono. Biol. Hassadeh, Israel 10, 202–204.

Rossetto, C.J., Igue, T., 1983. Inheritance of resistance to *Contarinia sorghicola* (Coq.) of sorghum variety AF 28. Bragantia 42, 211–219.

Rossetto, C.J., Goncalves, W., Diniz, J.L.M., 1975. Resistencia da variedade AF-28 a mosca do sorgho, na ausencia de outras variedades. Anais de Sociedade Entomologica do Brasil 4, 16.

Rossetto, C.J., Nagai, V., Overman, J., 1984. Mechanism of resistance in sorghum variety AF-28 to *Contarinia sorghicola* (Diptera: Cecidomyiidae). J. Econ. Entomol. 77, 1439–1440.

Satish, K., Srinivas, G., Madhusudhana, R., Padmaja, P.G., Murali, M., Nagaraja, R., et al., 2009. Identification of quantitative trait loci (QTL) for resistance to shoot fly in sorghum (*Sorghum bicolour* (L.) Moench). Theor. Appl. Genet. 119, 1425–1439.

Saxena, K.N., 1990. Mechanisms of resistance/susceptibility of certain sorghum cultivars to stem bor er *Chilo partellus*: role of behaviour and development. Entomol. Exp. Appl. 55, 91–99.

Saxena, K.N., 1992. Larval development of *Chilo partellus* (Swinhoe) (Lepidoptera: Pyralidae) on artificial diet incorporating leaf tissues of sorghum lines in relation to their resistance or susceptibility. Appl. Entomol. Zool. 27, 325–330.

Schweissing, F.C., Wilde, G., 1978. Temperature influence on greenbug resistance of crops in the seedling stage. Environ. Entomol. 7, 831–834.

Schweissing, F.C., Wilde, G., 1979. Temperature and plant nutrient effects on resistance of seedling sorghum to the greenbug. J. Econ. Entomol. 72, 20–23.

Seetharama, N., Mythili, P.K., Rani, T.S., Harshavardhan, D., Ranjani, A., Sharma, H.C., 2001. Tissue culture and alien gene transfer in sorghum. In: Singh, R.P., Jaiwal, P.K. (Eds.), Improvement of Food Crops. Sci-Tech Publishing Company, Houstan, TX, pp. 235–266.

Seitz, L.M., 2004. Effect of plant type (purple vs. tan) and mold invasion on concentrations of 3 deoxyanthocyanidins in sorghum grain. A. A. C. C. Annual Meeting Abstracts. <http://www.aaccnet.org/meetings/2004/abstracts/a04ma384.html>.

Sharma, H.C., 1985. Screening for sorghum midge resistance and resistance mechanisms. In: Proceedings of the International Sorghum Entomol. Workshop, 15–21 July 1984, College Station, Texas, USA. pp 275–292.

Sharma, H.C., 1993. Host plant resistance to insects in sorghum and its role in integrated pest management. Crop Prot. 12, 11–34.

Sharma, H.C., Franzmann, B.A., 2001. Orientation of sorghum midge, *Stenodiplosis sorghicola*, females (Diptera: Cecidomyiidae) to color and host-odor stimuli. J. Agric. Urban Entomol. 18 (4), 237–248.

Sharma, H.C., Lopez, V.F., 1990. Mechanisms of resistance in sorghum to head bug, *Calocoris angustatus*. Entomol. Exp. Appl. 57 (3), 285–294.

Sharma, H.C., Nwanze, K.F., 1997. Mechanisms of Resistance to Insects in Sorghum and their Usefulness in Crop Improvement. Information Bulletin No 45. ICRISAT, Patancheru, India, p. 56.

Sharma, H.C., Vidyasagar, P., 1994. Antixenosis component of resistance to sorghum midge, *Contarinia sorghicola* Coq., in *Sorghum bicolor* (L.) Moench. Ann. Appl. Biol. 124 (3), 495–507.

Sharma, H.C., Vidyasagar, P., Leuschner, K., 1988. No-choice cage technique to screen for resistance to sorghum midge (Diptera: Cecidomyiidae). J. Econ. Entomol. 81, 415–422.

Sharma, G.C., Jotwani, M.G., Rana, B.S., Rao, N.G.P., 1977. Resistance to the sorghum shoot fly, (*Atherigona soccata* Rondani) and its genetic analysis. J. Entomol. Res. 1, 1–12.

Sharma, H.C., Lueschner, K., Vidyasagar, P., 1990a. Factors influencing oviposition behaviour of the sorghum midge, *Contarinia sorghicola* Coq. Ann. Appl. Biol. 116, 431–439.

Sharma, H.C., Vidyasagar, P., Leuschner, K., 1990b. Components of resistance to the sorghum midge, *Contarinia sorghicola* Coq. Ann. Appl. Biol. 116, 327–333.

Sharma, H.C., Taneja, S.L., Leuschner, K., Nwanze, K.F., 1992. Techniques to Screen Sorghums for Resistance to Insect Pests. Information Bulletin No. 32. ICRISAT, Patancheru, India, p. 48.

Sharma, H.C., Vidyasagar, P., Subramanian, V., 1993. Antibiotic component of resistance in sorghum to sorghum midge, *Contarinia sorghicola*. Ann. Appl. Biol. 123, 469–483.

Sharma, H.C., Abraham, C.V., Vidyasagar, P., Stenhouse, J.W., 1996. Gene action for resistance in sorghum to midge, *Contarinia sorghicola*. Crop Sci. 36, 259–265.

Sharma, H.C., Nwanze, K.F., Subramanian, V., 1997. Mechanisms of resistance to insects and their usefulness in sorghum improvement. In: Sharma, H.C., Faujdar Singh, Nwanze, K.F. (Eds.), Plant Resistance to Insects in Sorghum. ICRISAT, Patancheru, India, pp. 81–100.

Sharma, H.C., Mukuru, S.Z., Manyasa, E., Were, J., 1999. Breakdown of resistance to sorghum midge, *Stenodiplosis sorghicola*. Euphytica 109, 131–140.

Sharma, H.C., Mukuru, S.Z., Gugi, H., King, S.B., 2000. Inheritance of resistance to sorghum midge and leaf disease in sorghum in Kenya. ISMN 41, 37–42.

Sharma, H.C., Franzmann, B.A., Henzell, R.G., 2002. Mechanisms and diversity of resistance to sorghum midge, *Stenodiplosis sorghicola* in *Sorghum bicolor*. Euphytica 124, 1–12.

Sharma, H.C., Taneja, S.L., Kameshwara Nao, N., Prasada Rao, K.E., 2003. Evaluation of Sorghum Germplasm for Resistance to Insects. Information Bulletin No. 63. ICRISAT, Patancheru, India, p. 184.

Sharma, H.C., Reddy, B.V.S., Dhillon, M.K., Venkateswaran, K., Singh, B.U., Pampapathy, G., et al., 2005. Host Plant resistance to insects in sorghum: present status and need for future research. Int. Sorghum Millet Newsl. 46, 36–42.

Sharma, H.C., Dhillon, M.K., Reddy, B.V.S., 2006. Expression of resistance to *Atherigona soccata* in f1 hybrids involving shoot fly resistant and susceptible cytoplasmic male-sterile and restorer lines of sorghum. Plant Breed 125, 473–477.

Sharma, H.C., Dhillon, M.K., Pampapathy, G., Reddy, B.V.S., 2007. Inheritance of resistance to spotted stem borer, *Chilo partellus*, in sorghum, *Sorghum bicolor*. Euphytica 156, 117–128.

Shivankar, V.I., Ram, S., Gupta, M.P., 1989. Tolerance in some sorghum germplasm to shoot fly (*Atherigona soccata* Rondani). Indian J. Entomol. 51, 593–596.

Shivaramakrishnan, S., Soman, P., Nwanze, K.F., Reddy, Y.V.R., Butler, D.R., 1994. Resistance in sorghum to shoot fly, *Atherigona soccata*: evidence for the source of leaf surface wetness. Ann. Appl. Biol. 125, 93–96.

Simmonds, M.S.J., 2001. Importance of flavonoids in insect-plant interactions: feeding and oviposition. Phytochemistry 56, 245–252.

Singh, S.P., 1998. Field efficacy of some bio-pesticides against shoot fly and stem borer in forage sorghum. Forage Res. 24, 177–178.

Singh, S.P., Jotwani, M.G., 1980a. Mechanisms of resistance in sorghum to shoot fly. I. Ovipositional non preference. Indian J. Entomol. 42, 240–247.

Singh, S.P., Jotwani, M.G., 1980b. Mechanisms of resistance in sorghum to shoot fly. II. Antibiosis. Indian J. Entomol. 42, 5–60.

Singh, S.P., Jotwani, M.G., 1980c. Mechanism of resistance to shoot fly IV. Role of morphological characters of seedlings. Indian J. Entomol. 42, 806–808.

Singh, R., Narayana, K.L., 1978. Influence of different varieties of sorghum on the biology of sorghum shoot fly. Indian J. Agric. Sci. 48, 8–12.

Singh, B.U., Rana, B.S., 1984. Influence of varietal resistance on oviposition and larval development of stalk borer *Chilo partellus* Swin., and its relationship to field infestation in sorghum. Insect Sci. Appl. 5, 287–296.

Singh, B.U., Rana, B.S., 1986. Resistance in sorghum to the shoot fly, *Atherigona soccata* Rondani. Insect Sci. Appl. 7, 577–587.

Singh, B.U., Rana, B.S., 1989. Varietal resistance in sorghum to spotted stemborer, Chilo partellus (Swinhoe). Insect Sci. Appl. 10, 3–27.

Singh, B.U., Rana, B.S., 1992. Stability of resistance to corn planthopper, *Peregrinus maidis* (Ashmead) in sorghum germplasm. Int. J. Trop. Insect Sci. 13, 251–263.

Singh, S.P., Verma, A.N., 1988. Antibiosis mechanism of resistance to stem borer, *Chilo partellus* (Swinhoe) in sorghum. Insect Sci. Appl. 9, 579–582.

Singh, B.U., Padmaja, P.G., Seetharama, N., 2004. Stability of biochemical constituents and their relationships with resistance to shoot fly, *Atherigona soccata* (Rondani) in seedling sorghum. Euphytica 136 (3), 279–289.

Soman, P., Nwanze, K.F., Laryea, K.B., Butler, D.R., Reddy, Y.V.R., 1994. Leaf surface wetness in sorghum and resistance to shoot fly, *Atherigona soccata*: role of soil and plant water potentials. Ann. Appl. Biol. 124, 97–108.

Soto, P.E., 1974. Ovipositional preference and antibiosis in relation to resistance to sorghum shoot fly. J. Econ. Entomol. 67, 165–167.

Sree, P.S., Butler, D.R., Nwanze, K.F., 1992. Morphological factors associated with leaf surface wetness. Cereal Program Annual Report 1992. ICRISAT, Patancheru, India.

Sree, P.S., Nwanze, K.F., Butler, D.R., Reddy, D.D.R., Reddy, Y.V.R., 1994. Morphological factors of the central whorl leaf associate with leaf surface wetness and resistance in sorghum to shoot fly, *Atherigona soccata*. Ann. Appl. Biol. 125, 467–476.

Subba Rao, M.V.S.S.T., Muralikrishna, G., 2002. Evaluation of the antioxidant properties of free and bound phenolic acids from native and malted fingermillet (Ragi, *Eleusine coracana 15*). J. Agric. Food Chem. 50, 889–892.

Swarup, V., Chaugale, D.S., 1962. A preliminary study of resistance of stem borer, *Chilo zonellus* (Swinhoe) infestation in sorghum (*Sorghum vulgare* Pers.). Curr. Sci. 31, 163–164.

Taneja, S.L., Leuschner, K., 1985. Resistance screening and mechanisms of resistance in sorghum to shoot fly. Proceedings of the International Sorghum Entomol. Workshop, 15–21 July 1984. Texas A&M University, College Station, Texas, USA, pp. 115–129.

Taneja, S.L., Woodhead, S., 1989. Mechanisms of stemborer resistance in sorghum. International Workshop on Sorghum Stemborers. ICRISAT, Patancheru, India, pp. 137–145.

Tao, Y.Z., Hardy, A., Drenth, J., Henzell, R.G., Franzmann, B.A., Jordan, D.R., et al., 2003. Identifications of two different mechanisms for sorghum midge resistance through QTL mapping. Theor. Appl. Genet. 107, 116–122.

Teetes, G.L., Johnson, J.W., 1978. Insect resistance in sorghum. In: Proceedings of the 33rd Annual Corn Sorghum Research Conference, Chicago, pp. 167–189.

Torto, B., Hassanali, A., Saxena, K.N., 1990. Chemical aspects of *Chilo partellus* feeding on certain sorghum cultivars. Insect Sci. Appl. 11, 649–655.

Tuinstra, M.R., Wilde, G.E., Krieghauser, T., 2001. Genetic analysis of biotype I resistance in sorghum. Euphytica 121, 87–91.

Unnithan, G.C., Reddy, K.V.S., 1985. Oviposition and infestation of sorghum shoot fly, *Atherigona soccata* Rondani on certain sorghum cultivars in relation to their relative resistance and susceptibility. Insect Sci. Appl. 6, 409–412.

van den Berg, J., van der Westhuizena, M.C., 1997. *Chilo partellus* (Lepidoptera: Pyralidae) moth and larval response to levels of antixenosis and antibiosis in sorghum inbred lines under laboratory conditions. Bull. Entomol. Res. 87, 541–545.

Venkateswaran, 2003. Diversity Analysis and Identification of Sources of Resistance to Downy Mildew, Shoot Fly and Stemborer in Wild Sorghums (Ph.D. thesis). Department of Genetics, Osmania University, Hyderabad, Andhra Pradesh, India.

Verma, T., Singh, S.P., 2000. Morpho-physio plant characters associated with shoot fly resistance in sorghum. Haryana Agric. Univ. J. Res. 30, 41–43.

Verma, O.P., Bhanot, J.P., Verma, A.N., 1992. Development of *Chilo partellus* (Swinhoe) on pest resistant and susceptible sorghum cultivars. J. Insect Sci. 5, 181–182.

Vijayalakshmi, K., 1993. Study of the Interrelationship of Important Traits Contributing to the Resistance of Shoot Fly in *Sorghum bicolor* (L.) Moench (MSc. thesis). Andhra Pradesh Agricultural university, Hyderabad, p. 161.

Vinayan, M.T., 2010. Genetic Architecture of Spotted Stem Borer Resistance in Sorghum as Inferred From QTL Mapping and Synteny with the Maize Genome (Ph.D. thesis). Tamil Nadu Agricultural University, Tamil Nadu, India.

Visarada, K.B.R.S., Padmaja, P.G., Saikishore, N., Pashupatinath, E., Royer, M., Seetharama, N., et al., 2014. Production and evaluation of transgenic sorghum for resistance to stem borer. In Vitro Cell. Dev. Biol. Plant 50, 176–189.

Waniska, R.D., Poe, J.H., Bandyopadhyay, R., 1989. Effects of growth conditions on grain molding and phenols in sorghum caryopsis. J. Cereal Sci. 10, 217–225.

Waquil, J.M., Teetes, G.L., Peterson, G.C., 1986. Adult sorghum midge (Diptera: Cecidomyiidae) nonpreference for a resistant hybrid sorghum. J. Econ. Entomol. 79, 455–458.

Watterson, J.J., Butler, L.G., 1983. Occurrence of an unusual leucoanthocyanidin and absence of proanthocyanidins in sorghum leaves. J. Agric. Food Chem. 31, 41–45.

Weibel, D.E., Starks, K.J., Wood Jr.E.A., Morrison, R.D., 1972. Sorghum cultivars and progenies rated for resistance to greenbugs. Crop Sci. 12, 334–336.

Widstrom, N.W., Wiseman, B.R., McMillian, W.W., 1984. Patterns of resistance to sorghum midge. Crop Sci. 24, 791–793.

Woodhead, S., Bernays, E.A., 1978. The chemical basis of resistance of *Sorghum bicolor* to attack by *Locusta migratoria*. Entomol. Exp. Appl. 24, 123–144.

Woodhead, S., Cooper, D., 1979. Phenolic acid and resistance to insect attack in *Sorghum bicolor*. Biochem. Syst. Ecol. 7, 301–310.

Woodhead, S., Taneja, S.L., 1987. The importance of the behaviour of young larvae in sorghum resistance to *Chilo partellus*. Entomol. Exp. Appl. 45, 47–54.

Woodhead, S., Padgham, D.E., Bernays, E.A., 1980. Insect feeding on different sorghum cultivars in relation to cyanide and phenolic content. Ann. Appl. Biol. 95, 151–157.

Wu, Y., Huang, Y., 2008. Molecular mapping of QTLs for resistance to the greenbug *Schizaphis graminum* (Rondani) in *Sorghum bicolor* (Moench). Theor. Appl. Genet 117, 117–124.

Wu, X., Prior, R.L., 2005. Identification and characterization of anthocyanins by high-performance liquid chromatography electro sprayionization tandem massspectrometry in common foods in the United States: vegetables, nuts, and grains. J. Agric. Food Chem. 53, 3101–3113.

Wuensche, A.L., 1980. An Assessment of Plant Resistance to the Sorghum Midge, *Contarinia sorghicola*, in Selected Lines of *Sorghum bicolor* (Ph.D. thesis). Texas A & M University, College Station.

Zein el Abdin, A.M., 1981. Review of sorghum shoot fly research in the Sudan. Insect Sci. Appl. 2, 55–58.

Zhang, M., Tang, Q., Chen, Z., Liu, J., Cui, H., Shu, Q., et al., 2009. Gene transfer of Bt gene into sorghum (*Sorghum bicolor* L.) mediated by *Agrobacterium tumefacians*. Chinese J. Biotechnol. 25, 418–423.

Zhao, Z.Y., Cai, T., Tagliani, L., Miller, M., Wang, N., Pang, H., et al., 2000. Agrobacterium-mediatedsorghum transformation. Plant Mol. Biol. 44, 789–798.

Zhu-Salzman, K., Salzman, R.A., Ahn, J.-E., Koiwa, H., 2004. Transcriptional regulation of sorghum defense determinants against a phloem-feeding aphid. Plant Physiol. 134, 420–431.

INSECT PEST RESISTANCE IN PEARL MILLET AND SMALL MILLETS

5

G.S. Prasad and K.S. Babu

ICAR-Indian Institute of Millets Research, Hyderabad, India

5.1 INTRODUCTION

Millets are a group of highly variable small-seeded cereals grown around the world as fodder and human food. Pearl millet and small millets are important crops in the semiarid tropics of Asia and Africa, with 97% of millet production occurring in developing countries. In India, they are important as food and feed crops. A large part of their production is confined to drier areas of the country. The major pearl millet growing areas in India comprise Rajasthan, Gujarat, Maharashtra, Uttar Pradesh, Haryana, Andhra Pradesh, Karnataka, and Madhya Pradesh (Murthi and Harinarayana, 1989; Kishore, 1987). Incidence and damage by different pests vary with time and space and are reflected in low grain yield of these crops ($300-400\,\mathrm{kg\,ha^{-1}}$) in farmers' field. The current knowledge on the bionomics, behavior, vulnerable stage, and population dynamics of these pests and different pest control components form the core for evolving the future strategy on pest management in millets (Kishore, 1987; Kishore and Solomon, 1989). Of late, with the adoption of an improved package of practices, new pests are getting introduced or the alteration in pest status is gaining importance. Yield losses of $10-20\%$ have been reported in India (Gahukar and Jotwani, 1980). Due to the increased importance of millets in food and nutritional security the status of the pest situation is being looked into in order to advocate improved management practices. The available information on the pests of pearl millet and small millets is scanty and scattered in the literature. Owing to the fact that most of the pests of millets are common across the crops, the available information on pests has been presented in the following section as per the crop stages.

5.2 PEST BIOLOGY

5.2.1 SEEDLING PESTS

5.2.1.1 Shoot flies

Shoot flies belonging to the genus *Atherigona*, infest a range of millets in Africa and Asia (Deeming, 1971; Seshu Reddy and Davies, 1977; Bonzi and Gahukar, 1983; Gahukar, 1985). The species are named based on the crop they infest, that is, pearl millet shoot fly (*A. approximata* Malloch), foxtail millet shoot fly (*A. atripalpis* Malloch), barnyard millet shoot fly (*A. falcate* Thompson), proso millet shoot fly (*A. pulla* Wiedemann), and kodo millet shoot fly (*A. simplex* Thompson).

In India, *Atherigona approximata* Malloch is known to occur on pearl millet in several states causing 12−46% loss in grain and 57% loss in dry fodder yield (Natarajan et al., 1973; Singh and Jotwani, 1973; Sharma and Singh, 1974). Shoot fly infestation is less common in pearl millet, unlike the situation in sorghum. This insect also causes damage to earhead (Sharma and Davies, 1988). In Burkino Faso the highest infestation was reported during the month of Jul. and Aug. (ICRISAT, 1981). In North India its population was reported to be lowest on the 38th and highest on the 37th standard week (Mittal et al., 2006). Wild hosts such as *Cynodon dactylon* help the pest to survive during the off season (Talati and Upadhyay, 1978). Pupation occurs either on the plant or in the soil, lasting for about 6−11 days. The adult fly is identified by the presence of three pairs of black spots on the dorsum of the abdomen. The entire life cycle is completed in about 3 weeks (Sharma and Singh, 1974). Shoot fly also attacks other millets in India (Jotwani et al., 1969b), where a regular occurrence of *A. soccata* and *A. miliaceae* Mall is noted on finger millet (Singh and Dias, 1972).

Shoot fly is commonly controlled by presowing soil application of granular formulations of phorate 10 G @ 3 kg a.i. ha^{-1} (Singh and Jotwani, 1973) and culturally by increasing the seed rate and later thinning out the damaged seedlings.

5.2.1.2 Other seedling pests

Seedlings of pearl millet and small millets are also infested by various leaf beetles namely, *Menolepta* spp., *Chaetocnematibialis* Illiger, *Chaetocnema indica* Jacoby, *Podagris* spp., *Oulema downsei*, *Longitasus* spp., *Phyllotreata chotanica* Duvivier, *Lema planifrons* Weise, *L. armata* Fab., *Eryxia holoserica* Klug, *Pseudocolapsis setulosa* Lefevre, and *Monolepta senegalensis* Bryant (Gahukar and Jotwani, 1980; Gahuhar, 1984; Nwanze and Harris, 1992).

Outbreaks of gray weevils are common in dry areas in India, where infestations of several species including *Myllocerus discolor*, *M. undecimpunctulus maculosus*. *M. viridians*, *M. cardoni*, and *Tanymecus indicus* Faustare are reported (Pal, 1971; Singh and Singh, 1977; Jotwani and Butani, 1978; Chhillar and Verma, 1981). Grubs of these weevils feed on roots while the adults cause notching in the leaf margins. Stem fly (*Agromyza* spp.) is occasionally reported on pearl millet (Reddy and Puttaswamy, 1981). *Afidentula minima* Groham (Coccinellidae; Coleoptera), a phytophagous pest, was reported for the first time feeding on finger millet in Gujarat (Shah et al., 1990). The beetle *Cyaneolytta aceton* Laporte (Meloidae, Coleoptera) was recorded infesting finger millet in Tamil Nadu (Selvaraj and Chander, 2011). In the southeastern United States of America, the key insect pests on both forage and grain pearl millet are the chinch bug, *Blissus leucopterus leucopterus* Say (Heteroptera: Blissidae), and the false chinch bug, *Nysius raphanus* Howard (Heteroptera: Lygaeidae), which cause significant injury and loss of seedling stands (Hudson, 1995; Cunfer et al., 1997). Spittle bug (*Poophilus costalis* Walker) infestation causes yellowing and wilting of mature pearl millet seedling and other minor millets. Brownish gray adults and the grayish nymphs remain covered by foamy spittle (Bonzi, 1981).

5.2.2 FOLIAGE PESTS

5.2.2.1 Leaf caterpillars

High incidences of hairy caterpillars (*Amsacta albistriga* Walker, *Estigmene lactinea* Huebnerand, *Spilosoma obliqua* Walker) have been recorded on millets in the northern, southern, and western parts of India (Jotwani and Butani, 1978). Due to the gregarious habits and voracious feeding habits of this insect, a complete defoliation of the plants or destruction of seedlings may occur in a short time. The large size larvae have many blackish hairs on a reddish body. They pass the hot summer as diapaused pupae in the soil. Moths emerge about a fortnight after the receipt of the first showers. The pest can complete one or two generations depending on the rainfall and its distribution. As attack is sporadic, application of insecticide is suggested for their control (Jotwani and Butani, 1978; Verma, 1981). Plowing of the field during the off season exposes the pupae to desiccation and can have a positive effect on control.

5.2.2.2 Armyworms

The armyworm (*Mythimna separata* Walker) feeds on the foliage (Balasubramanian et al., 1975). Differences in varietal susceptibility in pearl millet indicated a nonpreference for oviposition by *Spodoptera frugiperda* Smith and resistance appeared to be related to the absence of leaf pubescence (Leuck et al., 1968, 1977). Outbreaks of the noctuids (*M. separata*, *M. albistigma* Mooreand, and *Mocis frugalis* Fab.) were noticed during the rainy season of 1987 in Bangalore and Kolar districts in Karnataka. They attacked and defoliated finger millet and various other grasses. The *M. albistigma* and *M. frugalis* larvae can migrate even during the daytime. The average numbers of pupae were estimated to be 26.75, 18.5, and 15.0 per m^2 in fields of finger millet, maize, and grasses, respectively (Rajagopal and Musthak Ali, 1992).

5.2.2.3 Grasshoppers

The grasshopper feeds on the foliage and destroys the photosynthetic machinery of the crop. In India *Hieroglyphus nigrorepletus* Bolivar, *H. banian* Fab., *Chrotogonus* spp., *Colemania sphener-oides* Bolivar are reported to be destructive to millet crops (Jotwani and Butani, 1978; Gahukar and Jotwani, 1980). Measures like deep plowing of the field and dusting of insecticide were found to be effective to manage these pests. Senegalese grasshopper, *Oedaleus senegalensis* Krauss, was the main pest of millet in the Sahel (Bal et al., 2015). Grasshopper infestation in finger millet ranged from 0% to 4.6% and the leaf damage ranged from 0% to 18.5% in different tested genotypes (Sasmal, 2015).

5.2.3 SUCKING PESTS

Both the nymph and the adult of the sucking pests like aphids (*Rhopalosiphum maidis* Fitch), shoot bug (*Peregrinus maidis* Ashmead), and plant bug (*Aspavia armigera* Fab., *Callidea* spp.,) suck the sap from young leaves causing distortion, yellowing and wilting of the plants leading to formation of the shriveled and chaffy grains (Jotwani and Butani, 1978; Gahuhar, 1984; Balikai, 2010). Some of the sucking pests act as vectors of the plant viruses. Common sucking pests recorded on pearl millet include *Balclutha* spp. (Khurana and Deshmukh, 1974), pyrilla (*Pyrilla perpusilla* Walker)

(Jotwani et al., 1969a), spotted aleyrodid, *Neomaskellia bergii* Signoret (Verma, 1980), and thrips (*Chirothrips mexicanus* Crawford) (Ananthakrishnan and Thirumalai, 1977). Other species like *Nezara viridula* Linn., and *Dolycoris indicus* Stal., were recorded as minor pests of pearl millet sucking sap from the milky grains (Balikai, 2010). On finger millet, infestation of the rusty plum aphid (*Hysteroneura setariae* Thomas) was reported to be common (Desan and Kolandaiswamy, 1974; Kishore, 1991; Prabhuraj and Jagadish, 2001). This aphid could complete its life cycle in about 7−8 days under laboratory conditions and could produce 2.84−4.55 nymph's per day. Infestation ranged from 0% to 7.4% on plant and 1.4% to 12.4% on panicles (Sasmal, 2015).

Application of the insecticides like dichlorvos, chlorpyrifos, monocrotophos, and dimethoate could control the aphid population only for a short period of time (up to the third day after spraying), and pest populations increased thereafter. Carbofuran was, however, effective in controlling the plants up to 20 days after application. There was no increase in aphid population when carbofuran was sprayed at 75 days after sowing but the population increased when it was applied at 25, 35, 45, 55, and 65 days after sowing (Prabhuraj and Jagadish, 2001).

5.2.4 STEMBORER

Stemborers are one of the major destructive groups of insects attacking the millets (Tams and Boweden, 1953; Jepson, 1954; Ingram, 1958; Harris, 1962; Breniere, 1971). Two species of stemborer, that is, *Chilo partellus* Swinhoe and *Sesamia inferens* Walker, are predominantly observed in all the millets in India (Gahukar and Jotwani, 1980). Other species include *Acigona ignefusalis* Hmps., *Sesamia calamistis* Hmps., *Busseola fusca* Fuller in Africa (Gahukar, 1981a,b), *Chilo infuscatella* Snellen in China (Yongfu and Huiling, 1982), and *Diatraea grandiosella* Dyar in North America (Starks et al., 1982). *Acigona ignefusalis* was a major pest on pearl millet in West Africa (Harris, 1962; Ndoye, 1979b; Gahuhar, 1984) and *S. inferens* on finger millet in India (Kundu and Kishore, 1971; Jotwani and Butani, 1978). The pests bore into the stem and cause deadheart symptom, which results in the development of profuse numbers of tillers with unproductive spikes. Late attack results in peduncle damage and plant breakage contributing to loss of grains and fodder. Pearl millet appears to be immune to borer attack at the initial stages of the crop growth but it becomes susceptible to internode-injury later (Ahmed and Young, 1969). Infestation of *C. partellus* on pearl millet was observed during 35−42 standard week, and its population was highest on the 37th week under Delhi conditions (Mittal et al., 2006).

The solitary predatory wasp (*Dasyproctus agilis* Smith) was recorded damaging pearl millet in the field in Delhi, India. The damage was caused by the nesting of females, which made circular holes. Varieties with tall and thin stems were preferred for nesting (Srivastava et al., 1990).

5.2.4.1 Spotted stemborer

The spotted stemborer (*Chilo partellus* Swinhoe) is one of the most destructive and cosmopolitan species. It occurs in all the pearl millet growing areas of the country but was found to be more predominant in Gujarat and Delhi regions of India. The female lays patches of flattened, overlapping yellowish oval white eggs on the underside of the leaves. The larvae on hatching feed on the tender leaf whorls and later bore into the stem. Feeding on the central shoot causes formation of deadheart and death of the plant. Life cycle is completed within 4−6 weeks. The pink borer larvae are distinctly pink in color, while the *Chilo* larvae are creamy in color with dark spots (Kishore, 1996b).

5.2.4.2 Pink stemborer

The pink borer (*Sesamia inferens* Walker) lays spherical, yellowish white eggs in batches of 30–100 eggs in two to three parallel lines on the leaf sheaths or the leaf surface. The larva is pinkish, smooth, and measures about 30 mm when full grown. The larva bores directly into the stem without leaf feeding during early instars. Pupation occurs in stem galleries and the life cycle is completed in 40–50 days. Moths have straw colored forewings with a marginal dark brown streak. The pest completes four to five generations in a year under tropical conditions.

Management of borer population is possible by practicing field sanitation—collection and destruction of stubble and weeds after crop harvest—and pulling up deadhearts (Jotwani and Butani, 1978; Sagnia, 1983). The long duration millets and the late planting crops are prone to increased borer attack (Harris, 1962; Ndoye, 1979b). Efforts have been made to determine integrated management strategies by the study of insect rearing, screening, resistance mechanisms, and economic injury levels (Lingappa and Channabasavanna, 1983). In Africa, destruction of stalks is difficult to recommend as they are used for fencing or making roofs of houses. Natural enemies have been surveyed in Africa and India (Sharma et al., 1966; Mohyuddin and Greathead, 1970; Singh and Sandhu, 1977; Gahukar, 1981a). Hymenopterous parasitoids (*Apanteles*: *Euvipio* spp., *Syzeuctus* spp., *Goniozus procerae* Risbec) seem to be active in West Africa (Harris, 1962; Ndoye, 1980; Quicke, 1983). The parasitoids and predators recorded on life stages of *Sesamia* spp., and *C. partellus* in India (Sharma et al., 1966; Singh and Sandhu, 1977; Gahukar and Jotwani, 1980; Sharma, 1985) may play a significant role in reducing pink borer incidence.

5.2.4.3 Millet stemborer

Millet stemborer [*Coniesta* (=*Acigona*) *ignefusalis* (Hampson)] is well known as a persistent and often damaging pest of pearl millet, especially in the Sahelian and sub-Sahelian zones of West Africa. It is, however, not known to be important on other cereals. It is restricted to mainland Africa, south of the Sahara, and has been most frequently recorded in West Africa. It has also been recorded in Sudan, Ethiopia, and Angola and is probably more widely distributed in tropical Africa (Youm et al., 1996). The female moth lays about 200 eggs in groups on the leaf or sheath or underside of leaves. The larva comes out after hatching of the egg after a week and penetrates the whorl and moves into the stem. Larvae move from tiller to tiller and often infest tillers in a hill. A single stem may harbor as many as 20 larvae. After completing development in 30–40 days the larva pupates in the stem galleries. A full grown larva measures 15–18 mm, has a whitish gray body with black oval spots on each segment. The pupal stage lasts about 7–13 days. Some second or third generation larvae enter a facultative diapause phase, in which they survive the hot and dry summer months. Whitish gray moths, with 20–30 mm wing span and dark brown longitudinal stripes on the forewings, emerge about a month after the first rain. Usually the life cycle is completed in 2 months and 2–3 generations have been observed in a year. There is close synchronization between the stage of insect and crop phenology, and second generation larvae are destructive to late millets (Ndoye, 1979b). Borer infestation on pearl millet does not seem to be altered by intercropping with sorghum or maize (Adesiyun, 1983).

5.2.5 EARHEAD PESTS

5.2.5.1 Spike worms

The spike worms [*Heliocheilus* (=Raghuva) *albipunctella* De Joannis (Lepidoptera: Noctuidae)] have posed a potential threat to pearl millet production in the Sudano-Sahelian zone of the West Africa since 1972 (Vercambre, 1978; Ndoye, 1979a; Ndiaye, 1985; Payne, 2006). Several species have been described from Africa by Seymour (1973) and Laporte (1977). Gahukar et al. (1986a) reviewed the literature on the major species, *Heliocheilus albipunctella* De Joannis, which caused up to 85% loss in grain yield. The female lays about 400 eggs on spikelets or floral peduncles. Eggs hatch in 3−5 days and the young larva feeds on the flowers, leaving empty glumes. Older larvae cut the floral peduncles and eat between rachis and flowers, pushing out the destroyed flowers, and making a characteristic spiral trace on the spike. The larva completes its development in 23−39 days and pupates in the soil. Pupae pass the dry season in diapause followed by quiescence and moths emerge about a month after the first rain. The pest produces only one generation a year.

Cultural practices such as deep plowing before the rainy season expose the diapausing pupae to predators and desiccation. Coincidence of moth abundance with millet flowering can be avoided by delayed plantings. However, these measures have limitations and cannot be recommended to farmers for some practical difficulties (Gahukar et al., 1986a). Similarly, the applications of several insecticides on the flowering heads failed to increase the grain yield significantly in short or medium duration millets (Gahukar et al., 1986a). Resistant cultivars seem ideal to reduce damage in present farming conditions (Nwanze, 1985; Gahukar, 1987). The effectiveness of the use of predators, parasitoids, and the pathogens on management of this pest has been studied partially. In some seasons, the parasitism level was in the order of 48% on eggs, 62% on larvae, and 8% on pupae. A hymenopterous larval ectoparasitoid (*Bracon hebetor* Say) was reported to kill up to 90% of larvae by the end of the season (Gahukar et al., 1986a). Therefore, a short duration resistant cultivar supported by a good biological control strategy should be incorporated for management of this pest.

5.2.5.2 Grain midges

Five species of midge are reported on millets in West Africa, which include *Geromyia penniseti* Felt., *Contarinia sorghii* Harris, *Lasioptera* spp., *Lestodiplosis* spp., and *Stenodiplosis* spp. (Coutin and Harris, 1968, 1974). *Geromyia penniseti* is economically important on late millets in savanna areas in Africa and South India (Coutin and Harris, 1968; Santharam, et al., 1976). Larval feeding prevents normal seed development and causes complete or partial abortion of spikes, resulting in up to 90% yield loss (Coutin and Harris, 1968). The small orange colored female fly deposits about 300 eggs within the flowering spikelet. After an incubation of about 2−3 days the orange larvae complete their development in about 7−12 days and pupate beneath the glumes. Adults emerge in 2−4 days and remain active for 1−2 days. Since the midge is a nocturnal insect, its presence can easily be overlooked. The completion of one generation in a fortnight helps the pest to complete four to five generations in a cropping season with overlapping generations especially when flowers are available. Some larvae pass the dry season by entering diapause, which may last about 8−9 months and is terminated by the appearance of warm and humid conditions during the month of Aug. or Sep.

Late planted or long duration millets are heavily infested by the midge. Midge development is also favored when cultivars with different flowering periods are planted in the same field or region.

Larval and pupal parasitoids, particularly *Tetrastichus* spp., are active later in the season and the anthocorid bugs (*Orius* spp.) feed on the ovipositing females. Their practical use in management combined with host-plant resistance and adjustment of planting period should be explored in areas endemic to this pest.

A new dipterous pest, *Dicraeus pennisetivora* Deeming, was reported in West Africa (Deeming, 1979). Since the maggot feeds on the developing seed, an early attack causes complete withering of the ovary and a late attack develops lesions on the seed.

5.2.5.3 Head beetles

Several species of beetles belonging to the genera *Anomala*, *Cetonia*, *Cylindrothorax*, *Mylabris*, and *Cyaneolytta* have been recorded on millets (Ramanamurthy et al., 1970; Dhaliwal et al., 1974; Jotwani and Butani, 1978; Balikai, 2010; Selvaraj and Chander, 2011). They feed on flowers and inhibit grain formation. Beetles like *Lytta tenuicollis* Pall., *Mylabris pustulata* Thunb., and *Psalydolytta rouxi* Laporte have been considered as key pests of pearl millet in India for a long time (Singh, 1967; Gahukar and Jotwani, 1980; Balikai, 2010). Lately, the species such as *P. fusca* Oliv. and *P. vestita* Duf. have become a problem in a few countries in West Africa (Gahukar et al., 1986b). Scarabaeids become serious occasionally, for example, *Rhinyptia infuscata* Burm. in Senegal (Gahukar and Pierrard, 1983) and *R. laeviceps* Arrow. (Yadava et al., 1973; Mishra et al., 1979), *R. meridionalis* var. *puncticollis* Arrow. (Pal and Sharma, 1973), *Rhinyptia* spp. (Verma, 1979), and *Chiloloba acuta* Wied. (Srivastava et al., 1971) in India. Dusting or spraying with insecticides is generally practiced by farmers. Dusting of malathion at 20 kg ha^{-1} and spraying of endosulfan at 2 mL L^{-1} were found to be effective at protecting 75% of the crop from blister beetles and plant bugs for up to 10 days, while no other treatments afforded more than 25% protection (Balikai and Guggari, 2006).

5.2.5.4 Head caterpillars

Grain yield is reduced due to larval feeding on developing grains. Webbing with frass is often present. If such grains are exposed to late rains, sooty mold develops and grain quality deteriorates. Common caterpillars are *Heliothis armigera* Hbn., *Eublemma gayneri* Roths., *Eublemma silicula* Swinh., *Pyroderces simplex* Wsm., *Cryptoblabes gnidiella* Mill., and *Celama* spp. (Jotwani et al., 1966; Jotwani and Butani, 1978; Patel and Dhagat, 1981; Gahukar, 1984; Sandhu et al., 1977). Of late, the severity of the noctuid *E. silicula* (*Autoba silicula*) has been increased on pearl millet. During the rainy season of 1995, a serious outbreak of *A. silicula* was observed on pearl millet in Delhi, India. Larvae fed on maturing grains and concealed themselves under a small dome-shaped or elongated gallery made of anthers and silk. The time between late-Aug. and early-Sep. was the most active period for the pest (Kishore, 1996a,b). The *H. armigera* infestation on pearl millet could be checked by the use of larval parasitoid, *Campoletis chlorideae* Uchida (Pawar et al., 1986).

Infestation of *Helicoverpa armigera* was reported in finger millet in Karnataka and incidence ranged from 8% to 37% on different genotypes (Chakravarthy et al., 2008). Twelve genotypes recorded less than 15% damage of panicles. The timing of flowering appeared to influence the incidence. Field studies showed that an increase in the larval population of *Cryptoblabes* spp. resulted in a decrease in the yield of finger millet when the number of larvae per earhead exceeded eight (Hegde et al., 1995).

5.2.5.5 Head bugs

Head bug suck the saps from the developing grains, causing distortion and shrinking. Grains are later infected by fungus, causing blackening. The major species include *Dysdercus* spp., *Spilostethus* spp., *Diploxys* spp., *Agonoscells* spp., *Lygaeus* spp., *Leptocorisa acuta* Thunb., and *Calocoris angustatus* Leth. (Jotwani and Butani, 1978; Gahukar and Jotwani, 1980; Gahuhar, 1984). A number of other bugs also damage millet spikes in India. They include the painted bug (*Bagrada cruciferarum* L.) (Sandhu et al., 1974a), *Nezara viridula* Linn. (Tayade et al., 1976), milk weed bug (*Spilostefhus pandurus* Scop.,) (Sandhu et al., 1974c), plant bug (*Aethus taticotlis* Wagn.) (Ghauri, 1977), and lygaeid bug (*Nystus ericae* Sch.) (Deol, 1985). Bugs generally appear at low populations and no measures are usually required for their control. However, in the case of epidemics foliar sprays of fenthion, malathion, fenitrothion, or monocrotophos generally give excellent control (Sandhu et al., 1974a,c).

5.2.5.6 Thrips

Two important species of thrips are *Haplothrips ganglebauri* Schum. and *Thrips hawaiiensis* Mor. Thrips are considered as minor pests which can be controlled by the spraying of malathion at 0.05% or carbaryl at 0.7% (Jotwani and Butani, 1978).

5.2.5.7 Earwigs

Earwigs (*Forficuta senegalensis* Serv.) damage millets by feeding superficially on the flowers and milky grains and are found in West Africa. They often remain within the leaf whorls and leaf sheaths (Gahuhar, 1984).

5.2.6 SOIL DWELLING INSECTS

5.2.6.1 White grubs

Several species of *Holotrichia* and *Anomala* have been reported on millets. *Holotrichia consanguinea* Blanch devastates pearl millet in large areas in central and northwestern India (Jotwani and Butani, 1978; Khairwal and Yadav, 2005). The "C" shaped grubs devour the roots resulting in withering and death of the plant. Loss in plant stand becomes visible in an infested field. Infested seedlings remain stunted and produce no seed. Even three to four grubs may attack the same plant (Lal et al., 1976). In general eggs hatch in 1−3 weeks and grubs develop in 8−22 weeks. The larvae of *H. longipennis* Blanch feed on live roots resulting in stunting, yellowing, wilting, and drying of the plant. The extent of damage varied from 2.28% to 65.15% at different altitudes in the hilly regions of Uttranchal. Damage in almost all the crops was greatest (5.67−65.16%) in the mid-hills followed by high-hills (4.96−62.92%). Crops grown in high-hills showed lower mean damage (Singh et al., 2004). After a pupal period of 1−8 weeks the beetle emerges by Nov. or Dec. if the climatic conditions are favorable. Otherwise, the pest overwinters and adult become active during May−Jul. of the following year (Jotwani and Butani, 1978). Severe infestation was reported on *Echinochloa frumentacea* Link and finger millet in Uttranchal (Singh et al., 2004).

Grubs can be controlled by soil applications of chlordane, hepltachlor, BHC, DDT, aldrin, or lindane at 2 kg a.i. ha^{-1} or phorate granules at 7.5 kg a.i. ha^{-1} (Rangarajan, 1966; Jotwani and Butani, 1978). Mixing of phorate 10 G or carbofuran 3 G or quinalphos 5 G @12 kg ha^{-1} with pearl millet seed and applying in furrows at sowing helps in reducing the damage. Application of endosulphan 4% dust @ 25 kg ha^{-1} is also quite effective against white grub. Pearl millet when intercropped with pulses like clusterbean (*Cyamopsis tragonoloba* L. Taubert) or cowpea (*Vigna ungiculata* Walp.) or greengram (*Vigna radiata* var *radiata* Wilczek) significantly reduces the damage (Khairwal and Yadav, 2005). Beetle predation by birds may occur (Pandey and Yadava, 1974). Insecticidal use in endemic areas and the adjustment of cultural practices appear to be suitable for pest management against white grubs.

5.3 HOST-PLANT RESISTANCE

For millets being crops grown under marginal soil with minimum inputs, host-plant resistance forms the most practicable backbone of pest management. A major advantage of host-plant resistance is that it requires practically no input of the farmer except seed. To this effect a variety possessing horizontal resistance is considered to be the most desirable. The presence of multiple pest resistance in a variety further reinforces the defense (Breniere, 1980; Nwanze, 1985; Gahukar, 1987, 1988).

5.3.1 SOURCES OF RESISTANCE

The identification of sources of resistance to major pests of millets is of the utmost importance, and will provide material for breeders for the development of resistant varieties and hybrids (Kishore, 1995, 1996a,b). The millet germplasm, varieties, and hybrids found resistant or promising against various pests under different experiments, trials, and nurseries worldwide are compiled and presented in Table 5.1 (pearl millet) and Table 5.2 (small millets).

5.3.2 GENETICS OF RESISTANCE

The studies on the genetics of insect resistance in millets are scarce. Pearl millet inbreds and hybrids were evaluated for resistance to chinch bug at Lincoln, NE, and Tifton, GA, USA (Rajewski et al., 2009). The inbreds 59464B and 59668M-1 were the most frequently identified resistant lines. Inbred Tift 99B was susceptible. When insect damage among hybrids made with Tift 454 was evaluated, resistance tended to be dominant or overdominant in expression. Inbred lines 03GH707 and Tift 454, developed at Tifton, were resistant only in some assessments at Tifton, but not at Lincoln. Location-specific resistance influenced by environmental conditions or genetic differences in the insect populations between the two locations was observed. The line 16RmR1, developed at Lincoln, was susceptible in both the Lincoln experiments, but not at Tifton. The line 03GH706, on the other hand, was susceptible in some Tifton assessments, but was not among the most susceptible inbreds in the Lincoln experiments. Data suggested multilocation evaluations to effectively identify resistance to chinch bug feeding in pearl millet (Rajewski et al., 2009).

Wilson et al. (2000) suggested that expression of resistance is a quantitative trait, and can vary across locations and seasons. Resistance is not always fully dominant; both positive and negative general combining ability for plant damage were observed in diallel crosses of pearl millets from Africa. Generally hybrids were found to be more resistant than the parental inbreds.

Table 5.1 Resistant or Less Susceptible Genotypes of Pearl Millet Reported Against Various Insect Pests

Pest	Resistant/Promising Genotype	Reference
Shoot fly	23 D2A × H 403	Natarajan et al. (1973)
	5141 A × PT 1939, IP 241, PT 1939, MS 6317, PT 1522, PT 1930, IP 863, PT 1836, MS 6112	Appadurai et al. (1981)
	JFB 801, JFB 812	Pandey et al. (1985)
	MP 16, MP 19, MP 31, MP 53, MP 67, MH 49, MH 52, M 9, MH 82, MH 99, MH I05	Kishore (1996a)
	MP-19, MP-67, MH-91, HHB-67, Pusa-23, MH-855	Kishore (2000)
	MH 1616, MH 1600, MH 1642, MH 1650, MH 1671, MH 1685, MP 501, GHB 744	AICPMIP (2010)
	PB 106, MH 1684, MH 1663, MH 1697	AICPMIP (2011)
	MH 1720, MH 1759, MH 1751	AICPMIP (2012)
	MH 1785, MH 1828, MH 1852, MH 1864, MH 1880, MH 1900, MH 1904, GHB 744	AICPMIP (2013)
	MH 1975, MH 1828, MH 1957	AICPMIP (2014)
Spotted stem borer	A 10, A 21P1, A 63, A 66, A 163, A 280, A 281	Sandhu et al. (1976)
	MP 19, MP 2I, MP 31, MP 39, MP 47, MP 53, MP 56, MP 60, MP 63, MH 31, MH 9I, ICMS 7703, ICMS 7704, ICH 241, WCC 75, IVPS 77	Kishore (1996a)
	MH 1246, MH 1249, MH 1321, MP 473, MH 1399, MH 1405	Kishore et al. (2005)
	PUSA 23, PUSA 383, MH 1600, MH 1650, MH 1663, MH 1690, MP 489	AICPMIP (2010)
	MH 1695, MH 1754, MP 508	AICPMIP (2011)
	MH 1795, RAJ 171	AICPMIP (2012)
	MH 1700, MH 1771, MH 1777, MH 1785, MH 1792	AICPMIP (2013)
	MH 1888, MH 1852, HHB67 Imp, MH 1901, MH 1957, MH 1969, MH 1977, MP 545	AICPMIP (2014)
Millet stem borer	CIVT, Sadore local	ICRISAT (1983)
	Zongo	Gahuhar (1984)
	INMB 106, INMB 218, INMB 155	Ndoye et al. (1986)
Oriental armyworm	Souga Local 4, 700112, PIB 228, D 1051	Sharma and Davies (1982)
	IP 6577, PIB 228, IP 6069, IP 6251, IP 5836	Sharma and Sullivan (2000)
Spike worm	Ex-Bornu and Souna, HKP, Zongo 3, Nieluve, Bou	Vercambre (1976, 1978)
	Souna, 314 HK 78, ICMS 7819, ICMS 7838, IBV 8001, M 24-38, Nigerian Composite, HKB Tif, CIVT, HKP, Zongo, Nieluva, Boudouma, IBMV 8392, INMG 52, INMV 5001, SRM-Dori, P3 Kolo, ITV 8001, Kassblaqa, Yolusee-Nial, Tara Yombo	Ndoye et al. (1986)
	Souna, KH-78, IBV 8001, ICMS 7819	Ndoye and Gahukar (1987)

Table 5.1 Resistant or Less Susceptible Genotypes of Pearl Millet Reported Against Various Insect Pests *Continued*

Pest	Resistant/Promising Genotype	Reference
Spike worm	IBMV 8302, INMG-1, INMG-52, ITMV 8001	ICRISAT (1984)
Earhead caterpillar *Eublemmasilicula*	29 MD, 146, RSK, 268, Pusa 605, MLBH 104	Kishore (1996a,b)
Earhead caterpillar, *Helicoverpa armigera*	MH 1910, MH 1984, MP 533, HHB 67 Imp, Nandi 61, 86M64	AICPMIP (2014)
Shoot bug	26 J, 78 J, 98, 103, 107TD, RSJ, RSK, 13073, 6D, 29MD, 146	Kishore (1996a)
Pyrilla	IP Nos. 22B, 36D, 44, 79, 214, 263, 1266, 1301, 1345, 1395, 1402	Pradhan (1971)
	79, 1395, 263, 1307, j-98, 1301, 1402, 44265, 23B, 1362	Jotwani (1978)
	36 D, IP Nos. 44, 79, 214, 263, 1266, 1301,1307, 1345, 1395, 1402, 1708	Kishore (1996a)
	MH 1236, MH 1317, MH 1234, MP 475	Kishore et al. (2005)
White grub	23 × KS30, RSK Nos. 1086, 213, 315,1826, 833, 23 A × K 230	Kishore (1996a)
Gray weevil	NHB 5	Singh and Singh (1977)
	MP 17, MP 3I, MP 38, MH 34, MH 36, MH62, MH 80, MH 82, MH 88, MH 90, MH 93, MH 94	Kishore (1996a)
	MH 1580, MH 1610	AICPMIP (2010)
	MH 1816, MH 1887, MH 1900	AICPMIP (2013)
Leaf roller	36 D, 29 MD, 146, MP 31	Kishore (1996a)
	MH 1964, MH 1969, 86M86, GHB 732, 86 M 64, ICTP 8203	AICPMIP (2014)
Greenbug	GAHI 1	Stegmeier and Harvey (1976)
	C-591, Pak-75211, Pak-75212, Pak-75219, Pak-75194, Pak-75227, Pak-75238, Pak-75272, Pak-75276, WCA-78, C-47, Pak-75322, Pak-75323, Pak-75329, Pak-75331, Pak-75334, Pak-75337, Pak-75338, Pak-75339, Pak-75353, Pak-75359	Akhtar et al. (2012)
Chinch bug	Variety: TifGrain 102, Hybrids: 59668 A × NM-5B, 59668 M × 9Rm/4Rm, 02GH973 × Tift 454, NM-5A1 × NM-7R1R5	Ni et al. (2007)
	Varieties: 04-7049, 05-5212a, 05-5206a, 04-7041, 02-7978, 02-7747, 04-7040	Maas and Ni (2009)
	Hybrids: 6017, 6059, 6064, 7017, 7018, 7021, 7028, 7030	
	07F-1226, 07F-1229, 07F-1231, 07F-1235, 07F-1238, 07F-1239, 07F-1240	Xinzhi et al. (2009)
	59464B and 59668M-1	Rajewski et al. (2009)

Table 5.2 Resistant or Less Susceptible Genotypes of Small Millets Reported Against Various Insect Pests

Crop/Pest	Resistant/Promising Genotype	Reference
Finger millet		
Pink borer	VR 94, C 180, PR 722, S 81-10	Jotwani (1978)
	IE 932, IE 982, and IE 1037	Lingappa (1979)
Gray weevil	PES 9, 144, 224, KM 1, 14, HR 228, JNR 1008, T36-B	Kishore and Jotwani (1980)
	KM 1, RAU 1, RAU 3, INNDAF 7, INDAF 8, HR 374, HR 1523, HR 154, PES 110, PES 400, WR 9, VL 110	Murthi and Harinarayana (1989)
	HR-154, PES-176, JNR-852	Kishore (1991)
	IGRFM 08-4,VL 352, GPU 88, TNEC 1234, KMR 344, DHFM V 10-2-1, GK 1, VL 376, GPU 89, PPR 1040, GK 2	Sasmal (2015)
Chilo partellus	PES 172, KM 1, PR 202, LES 224, IE 169	Kundu et al. (1980)
Earhead worm	Indaf 7, Indaf 8, PR 202, PR 202, PR 177, HR 374, HR 1523, PES 110, PES 1877, TNAU 1877, TNAU 294, VL 110	Murthi and Harinarayana (1989)
	HR 174, JAN 852, B7-43, PR 1044, PES 8, PES 176, INDAF 5, T 20-1, PES 144, CO-10, KM 14	Kishore and Jotwani (1980)
Aphid	PES 176, RAU 1, HR 374	Murthi and Harinarayana (1989)
Kodo millet		
Shoot fly	Accession nos. 6, 10, 12, 21p, 22, 44, 48, 221, 227, 232, 278, Bulk, 291, 296	Sandhu et al. (1977)
	Germplasm: 6, 11, 20, 21, 29, 32, 39, 42, 45, 50, 60, 106, 110, 113, 117, 119, 120, 121, 131, 142, 155, 158, 160, 170, 172, 173, 178, 180, 185	Murthi and Harinarayana (1989)
	Varieties: RPS 40-1, RPS 40-2, RPS 62-3, RPS 61-1, RPS 69-2, RPS 72-2, RPS 75-1, RPS102-2, RPS 107-1, RPS 114-1, RPS 120-1, IQS 147-1, CO 2, Keharpur	
	RPS 811, 902, 904, 905, 929, 941, 946, 967, 968	Jain et al. (2014)
Foxtail millet		
Shoot fly	GS No. 101, 107, 110, 112, 119, 124, 128, 129, 132, 142, 150, 151, 155, 156, 157, 160, 167, 170, 172, 174, 175	Murthi and Harinarayana (1989)
	Varieties: RAU 1, 2, 6 ISe 119, 185, 358, 700, 700, 702, 703, SIA 5, 36, 67, 242, 326, 395, SE 21-1, SIC 1, 2 CO 3	
Flea beetle	Germplasm: 2, 12, 33, 47, 62, 64, 73, 89, 101, 111, 116, 117, 118, 123, 125, 129, 157, 167, 168, 170, 179, 182, 201, 213, 219	Murthi and Harinarayana (1989)
	Varieties: SIA 1432, 1557, 1583, 1720, 2423, 2424, 2425, SE 21.1, TNAU 18, TNAU 82, Chitra	

Table 5.2 Resistant or Less Susceptible Genotypes of Small Millets Reported Against Various Insect Pests *Continued*

Crop/Pest	Resistant/Promising Genotype	Reference
Armyworm	Germplasm: 12, 29, 39, 102, 103, 104, 116, 117, 123, 125, 138, 157, 167, 168, 169, 198, 201, 219 Varieties: SIA 1557, 1583, 1720, 2423, 2424, 2425, 2425, SS 21-1, ITS 69, SIC 31	Murthi and Harinarayana (1989)
Leaf roller	Germplasm: 26, 39, 73, 101, 121, 123, 126, 128, 137, 144, 170 Varieties: SIA 1432, 2423, 2424, 2425, SE 21-1, SIC 28	Murthi and Harinarayana (1989)
Little millet		
Shoot fly	GPMR No. 7, 17, 18, 20, 22, 26, 46, 53, 78, 84, 92, 98, 101, 104, 106, 107, 112, 114, 115, 116, 117, 124, 132, 134, 136, 141, 148, 149, 163, 169, 170, 171, 172, 175 Varieties: PRC 2, 3, 7, 8, 9, 10, 11, 12 RPM 1-1, 8-1, 12-1, 41-1, RAU 1, 2, K 1, CO 2, Dindori 2-1	Murthi and Harinarayana (1989)
Proso millet		
Shoot fly	GPMS No. 101, 102, 105, 108, 112, 114, 115, 117, 122, 123, 124, 125, 126, 135, 136, 138, 148, 152, 153, 155, 156, 157, 159, 164 Variety: RAUm1, 2, 3, MS 1307, 1316, 1437, 1595, 4872, PM 29-1, BR 6, CO 1	Murthi and Harinarayana (1989)
Barnyard millet		
Shoot fly	GECH No. Variety 102, 106, 108, 111, 120, 123, 127, 142, 149, 151, 157, 180, 205, 210, 218, 224, 226, 227, 230, 235, 236, 240, 241, 246, 247, 248, 250, 260, 276, 288, Bhageshwar Local-2 Variety: VL 8, 13, 21, 24, 30, 31, 32, ECC 19, 18, 20, 21, RAU 7, KE 16, K 1, PUNE 2386	Murthi and Harinarayana (1989)

Presence of chinch bug resistance in elite US grain pearl millet lines and their high inheritance provides an opportunity for effective selection for this trait in a pearl millet breeding program (Maas and Ni, 2009).

5.3.3 MECHANISMS OF RESISTANCE

In general, host-plant resistance to insects is based on direct or indirect defense mechanisms, which are inherently present or induced upon herbivore attack (Schoonhoven et al., 2005). Direct defense mechanisms involve physical or chemical plant traits that by themselves interfere with the physiology or behavior of the herbivore and are the main determinant of plant resistance. Morphological characteristics are known to contribute to plant resistance to insect pest (Norris and Kogan, 1980). Studies on the mechanisms of resistance in millets against insect have been scanty. Most of the statements made are based on the field observations recorded in routine screenings with meager data to support the statements.

5.3.3.1 Plant defense traits

5.3.3.1.1 Antibiosis

Antibiosis in plant resistance is expressed as prolonged or incomplete life cycle, reduced progeny production, reduced fecundity, viability, and mortality of immature stages. These may arise from physical factors and may result in the inability of the insect to utilize the host for feeding or shelter leading to starvation.

In stemborers, differences exist in the initial levels of infestation between genotypes and infestation shifted with crop age and phenology. Such changes were due to differences in the biophysical and chemical constituents between varieties at various physiological growth stages, which play a role in affecting pest populations. It was suggested that traits like size, thickness, and hardness of stem may also affect progeny development. Ndoye (1977) also suggested that in local pearl millet cultivar Zongo a secretion in the galleries where the larvae are lodged may be a resistance mechanism. Some pearl millets were found associated with *Heliocheilus albipunctella* attack. Low level of damage on long and compact panicles was observed and was not affected by number and length of floral peduncles (Vercambre, 1978). Gahuhar (1984) investigated the relationship between *Heliocheilus albipunctella* damage and bristle length, position, panicle length, compactness, and diameter and found that a relationship existed between compactness and damage. Resistance was expressed by a slower rate of plant damage due to chinch bug on resistant pearl millet genotypes as compared to the susceptible ones (Rajewski et al., 2009). The chinch bug-infested plants had lower photosynthetic rates than the noninfested control plants.

5.3.3.1.2 Nonpreference

In pearl millet, *Heliocheilus* emergence although coinciding with panicle exertion, showed low panicle damage (ICRISAT, 1983, 1984), which was attributed to oviposisional nonpreference or antibiosis against larval feeding. Nonpreference to oviposition may be due to the presence of involucral bristles, their density, length, and orientation. Bristle length was one of the few characters found associated with *Heliocheilus* infestation. Bristles on panicles of pearl millet also caused reduced damage by the blister beetle, *Psalydolytta fusca* (Gahukar, 1988, 1991). Long duration cultivars (Sanio, NKK, Sadore, Torini, and Haini-Kiei) escaped pest attack. Compact spikes were less preferred for oviposition. Incorporation of these characters in high yielding cultivars was suggested (Gahukar, 1987). Long awn on the spike of pearl millet and lack of covering by the flag leaf were found to be associated with resistance to *Pyroderces simplex* (Sandhu et al., 1977).

In finger millet a number of lepidopterous larvae infest the earhead at maturity. The total damage varies considerably with the variety, the season, and other factors. The more compact or tightfisted the panicles, then the greater is the susceptibility to attack, as such panicles provide a congenial microclimate for the worms to multiply or to hide within the closed head (Murthi and Harinarayana, 1989; Sharma et al., 1998). The presence of a high number of vascular bundles was linked to susceptibility to pink borer (Kishore and Jotwani, 1980).

High trichome length and density in little millet (*Panicum miliaceae* L.,) induced nonpreference for ovipoistion by the shoot fly. The susceptible genotypes were more vigorous in growth (higher plumule, coleoptile, and radical length) than the resistant genotypes (Gowda et al., 1996).

5.3.3.1.3 Tillering capacity

Tillering capacity is an adaptive form of tolerance of the native grasses to stem injury and may result in an overall increase in head production and yield (Nwanze, 1985). Local genotypes of pearl millet are reported to produce tillers profusely under moderate to low attack by borers and still produce reasonable yields. Harris (1962) and Nwanze (1989) indicated higher yields of millet under low borer infestation due to production of large number of tillers.

5.3.3.1.4 Pseudo-resistance

Infestation of *Heliocheilus* results in severe damage of panicle and yield loss when the peak of the moth emergence period coincides with the panicle exertion. Hence, the early as well as the late varieties of pearl millet evaded the pest infestation. It was shown that the extent of crop damage was directly related to the period of crop maturity and head exertion (ICRISAT, 1984). The short-cycle pearl millet cultivar, Souna millet, was reported to have an escape mechanism from damage by blister beetle (Gahukar, 1991). Coop et al. (1993) reported that millet grains compensate for damage by the meloids through enlargement of grains in neighboring undamaged glumes. Typically hybrid pearl millet plants grow so vigorously that severe damage by chinch bug and yield loss are not observed (Maas and Ni, 2009).

Jotwani (1978) opined that early maturing lines of finger millet were less susceptible to earhead caterpillars. Late sown millets generally evade attack by white grubs, but crops may be infested severely later in the season (Singh et al., 2004). Late maturing finger millet varieties had more severe incidence of pink borer and gray weevils than the early and mid-late varieties (Lingappa, 1979). The rusty plum aphid (*Hysteroneura setariae*) is often found infesting the leave, stem, and shoot of finger millet in large numbers. Aphids were found in higher frequency on mid-late than early and late varieties (Nageshchandra, 1981).

5.3.4 UTILIZATION OF RESISTANCE

Incorporation of identified resistance into a variety or hybrid seems to be the real bottleneck in the pathway to the utilization of pest resistance. Most of the resistant sources are from germplasm which possess traits often not preferred by the farmers. This has been, however, amply exploited in the case of pearl millet against chinch bug. The availability of sources of resistance to certain pests has now broadened the base for developing varieties to suit endemic areas, hot spots, or location-specific pest problems.

5.4 CONCLUSIONS

The available evidences suggest that pearl millet and small millets are relatively less subject to many pests compared with other graminaceous crops. The perusal of literature reveals that in pearl millet the major pests are shoot fly, stemborer, and white grub, while in finger millet it is pink borer. Information on the key pests of pearl millet and small millets with respect to data on economic injury levels, yield loss, effectiveness of natural enemies, and their use in existing ecosystems are scanty. Standardized screening techniques for shoot flies and borer in pearl and finger millet are lacking and the information on the resistance mechanism is rare and needs to be generated. Little progress has been made on the incorporation of the resistant-traits into agronomically suitable cultivars. Pest monitoring systems in these crops are in their infancy and warrant special emphasis under the changing global climate scenario.

5.5 FUTURE PRIORITIES

- The biology of shoot flies should be studied and their population dynamics throughout the annual cycle should also be determined. The specificity of shoot fly and the occurrence of alternate or collateral hosts need critical investigations.
- The seasonal occurrence of polyphagous pests like spotted and pink borer should be studied through nationwide monitoring. There is a need to generate data on crop loss due to pests and identify economic injury levels.
- The available germplasm should be screened under artificial infestation to identify stable resistance sources for shoot fly and borers and confirm field resistance in developed entries.
- Attention is needed for standardization of the screening techniques and continued search for the sources of resistance.
- The traits responsible for conferring resistance must be identified so that traits of interest can be incorporated into agronomically suitable cultivars.
- Information needs to be generated on the natural enemies of millet pests for their use in complementing the host-plant resistance.
- There is a need to identify suitable insecticides to get the desired level of protection of millets in the event of pest outbreak.
- In view of climate change the pests on these crops should be monitored through network programs as minor or occasional pests are gaining the status of key pests.

REFERENCES

Adesiyun, A.A., 1983. Some effects of intercropping sorghum, millet and maize on infestation by lepidopterous stalk borers, particularly *Busseola fusca*. Insect Sci. Appl. 4, 387−391.

Ahmed, S., Young, W.R., 1969. Field observations on the susceptibility of sorghum, maize, and bajra to *Chilo zonellus* (Swinhoe). Indian J. Entomol. 31, 32−35.

AICPMIP, 2010. Annual Progress Report of All India Coordinated Pearl Millet Improvement Project, ICAR, Jodhpur 342 304, Rajasthan, India.

AICPMIP, 2011. Annual Progress Report of All India Coordinated Pearl Millet Improvement Project, ICAR, Jodhpur 342 304, Rajasthan, India.

AICPMIP, 2012. Annual Progress Report of All India Coordinated Pearl Millet Improvement Project, ICAR, Jodhpur 342 304, Rajasthan, India.

AICPMIP, 2013. Annual Progress Report of All India Coordinated Pearl Millet Improvement Project, ICAR, Jodhpur 342 304, Rajasthan, India.

AICPMIP, 2014. Annual Progress Report of All India Coordinated Pearl Millet Improvement Project, ICAR, Jodhpur 342 304, Rajasthan, India.

Akhtar, N., Ahmad, Y., Shakeel, M., Gillani, W.A., Khan, J., Yasmin, T., et al., 2012. Resistance in pearl millet germplasm to greenbug, *Schizaphis graminum* (Rondani). Pak. J. Agric. Res. 25, 228–232.

Ananthakrishnan, T.N., Thirumalai, G., 1977. The grass seed infesting thrips, *Chirothrips mexicanus* Crawford on *Pennisetum typhoides* and its principal alternate host. *Chloris barbata*. Curr. Sci. 46, 193–194.

Appadurai, R., Natarajan, U.S., Raveendran, T.S., Regupathy, A., 1981. Combining ability for shoot fly (*Atherigona approximate* Malloch) resistance in pearl millet (*Pennisetum americanum* (L.) Leeke). Madras Agric. J. 68, 491–495.

Bal, A.B., Ouambama, Z., Moumouni, A., Dieng, I., Maiga, I.H., Gagare, S., et al., 2015. A simple tentative model of the losses caused by the Senegalese grasshopper, *Oedaleus senegalensis* (Krauss 1877) to millet in the Sahel. Int. J. Pest Manage. 61 (3), 198–203.

Balasubramanian, R., Seshu Reddy, K.V., Govinda, R., Deviah, M.A., 1975. New record armyworm, *Pseudaletia separate* Walker (Lepidoptera: Noctuidae) as a pest of ragi in India. J. Bomb. Nat. Hist. Soc. 72, 588–589.

Balikai, R.A., 2010. Insect pest status of pearl millet (*Pennisetum glaucum* (L.) R. Br.) in Karnataka. Int. J. Plant Prot. 2 (2), 189–190.

Balikai, R.A., Guggari, A.K., 2006. Evaluation of indigenous technologies for the management of insect pests in pearl millet. J. Entomol. Res. 30 (4), 321–324.

Bonzi, M., 1981. Note sur la cicadelle du sorgho, *Poophilus costalis* Walker. Agron. Trop. 36, 185–187.

Bonzi, M., Gahukar, R.T., 1983. Repartition de la population d'*Atherigona soccata* Rondani (Diptera: Muscidae) et des éspeces alliés pendant la saison pluvieuse en Haute Volta. Agron. Trop. 38, 331–334.

Breniere, J., 1971. Les problems de lepidopteres foreurs des graminees en afrique de l'ouest. Ann. Zool. Ecol. Anim. 3, 287–296.

Breniere, J., 1980. Varietal resistance to insects in Africa: a case study of millet, Sorghum and maize. Afr. J. Plant Prot. 2, 135–148.

Chakravarthy, A.K., Hegde, J.N., Shivakumar, M.S., 2008. Unusual infestation of pod borer (*Helicoverpa armigera* Hubner) on finger millet (*Eleusine coracana* Amheric) ears in Karnataka, South India. Insect Environ. 14 (3), 135–138.

Chhillar, R.B.S., Verma, A.N., 1981. Grey weevil (*Myllocerus bicolor* Boh) grubs cause serious damage to bajra seedlings. Haryana Agric. Univ. J. Res. 11, 550.

Coop, L.B., Dively, G.P., Dreves, A.J., Jago, N.D., 1993. Damage recognition. In: Jago, N. (Ed.), Millet Crop Loss Assessment Methods. Bull. 62. Natural Resources Institute [Overseas Development Administration], Chatham, England, pp. 48–61.

Coutin, R., Harris, K.M., 1968. The taxonomy, distribution, biology and economic importance of the millet grain midge, *Geromyia penniseti* (Felt.) gen. n. comb. n. (Diptera, Cecidomyiidae). Bull. Entomol. Res. 59, 259–273.

Coutin, R., Harris, K.M., 1974. Biologie de *Contarinia sorghi* (Harris) comb. nov sur le mil au Sénégal. (Dipt., Cecidomyiidae). Ann. de Soc. Entomol. de France 10, 457–465.

Cunfer, B.M., Allison, J.R., Buntin, G.D., Phillips, D.V., Jones, S.R., Lee, R.D., et al., 1997. Disease and insect management using new crop rotations for sustainable production of row crops in the southeastern United States. In: Sustainable Agriculture Research and Education, Southern Region 1997 Annual Report, Griffin, GA, p. 13.

Deeming, J.C., 1971. Some species of *Atherigona Rondani* (Diptera: Muscidae) from northern Nigeria, with special reference to those injurious to cereal crops. Bull. Entomol. Res. 61, 133–190.

Deeming, J.C., 1979. A new species of *Dicraeus* Loew. (Diptera: Chloropidae) attacking millet in West Africa and its related species. Bull. Entomol. Res. 69, 541–552.

Deol, G.S., 1985. New record of *Nysius ericae* Schilling (Lygaeidae: Hemiptera) on pearl millet and efficacy of some insecticides for its control. Indian J. Plant Prot. 13, 51−52.

Desan, A.A., Kolandaiswamy, S., 1974. Effect of 2-chloroethyl-trimethyl ammonium chloride on the control of rusty plum aphid (*Hysteroneura setariae* T.) on ragi with graded doses of nitrogen. Madras Agric. J. 61, 780−782.

Dhaliwal, T.S., Singh, B.P., Atwal, A.S., 1974. *Cyaneolytta aceton* (Laporate) (Coleoptera: Meloidae) a new pest of maize and bajra. Curr. Sci. 43, 765.

Gahukar, R.T., 1981a. Biological control of insect pests of sorghum and pearl millet in West Africa. Proc. Conf. Biol. Control of Pests: Its Potential in West Africa. USAID, Dakar, Senegal, pp. 69−91.

Gahukar, R.T., 1981b. Control strategies for the pests of sorghum and millet in West Africa. In: Proc. Conf. Food Crisis and Agric. Res. Prod. in Afr. Ibadan, Nigeria: Assoc. Adv. Agric. Sci. Afr., pp. 169−189.

Gahukar, R.T., 1984. Insect pests of pearl millet in West Africa − a review. Trop. Pest Manage. 30, 142−147.

Gahukar, R.T., 1985. Some species of *Atherigona* (Diptera: Muscidae) reared from Gramineae in Senegal. Ann. Appl. Biol. 106, 399−403.

Gahukar, R.T., 1987. Relationship between spike worm (*Raghuva albipunctella*) infestation and flowering of pearl millet and some sources of resistance. Agronomie 7, 595−598.

Gahukar, R.T., 1988. Problems and perspectives of pest management in the Shel: a case study of pearl millet. Trop. Pest Manage. 34, 35−48.

Gahukar, R.T., 1991. Pest status and control of blister beetles in West Africa. Int. J. Pest Manage. 37 (4), 415−420.

Gahukar, R.T., Jotwani, M.G., 1980. Present status of field pests of sorghum and millets in India. Trop. Pest Manage. 26, 138−151.

Gahukar, R.T., Pierrard, G., 1983. Chafer beetles as a pest of sorghum and pearl millet. FAO Plant Prot. Bull. 31, 168−169.

Gahukar, R.T., Guevremont, H., Bhatnagar, V.S., Doumbia, Y.O., Ndoye, M., Pierrared, G., 1986a. A review of the pest status of millet spike worm, *Raghuva albipunctella* De joannis (Noctuidae: Lepidoptera) and its management in the Sahel. Insect Sci. Appl. 7, 457−463.

Gahukar, R.T., Sagnia, S.B., Pierrard. G., 1986b. Rapport du séminaire Régional sur les Méloides. Project CILSS de lutte intégrée, Ouagadougou, Burkina Faso.

Ghauri, M.S.K., 1977. On a new subspecies of *Aethus laticollis* Wagner (Hemiptera: Heteroptera: Cydnidae) as a serious pest of *Pennisetum typhoides* (Burm.) in India. J. Bomb. Nat. Hist. Soc. 72, 226−229.

Gowda, K., Jagadish, P.S., Ramesh, S., Muniswamy Gowda, K.N., 1996. Morphological characters associated with resistance to shoot fly in little millet. Kar. J. Agric. Sci. 9 (1), 63−66.

Harris, K.M., 1962. Lepidopterous stem bores of cereals in Nigeria. Bull. Entomol. Res. 53, 139−171.

Hegde, S.K., Rajagopal, B.K., Gowda, K.N.M., 1995. Assessment of loss in ragi caused by earhead caterpillar. Curr. Res. 24 (11), 214.

Hudson, R., 1995. Insects of pearl millet and their control. In: Teare, I.D. (Ed.), Proc. First Natl. Grain Pearl Millet Symp., Tifton, GA, 17−18 January 1995. Univ. of Georgia and U.S. Dept. Agric. Special Publication, pp. 72−74.

ICRISAT, 1981. Upper Volta Cooperative Program, Annual Report 1980, Section F Entomology. ICRISAT, Ouagadougou, Burkino Faso, p. 36.

ICRISAT, 1983. Annual Reports 1982. ICRISAT, Patancheru, India.

ICRISAT, 1984. Entomology. ICRISAT Sahelian Centre Annual Report, 1983. ICRISAT Sahelian Centre, Niamey, Niger, pp. 31−37.

Ingram, W.R., 1958. The lepidopterous stemborers associated with *Gramineae* in Uganda. Bull. Entomol. Res. 49, 367−383.

Jain, A.K., Dhingra, M.R., Joshi, R.P., 2014. Integrated approach for management of head smut and shoot fly in Kodo millet (*Paspalum scrobiculatum* L.) under rainfed ecosystem. Ann. Plant Prot. Sci. 22 (1), 116−121.

Jepson, W.F., 1954. A Critical Review of the World Literature on the Lepidopterous Stalk Bores of Tropical Graminaceous Crops. Commonwealth Institute of Entomology, London.

Jotwani, M.G., 1978. Investigations on insect pests of sorghum and millets with special reference to host plant resistance. Research Bulletin, Indian Agricultural Research Institute, Div. Entomol., New Delhi, India, p. 114.

Jotwani, M.G., Butani, D.K., 1978. Crop and their control: pearl millet. Pesticides 12, 20−30.

Jotwani, M.G., Dang, K., Saxeena, P.N., Young, W.R., 1966. Occurrence of *Eublemma silicula* Swinhoe as a pest of bajra in Delhi. Indian J. Entomol. 28, 407−408.

Jotwani, M.G., Srivastava, B.G., Dang, K., 1969a. Occurrence of *Pyrilla perpusilla* Walker as a serious pest of some lines of bajra (*Pennisteum typhoides*). Indian J. Entomol. 31, 368−369.

Jotwani, M.G., Verma, K.K., Young, W.R., 1969b. Observations on shoot flies (*Atherigona* spp.) damaging different minor millets. Indian J. Entomol. 31, 291−294.

Khairwal, I.S., Yadav, O.P., 2005. Pearl millet (*Pennisetum glaucum*) improvement in India − retrospect and prospects. Indian J. Agric. Sci. 75 (4), 183−191.

Khurana, S.M.P., Deshmukh, U., 1974. Serious infestation of bajra by the jassid, *Balclutha* sp. Labdev J. Sci. Technol. 12, 34.

Kishore, P., 1987. Host-plant resistance to pearl millet insect pests in India. In: International Pearl Millet Workshop, 7−11 April 1986. ICRISAT, Patancheru, India.

Kishore, P., 1991. Biology of aphid, *Hysteroneura setariae* (Thomas) and its host preference amongst ragi, *Eleusine coracana* Gaertn. genotypes. J. Entomol. Res. 15 (1), 40−42.

Kishore, P., 1995. Search for new sources of resistance in newly developed genotypes of pearl millet. J. Entomol. Res. 19 (2), 187−190.

Kishore, P., 1996a. Changing pest status of earhead caterpillar, *Eublemma silicula* Swinh. on pearl millet, *Pennisetum typhoides* (Burm.). J. Entomol. Res. 20 (3), 277−279.

Kishore, P., 1996b. Evolving management strategies for pests of millets in India. J. Entomol. Res. 20 (4), 287−297.

Kishore, P., 2000. Eco-friendly viable options for formulating management strategy for insect pests of sorghum and pearl millet. J. Entomol. Res. 24 (1), 63−72.

Kishore, P., Jotwani, M.G., 1980. Screening of some improved genotypes of the finger millet (*Eleusine coracana* Gaertn.) for resistance to *Sesamia inferens* Walker and *Myllocerus maculosus* Desbrochers. J. Entomol. Res. 4, 221−223.

Kishore, P., Solomon, S., 1989. Research needs and future strategy for controlling insect pest problems on bajra based on cropping system. Seeds Farms 15, 7−8.

Kishore, P., Mittal, V., Rai, G., Raghavani, K.L., Ram, C., Khairwal, I.S., 2005. Pest spectrum in pearl millet genotypes in India. J. Entomol. Res. 29 (4), 323−327.

Kundu, G.,G., Kishore, P., 1971. New record of parasites of *Sesamia inferens* (Walker) and *Atherigona nudiseta* (Rondani) infesting minor millets. Indian J. Entomol. 33, 466−467.

Kundu, G.G., Jotwani, M.G., Verma, K.K., Srivastava, K.P., 1980. Screening of some high yielding genotypes of ragi (*Eleusine coracana* Geartn.) to the pink borer, *Sesamia inferens* (Walker) (Lepidoptera: Noctuidae). J. Entomol. Res. 4, 97−100.

Lal, R., Bose, B.N., Katiyar, R.N., 1976. Incidence of white grubs on bajra around Delhi. Entomol. Newsl. 6, 41−42.

Laporte, B., 1977. Note concernant des noctuidae (Melicleptriinae) dont les chenilles sont mineuses des chandells de mil au Sénégal. Agron. Trop. 32, 429−432.

Leuck, D.B., Taliafero, C.M., Burton, G.W., Bowman, M.C., 1968. Fall armyworm resistance in pearl millet. J. Econ. Entomol. 61, 693–695.

Leuck, D.B., Burton, G.W., Windstrom, N.W., 1977. Insect oviposition and foliage feeding resistance in pearl millet. J. Georgia Entomol. 12, 138–140.

Lingappa, S., 1979. Development of artificial diet for mass rearing *Sesamia inferens* Walker (Lepidoptera: Noctuidae) and screening for resistance to finger millet germplasm. Mysore J. Agric. Sci. 8 (3), 353.

Lingappa, S., Channabasavanna, G.P., 1983. Standardization of techniques for mass multiplication of pink stemborer, *Sesamia inferens* Walker (Lepidopetera: Noctuidae) and for screening for resistance in ragi (*Eleusine coracana* Gaertn). Proc. Natl. Seminar on Breed. Crop plants for Resistance to pests and Diseases. Tamil Nadu Agricultural University, Coimbatore, India, pp. 56–57.

Maas, A., Ni, X., 2009. Inheritance of chinch bug resistance in grain pearl millet. J. SAT Agric. Res. 7, 8.

Mishra, R.K., Saxena, R.C., Yadava, C.P.S., Dadheech, L.N., 1979. Laboratory evaluation of some insecticides against adults *Rhinyptia laeviceps* Arrow. Pesticides 13, 50.

Mittal, V., Rai, G., Kishore, P., 2006. Monitoring of insect pest problems on pearl millet. J. Entomol. Res. 30 (1), 33–34.

Mohyuddin, A.I., Greathead, D.J., 1970. An annotated list of the parasites of graminaceous stems borers in East Africa, with a discussion on their potential in biological control. Entomophaga 15, 241–274.

Murthi, T.K., Harinarayana, G., 1989. Insect pests of small millets and their management in India. In: Seetharam, A., Riley, K.W., Harinarayana, G. (Eds.), Small Millets in Global Agriculture, Proceedings of the First International Small Millets Workshop Bangalore, India, October 29–November 2, 1986, 255–270pp.

Nageshchandra, B.K., 1981. Insects and ragi production. Paper presented at All India Coordinated Millets Workshop, April 26–28 held at College of Agriculture, Tamil Nadu Agrl. Univ., Coimbatore.

Natarajan, V.S., Raja, V.D.G., Selvaraj, S., Anavardham, L., 1973. Extent of damage caused by shoot fly (*Atherigona approximate*) on bajra hybrids. Madras Agric. J. 60, 584–585.

Ndiaye, A., 1985. Etude de biologie et du cycle vital de Raghuva albipunctella De joannis (Lepidoptea: Noctuidae). ravageur de pennisetum typhoides Burn. (mil penicillaire) au Niger MSc Dissertation, University of Quebec, Montreal, Canada.

Ndoye, M., 1977. Synthese de Quelques Resultants sur les Insects Forcers des mil et sorgo au Senegal. Centre National de Recherches Agronomidues, Bambey, Senegal, p. 15.

Ndoye, M., 1979a. New millet spike pests in Senegal and the Sahelian zone. FAO Plant Prot. Bull. 27, 7–8.

Ndoye, M., 1979b. L' entomofaune nuisible au mil a chandelle (*pennisetum typhoides*) Sénégal. Compte Rendu du Congrés sur la Lutte Contre les Insects en Compte Rendu Milieu Tropical. Chambre de commerce, Marselle, pp. 515–530.

Ndoye, M., 1980. Goniozus Procerae Risbec (Hymenoptera, Bethylidoe) Ectoparasite Larvaire d'Acigona Imefusalis (Lepid, Pyralidae, Crambinae). CNRA, Barnbey, Multigr, p. 6.

Ndoye, M., Gahukar, R., 1987. Insect pests of pearl millet in West Africa and their control – a review article. In: International Pearl Millet Workshop, 7–11 April 1986. ICRISAT, Patancheru, India.

Ndoye, M., Nwanze, K., Gahukar, R.T., 1986. Insect pests of pearl millet in West Africa and their control. Paper presented at the International Pearl Millet Workshop, 7–11 April, 1986. ICRISAT, Patancheru 502 324, India.

Ni, X.J.P., Wilson, J.A., Rajewski, G.D., Buntin, Dweikat, I., 2007. Field screening of pearl millet for chinch bug (Heteroptera: Blissidae) resistance. J. Entomol. Sci. 42, 467–480.

Norris, D.M., Kogan, M., 1980. Bio chemical and morphological basis of resistance. In: Max Well, F.G., Jennings, P.R. (Eds.), Breeding Plants Resistant to Insects. John Wiley and Sons, Inc, New York, NY, p. 683.

Nwanze, K.F., 1985. Some aspects of pest management and host plant resistance in pearl millet in the Sahel. Int. J. Trop. Insect Sci. 6, 461–465.

Nwanze, K.F., 1989. Insect pests of pearl millet in Sahelian West Africa I. *Acigona ignefusalis* (Pyralidae, Lepidoptera): Distribution, population dynamics and assessment of crop damage. Int. J. Pest Manage. 35 (2), 137–142.

Nwanze, K.F., Harris, K.M., 1992. Insect pests of millet in West Africa. Rev. Agric. Entomol. 8, 1134–1154.

Pal, S.K., 1971. A new record of leaf weevil, *Myllocerus cardoni* Marshall (Coleoptera: Curculionidae) on bajra, *pennisetum typhoides*. Labdev J. Sci. Technol. (B) 9, 138–139.

Pal, S.K., Sharma, V.P., 1973. Occurrence of *Rhinyptia meridinalis var. puncticollis* Arr. (Scarabaeidea: Coleoptera) as pest on bajra in western Rajasthan. J. Bomb. Nat. Hist. Soc. 70, 574–575.

Pandey, K.C., Faruqui, S.A., Gupta, S.K., 1985. Resistance of pearl millet varieties to shoot fly. Indian J. Agric. Sci. 55, 201–202.

Pandey, S.N., Yadava, C.P.S., 1974. King crow (*Dicrurus macrocerus*), a predator of the adult of *Holotrichia Consan guinea* Blanch. Entomol. Newsl. 4, 52.

Patel, R.K., Dhagat, N.K., 1981. Severe infestation of webworm, *Cryptoblabes gnidiella* (Mill.) on ragi (*Eleusine coracana* G.) in Madhya Pradesh. Jawaharlal Nehru Krishi Vidhyapith Res. J. 15, 135.

Pawar, C.S., Bhatnagar, V.S., Jadhav, D.R., 1986. *Heliothis* species and their natural enemies, with their potential for biological control. Proc. Indian Acad. Sci. (Animal Science) 95, 695–703.

Payne, W.A., 2006. Dryland cropping systems of West and East Africa. In: Peterson, G.A., Payne, W.A., Unger, P.W. (Eds.), Dryland Agriculture Monograph. American Society of Agronomy, Madison, WI, pp. 733–768.

Prabhuraj, A., Jagadish, P.S., 2001. Insecticidal control of leaf aphid, *Hysteroneura setaria* (Thomas) (Aphididae: Homoptera) on finger millet, *Eleusine coracana* (L.) Gaertn. Kar. J. Agric. Sci. 14 (2), 470–473.

Pradhan, S., 1971. Investigations on Insect Pests of Sorghm and Millets (1965–1970). Final Technical Report. Div. Entomol. Indian Agricultural Research Institute, New Delhi, India.

Quicke, D.L.J., 1983. The true identity of reported braconid (Hym.) parasites of *Coniseta ignefusalis* Hmps. (Lep. Pyralidae) In Nigeria. Entomol. Month. Mag. 119, 94.

Rajagopal, D., Musthak Ali, T.M., 1992. Outbreak of cutworms under dryland conditions. Mysore J. Agric. Sci. 26 (1), 46–50.

Rajewski, J.A., Ni, X., Wilson, J.P., Dweikat, I., Buntin, G.D., 2009. Evaluation of resistance to chinch bug in pearl millet in temperate and subtropical environments. Plant Health Prog. (online). Available from: http://dx.doi.org/10.1094/PHP-2009-0112-01-RS.

Ramanamurthy, T.G., Selvaraj, S., Raghupathy, A., 1970. Occurrence of blister beetle, *Mylabris pustulata* Thumb. (Meloidae: Coleoptera) on Cumbu, *Pennisetum typhoides* Stapf. & Hubb. and its damage. Madras Agric. J. 57, 134–135.

Rangarajan, A.V., 1966. A note on the occurrence of *Holotrichia* sp. (Melolonthidae: Coleoptera) on ragi (*Eleusine coracana* Gaertn.) and its control. Sci. Cult. 32, 604–605.

Reddy, D.N.R., Puttaswamy, R., 1981. Record of agromyza (Diptera: agromyzidae) as a pest bajra and its parasite, *Bracon* sp. (Hymenoptera: Braconidae). Curr. Res. 10, 23–24.

Sagnia, S.B., 1983. Possible integrated pest management tools for the effective control of cereal stemborers in the Gambia. Insect Sci. Appl. 4, 217–219.

Sandhu, G.S., Singh, B., Bhalla, J.S., 1974a. Note on the relative efficacy of different insecticides for the painted bug. *Bagrada cruciferum* Kirk. (Hemiptera: Pentatomidae) on pearl millet. Indian J. Agric. Sci. 44, 156–166.

Sandhu, G.S., Singh, B., Bhalla, J.S., 1974c. Note on chemical control of milk weed bug on pearl millet with different low volume concentrate insecticides in Punjab. Indian J. Entomol. 44, 558–559.

Sandhu, G.S., Luthra, R.C., Singh, J., 1976. Preliminary studies on the resistance of pearl millet to Chilo p'artellus (Swinhoe) (Pyralide: Lepidoptera). Sci. Cult. 42 (4), 222–223.

Sandhu, G.S., Luthra, R.C., Singh, J., 1977. Note on the comparative infestation of *Pyrodercessimplex* Wlsm. On pearl millet inbreds in Punjab. Indian J. Entomol. 39 (4), 385–387.

Santharam, G., Mohanasundaram, M., Jayaraj, S., 1976. Control of bajra grain midge, *Geromyia Penniseti* Felt. with insecticides. Pesticides 10, 45–46.

Sasmal, A., 2015. Screening of finger millet varieties against major insect pests at Odisha. J. Crop Weed 11 (1), 227–228.

Schoonhoven, L.M., Van Loon, J.J.A., Dicke, M., 2005. Insect Plant Biology. Oxford University Press, New York, NY.

Selvaraj, K., Chander, S., 2011. *Cyaneolytta acteon* (Laporte): a new pest of finger millet in Tamil Nadu. Ann. Plant Prot. Sci. 19 (1), 230–231.

Seshu Reddy, K.V., Davies, J.C., 1977. Species of *Atherigona* in Andhra Pradesh. Pest Art. News Summ. 23, 379–383.

Seymour, P.R., 1973. A revison of the genus *Masalia* (Lepidoptera: Heliothidinae). Bull. British Museum (Natural History) 27, 1–100.

Shah, A.H., Purohit, M.S., Patel, G.M., Patel, C.B., 1990. New record of *Afidentula minima* (Groham) (Coccinellidae: Coleoptera) as a pest of finger millet (*Eleusine coracana* Gaertn) in Gujarat. Indian J. Entomol. 52 (1), 160–161.

Sharma, A.K., Saxena, J.D., Subba Rao, B.R., 1966. A catalogue of the hymenopterous and dipterous parasites of *Chilo zonellus* (Swinhoe) (Crambidae; Lepidoptera). Indian J. Entomol. 28, 510–542.

Sharma, H.C., 1985. Strategies of Pest control in sorghum in India. Trop. Pest Manage. 31, 167–185.

Sharma, H.C., Davies, J.C., 1982. Studies on Pearl Millet Insects. Sorghum Entomology Progress Report-7. ICRISAT, Patancheru, India, p. 82.

Sharma, H.C., Davies, J.C., 1988. Insect and Other Animal Pests of Millets. ICRISAT, Patancheru, India, p. 86.

Sharma, H.C., Sullivan, D.J., 2000. Screening for plant resistance to Oriental armyworm, *Mythimna separate* (Lepidoptera: Noctuidae) in pearl millet, *Pennisetum glaucum*. J. Agric. Urb. Entomol. 17, 125–134.

Sharma, H.C., Mukuru, S.Z., Kibuka, J., 1998. *Helicoverpa armigera* incidence in finger millet (*Eleusine coracana* Gaertn.) at Kiboko, Kenya. ISMN 39, 147–149.

Sharma, S.K., Singh, B., 1974. *Atherigona* sp. nr. *Approximate* Malloch (Anthomyiidae: Diptera) as a pest of bajra in Rajasthan. Indian J. Entomol. 36, 246–247.

Singh, J.P., 1967. Record of blister beetle, *Epicauta tenicollis* Budaun, Pall. (Coleoptera: Meloidae) on the hybrid bajra in district Budaun (U.P.). Madras Agric. J. 54, 251–252.

Singh, K.M., Sandhu, G.S., 1977. New record of predatory beetles on *Chilo partellus* (Swinhoe). Curr. Sci. 46, 422.

Singh, K.M., Singh, R.N., 1977. The upsurge of *Myllocerus undecimpustulatus maculosus* Desb. on pearl-millet under dryland condition at Delhi. Indian J. Ent. 39 (3), 300.

Singh, M.P., Mishra, P.N., Bisht, R.S., 2004. Nature and extent of damage of white grub *Lachnosterna longipennis* (*Holotrichia longipennis* Blanch.) under various farming situations of Uttaranchal hills. Indian J. Entomol. 66 (3), 277.

Singh, V.S., Dias, C.A.R., 1972. Occurrence of different species of *Atherigona* attacking some minor millets at Kanpur. Entomol. Newsl. 2, 38.

Singh, V.S., Jotwani, M.G., 1973. Efficacy of some systematic insecticides against the bajra shoot fly, *Atherigona approximate* Malloch. Indian J. Entomol. 35, 130–133.

Srivastava, A.S., Srivastava, K.M., Katiyar, S.S., Bhadauria, A.S., 1971. New records of *Chiloloba acuta* Wied. (Scarabaeidae: Coleoptera) causing damage to hybrid bajra in Uttar Pradesh. Indian J. Entomol. 33, 97.

Srivastava, K.P., Bhatnagar, R.K., Sukhani, T.R., 1990. Occurrence of the predatory wasp, *Dasyproctus agilis* (Smith) as a pest of pearl millet. Indian J. Entomol. 52 (1), 155−157.

Starks, K.J., Burton, G.W., Wilson, R.L., Davies, F.M., 1982. Southwestern corn borer: influence of planting dates and time of infestation on damage to corn, pearl millet and sorghum. J. Econ. Entomol. 75, 57−60.

Stegmeier, W.D., Harvey, T.L., 1976. Resistance to greenbug in pearl millet. Sorghum Newsl. 19, 105.

Talati, G.M., Upadhyay, V.R., 1978. Status of shoot fly, *Atherigona approximate* (Malloch) as a pest of bajra (*pennisetum typhoides*) crops Gujarat state. Guj. Agric. Univ. Res. J. 4, 30−35.

Tams, W.H.T., Boweden, J., 1953. A revision of African species of *Sesamia* Guenée and related genera (Agrotidae: Lepidoptera). Bull. Entomol. Res. 43, 645−648.

Tayade, D.S., Payar, V.M., Wadnerkar, D.W., 1976. Painted bug on bajra in Maharashtra. Entomol. Newsl. 6 (8−9), 52.

Vercambre, B., 1976. Millet Earhead Caterpillar in Senegal. (Fr). Institut Senegalais de Recherches Agricoles, Bambey, Senegal, p. 8.

Vercambre, B., 1978. *Raghuva* spp., *Masalia* sp., chenilles des chandelles du mil en zone. Agron. Trop. 33, 62−79.

Verma, S.K., 1979. Field control of *Rhinyptia* spp. (Coleoptera: Rutelidae) adults using fenin compounds. Ann. Arid Zone 18, 274−275.

Verma, S.K., 1980. Occurrence of the spotted aleyrodid, *Neomeskellia bergii* (Singh) on pearl millet. Ann. Arid Zone 19, 171−172.

Verma, S.K., 1981. Field efficacy of insecticides and antifeedants against advanced stage larvae of *Amsacta moorei* Butler. Ann. Arid Zone 20, 253−258.

Wilson, J.P., Ouendeba, B., Hanna, W.W., 2000. Diallel analysis of chinch bug damage to pearl millet. ISMN 41, 78−79.

Xinzhi, N.I., Wilson, J.P., Buntin, G.D., 2009. Differential responses of forage pearl millet genotypes to chinch bug (Heteroptera: Blissidae) feeding. J. Econ. Entomol. 102 (5), 1960−1969.

Yadava, C.P.S., Pandey, S.N., Bharadwaj, S.C., Misra, R.K., 1973. Record of *Rhinyptia laeviceps* arrow. (Coleoptera: Scarabaeidae: Rutelidae) as a pest of bajra (*Pennisetum typhoides* S. & H.) from Rajasthan. Indian J. Entomol. 35, 271.

Yongfu, L., Huiling, C., 1982. The geographical distribution and control of millet borer, *Chilo infuscatella* (Snellen) in Luliang mountainous region. Acta Phytophylactica Sinica 9, 203.

Youm, O., Harris, K.M., Nwanze, K.F., 1996. Coniesta ignefusalis (Hompson), the millet stem borer: a handbook of information. Info. Bull. no. 46. ICRISAT, Patancheru 502 324, India, 60 pp.

STRIGA AND WEEDS

STRIGA: A PERSISTENT PROBLEM ON MILLETS

6

B.A. Kountche[1], S. Al-Babili[1] and B.I.G. Haussmann[2]

[1]*King Abdullah University of Science and Technology (KAUST), Thuwal, Kingdom of Saudi Arabia*
[2]*University of Hohenheim, Stuttgart, Germany*

6.1 INTRODUCTION

Striga spp., *S. hermonthica* (Del.) Benth. and *S. asiatica* (L.) Kuntze, known as witch weeds, are one of the most troublesome and damaging weeds in the world (Parker, 2009). They belong to the *Orobanchaceae* family (ex *Scrophulariaceae*) and are obligate root hemiparasitic plants (Olmstead et al., 2001; Tank et al., 2006). The genus *Striga* comprises of 42 species worldwide, out of which *S. hermonthica* (Del.) Benth. and *S. asiatica* (L.) Kuntze (henceforth denoted as *Striga*) are deemed to be the most economically important weeds parasitizing cereals (Berner et al., 1995; Cochrane and Press, 1997; Haussmann et al., 2000; Gressel et al., 2004; Gethi et al., 2005; Aly, 2007; Spallek et al., 2013). *Striga* infest millets, such as sorghum, pearl millet, finger millet, foxtail millet, little millet, proso millet, fonio, teff, and barnyard millet (Chemisquy et al., 2010; Olivier, 1995; Parker, 2012; Atera et al., 2012), which represent the staple food and fodder crops of millions of poor rural families in the tropical and semiarid regions of their cultivation, and therefore are of utmost significance to subsistence farmers. Millets thus play a critical role in ensuring food security in these regions (Hash et al., 2000; Senthilvel et al., 2008). *Striga* also attack other crops including maize (*Zea mays* L.), upland rice (*Oryza sativa* L.), sugar cane (*Saccharum officinarum* L.), and wheat (*Triticum aestivum* L.) (Gurney et al., 2003, 2006; Elzein and Kroschel, 2004; Vasey et al., 2005; Amusan et al., 2008). Indeed, *Striga* are the major and most persistent biotic threat to production of these crops mostly grown on the hottest and driest marginal regions of sub-Saharan Africa, Middle East, and large parts of Asia (Gurney et al., 2002; Gressel et al., 2004; Ejeta, 2007; Rispail et al., 2007; Scholes and Press, 2008; Parker, 2012). At present, over 50 million ha of the arable farmland under cereals in sub-Saharan Africa are infested by *Striga*. Crop yield losses due to *Striga* attacks range from a few percent to complete crop failure and depend largely on the cereal host species and variety grown, rainfall distribution, soil fertility, and the *Striga* seed density in soil (Haussmann et al., 2000; Parker and Riches, 1993). The annual yield loss has been estimated to exceed US$10 billion (Gressel et al., 2004; Venne et al., 2009; Westwood et al., 2012; Pennisi, 2015). More recently, *S. hermonthica* has been identified as one of the seven most severe biological constraints to food production. Thus, *Striga* present a worrying problem to subsistence farmers with small land holdings (Pennisi, 2010).

Biotic Stress Resistance in Millets. DOI: http://dx.doi.org/10.1016/B978-0-12-804549-7.00006-8

Deployment of several management strategies, including quarantine imposed on infested areas, control of movement of farm equipment between infected and uninfected areas, intensive herbicide application, and depletion of *Striga* seed bank through injection of ethylene gas, has resulted in the control of the witchweed, *S. asiatica*, in the United States, where the parasite was accidentally introduced (Berner et al., 1995, 1999; Van Mourik, 2007). However, these strategies are expensive and are not generally available to small farmers in *Striga*-prone zones of Africa and Asia. In these regions, the *Striga* problem is, in general, associated with low economic resources, low soil fertility, marginal environments with continued cereal monoculture, and newly infested areas regrettably due to various human and agricultural activities (Oswald, 2005; Rodenburg et al., 2005; Parker, 2009).

Over the years, efforts have been made to develop effective measures to combat *Striga*. A number of control strategies that aim at improving soil fertility or directly affecting the parasite life have been suggested to farmers (Haussmann et al., 2000; Rector, 2009). Genetic control of the witchweed through deployment of resistant-crops has been also extensively explored in millets, especially sorghum, and to a lesser extent in pearl millet (Ejeta, 2007; Kountche et al., 2013a). However, the complexity of the host–parasite relationship, the life cycle of *Striga* and the ability of the parasite to adapt to diverse environments, and the diversity of millets farming systems and that of the parasite (case of *S. hermonthica*) have made the use of a single control approach ineffective. Considering this, there has been an increased effort to combine two or more methods to address the long-term management of the root-parasitic weeds (Joel, 2000; Oswald, 2005). Although these efforts have helped in reducing the *Striga* damage on the host crops, they have not resulted in the complete control of the parasites, as this requires the destruction of *Striga* seed bank in infested fields.

In this chapter, we present the latest knowledge on *Striga* research and control in millet crops, highlighting the main gaps in our knowledge and suggesting how these gaps can be filled. We will split this chapter into three sections; the first section deals with the complex life cycle and *Striga* impact on millets. This section would provide brief overview on the biology of *Striga* and how this relates to current and important future control strategies. The second section describes the host plant–*Striga* pathosystem, focusing on the events up to the recognition of the host by the parasite and the genetic basis of host resistance and defense mechanisms that have evolved as part of that interaction. The literature about *Striga* research in millet crops particularly in small millets is very limited. This makes it difficult to write this chapter exclusively about the sources and genetic resistance to *Striga*, and subsequent sections. In the third section, we discuss the deployment of genetically resistant cultivars as an important component of an integrated approach toward tackling the parasitic weeds *Striga* in millets.

6.2 IMPORTANCE AND BIOLOGY

6.2.1 IMPORTANCE

Striga parasitize important food and forage grain grasses (*Poaceae*) and are therefore among the most agronomically destructive parasitic plants globally. Geographically, *Striga* are widely distributed in tropical and semi-arid regions of Africa, Middle East, Asia, and Australia (Cochrane and Press, 1997; Teka, 2014), and have been reported in more than 40 countries globally. Occurrence of the parasites in 25 African countries has been documented; the most severely affected being located in sub-Saharan Africa and India (Fig. 6.1) (Vasey et al., 2005; Ejeta, 2007; De Groote et al., 2008; Parker, 2012). Mohamed et al. (2001) suggested *Striga* to be originated from the Nuba hills of Sudan and the Simien Mountains of Ethiopia. This region has been also reported to be the native place of domesticated sorghum (*Sorghum bicolor* L.). The distribution of *Striga* in relation to ecological zones in Africa and India indicates that sorghum and pearl millet

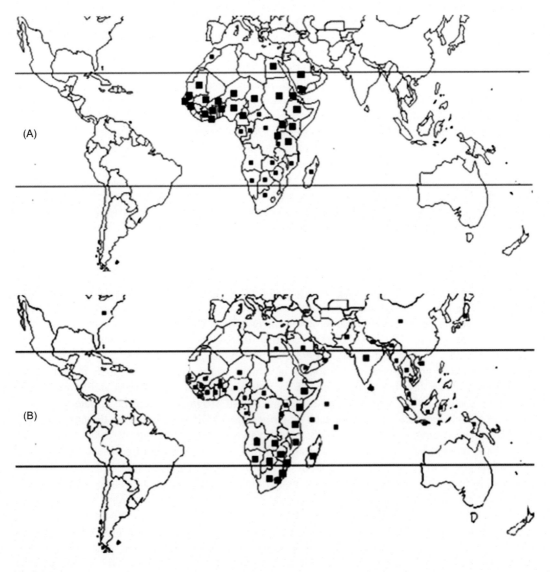

FIGURE 6.1

World distribution of *S. hermonthica* (A) and *S. asiatica* (B).

Adapted from Parker, C., 2012. Parasitic weeds: a world challenge. Weed Sci. 60, 269–276.

are the most parasitized hosts (Ramaiah, 1984; Parker, 2009, 2012). Thus, over 50 million ha of the arable farmland under cereals in sub-Saharan Africa are reported to be infested with *Striga* (Westwood et al., 2010). In West Africa, *Striga* has been estimated to infest 17.2 million ha, covering about 64% of the total area of the major millets like sorghum and pearl millet (Gressel et al., 2004), and the parasites have been reported to have expanded their infection range (Ejeta, 2007; Parker, 2012). Crop yield loss due to *Striga* infection has been estimated to range from about 10−31% in pearl millet experimental trial (Ramaiah, 1984). Gressel et al. (2004) reported an estimated yield loss of 26% on average in pearl millet and sorghum in sub-Saharan Africa. In areas of heavy *Striga* infestation, yield loss may even reach 90−100% (total crop failure) in some years (Ramaiah, 1984; Wilson et al., 2000; Gressel et al., 2004). As a result, farmers have been reported to be eventually forced to abandon highly *Striga*-infested fields (Atera et al., 2011). For these poor farmers, millets are the major staple food providing them with carbohydrates and are the main source of vitamins and minerals including zinc and iron (Andrews and Kumar, 1992; Rai et al., 2012; Bangoura et al., 2011; Mannuramath et al., 2015; Mishra et al., 2014). Hence, yield losses lead to significant negative socioeconomic problems. *Striga* affect the life of more than 300 million people in Africa and cause economic damage equivalent or even more than US$10 billion annually (Obilana and Ramaiah, 1992; Gurney et al., 2002; Rodenburg et al., 2005; Ejeta, 2007; Scholes and Press, 2008; Westwood et al., 2012). More recently, as a consequence, sub-Saharan Africa has been reported to be the region with the highest prevalence of poverty and undernourishment, with one in four people (24.8%) estimated to be hungry (FAO, IFAD and WFP, 2013).

6.2.2 BIOLOGY

Striga are annual, chlorophyll-bearing, root-parasitic plants that need a host plant to complete their life cycle. The latter is complex, intimately associated with that of the host and to the climate, particularly during postripening (Hearne, 2009). *Striga* plants have a high reproductive capacity: a single plant can produce 100,000−200,000 very tiny (0.15 ∼ 0.30 mm in diameter) seeds, which are easily dispersible (Parker and Riches, 1993; Gurney et al., 2006; Hearne, 2009). *Striga* seeds require a period of pretreatment, conditioning in a moist warm environment (30°C in germination bioassays) for 2−16 days before they acquire the potential to germinate (Logan and Stewart, 1991; Parker and Riches, 1993). After this phase, germination of *Striga* seeds will be initiated only upon induction by some specific chemicals, such as *strigolactones* (SLs), released by the host roots into the rhizosphere (Bouwmeester et al., 2003; Matusova et al., 2005; Shen et al., 2006; Zwanenburg et al., 2009; Yoneyama et al., 2010; Xie et al., 2010). The concentration of such chemicals is very low and ranges from 10^{-10} to 10^{-15} mole m^{-3} (Hearne, 2009). The root system in *Striga* is vestigial. Instead of a usual angiosperm root system, germinating seeds establish a sticky radical, which, in response to haustorial initiation factors derived from the host roots, develops to a haustorium. The haustorium connects the host and its parasite. Indeed, upon coming in contact with a host root, the haustorium develops a wedge-shaped group of cells that penetrates the host root cortex and endodermis to establish parasite−host xylem−xylem connections (Albrecht et al., 1999). This allows the direct uptake of water, assimilates, and nutrients from the host plant to the parasite. Following the establishment of the host−parasite connection, *Striga* depend entirely on the host before emergence from the soil. During this holoparasitic stage of development, the parasite inflicts a severe damage to the host. Parker and Riches (1993) reported that *Striga* parasitism resulted in chlorosis and wilting and therefore, in drastic reduction of host-plant growth and development and even in plant death under a severe infection. Subsequently, the parasites grow toward the soil surface and emerge above the ground, develop chlorophyllous shoots (hemiparasitic stage), and produce flowers and seeds which remain viable in the soil for 20 years or even more (Parker and Riches, 1993; Berner et al., 1997). However, in spite of being capable of photosynthesis, *Striga* cannot survive independently of a host in the postemergence stage. Joel et al. (2007) indicated that subsequent haustoria development, attachment and penetration, as well as further growth and development of the parasite also require signals or resource commitment from the host plant. A schematic version of the intricate life cycle of *Striga* is provided in Fig. 6.2.

FIGURE 6.2

Schematic illustration of the *Striga* life cycle. *Striga* seeds conditioning consists of favorable moisture and warm stratification that enables *Striga* seeds to become responsive to germination stimuli secreted by the host roots and some nonhost plants. *Striga* seed germinates on exposure to specific chemicals, SLs, released by the host-plant roots into the rhizosphere. Development of the haustorium by germinated *Striga* seed in response to haustorial initiation factors released by the host-plant roots. The haustorium enables parasite−host xylem−xylem connections in contact to the host-plant root, thus allowing uptake of water, mineral nutrients, and assimilates. The *Striga* plants grow toward the soil surface, emerge above the ground, and flower to produce thousands of tiny seeds. After maturity, *Striga* seeds remain dormant for several months (after ripening).

Although the parasitic plants exhibit a common lifestyle (cycle), they show differences in the reproductive phase. From a geneticist's point of view, these differences are expected to have some implications on the genetic diversity and the aggressiveness of the parasites and hence, on the breeding strategies toward development of resistant hosts. *Striga asiatica* is an autogamous (self-pollinating) species and genetic diversity analyses have shown distinct races of that species across their ranges (Botanga et al., 2002). In contrast, *S. hermonthica* is a highly out-crossing species, thus it is expected to show greater diversity within a population than seen in related autogamous species (Hamrick, 1982; Koyama, 2000). This mode of pollination has contributed to the genetic variation in *S. hermonthica* plants and also restricted the geographical distribution of this species depending on the availability of pollinators (Berner et al., 1997; Mohamed et al., 2007). Genetic diversity in parasitic weed populations has been reported to impinge on host-plant reaction (resistance/susceptibility) (Awad et al., 2006). More recently, Huang et al. (2011) reported the presence of genetic variation for host range specificity within *S. hermonthica* populations. This resulted in each *S. hermonthica* plant in a natural population having a different genotype and therefore carrying potentially different alleles for virulence (Kountche et al., 2013a). This intraspecific variation in the parasite implies that an acceptable level of resistance to one ecotype of *S. hermonthica*, for instance, may not hold when a cultivar is grown in the presence of different *S. hermonthica* populations. This nature of *Striga* is certainly one of the main reasons why the parasites cause significant damage on the host plant.

6.3 HOST-PLANT RESISTANCE AND HEREDITY

6.3.1 HOST FINDING AND ORIENTATION: THE KEY ROLE OF STRIGOLACTONES

In the course of evolution, many flowering plants have lost their autotrophic way of life and parasitized other plants in order to feed themselves directly from the host plants (Rubiales, 2003). Such parasitic plants pose a tremendous threat to today's agriculture and provide an intriguing case of pathogenesis between species of relatively close evolutionary ancestry. Almost all crop species are potential hosts for parasitic plants. So one can ask: which evolutionary conditions have favored the development of parasitism in plants? How has a crop plant become a host to the parasitic species? The evolutionary strategy of exchanging autotrophy for dependence on host plants (parasitism) may seem odd, but it has been proven to be evolutionarily successful for several plant species (Westwood et al., 2010). It has been hypothesized that during the evolution of parasitism, the parasitic plants have significantly reduced or even lost their capability for photosynthesis and, hence, cannot survive without a host plant (Xie et al., 2010). Depending on which host organ is infected, parasitic plants are grouped into stem or root parasites (Mayer, 2006).

Research on the root-parasitic plants, their interaction with the host, and the environmental cues started more than 50 years ago. One of the major questions to be answered was how the invasive root-parasitic plants, including *Striga*, find and recognize their hosts. It has been reported that plants produce and release into the rhizosphere via their roots a multitude of compounds that play roles in plant–plant communication (Bertin et al., 2003; Hirsch et al., 2003; Steinkellner et al., 2007). Like other parasitic plants from the *Orobanchaceae* family, *Striga* seeds contain limited energy reserves, and therefore, germinating seedlings must quickly locate and attach to suitable hosts upon germination in order to survive (Bouwmeester et al., 2003). Essentially, a bulk of work on host—parasite communication has focused on the parasitic partner and demonstrated that *Striga* recognizes and invades the host via a series of developmental processes that depend on the presence of a host and are triggered by host signals: for example, seed germination, haustorium formation, attachment to the host root, establishing vascular connections, and penetration. Each of these complex processes has been reported to be stimulated independently by different specific host signals (Yoshida and Shirasu, 2009; Estep et al., 2011). Thus, the high selection pressure associated with host coevolution has prompted the parasite to employ efficient strategies for host recognition. To prevent the seeds from germinating too far from a host root, *Striga* have evolved a requirement for so-called germination stimulants that are produced by the plant via their roots (Hirsch et al., 2003; Bouwmeester et al., 2003). Such a strategy ensures the parasite seed germinates when a host plant is within reach. Extensive work has been directed toward the characterization of these compounds, which are of economical and scientific importance for the biology and management of parasitic weeds. Cook et al. (1966) reported the first *Striga* germination stimulant, strigol, which was isolated from the roots of cotton (*Gossypium hirsutum* L.), a nonhost of *Striga*. Later, the same group discovered a new compound, strigyl acetate, from root exudates of cotton (Cook et al., 1972). The first *Striga* germination stimulant isolated from a host plant was sorgoleone, from sorghum (Chang et al., 1986). Hauck et al. (1992) reported the extraction and identification of a sorgolactone as the major *Striga* germination stimulant exuded by sorghum roots. These authors reported that strigol also occurs in root exudates of several host species of millets such as sorghum and proso millet, in addition to the related and more active sorgolactone (Hauck et al., 1992; Siame et al., 1993). After the discovery of the first root-derived germination stimulants, Butler (1995) coined the name strigolactones (SLs) for these strigol-related compounds. So far, a variety of different SLs have been isolated from a range of plant species, and it has been shown that exudates contain more than one SL and differences in the composition exist even between varieties of one species (Awad et al., 2006; Xie et al., 2010; Cardoso et al., 2014).

The development of highly advanced analytical methods, for example, the high-performance liquid chromatography (HPLC) connected to tandem mass spectrometry (LC-MS/MS) (Sato et al., 2003), has allowed new SLs to be surveyed in the root exudates of many different plant species, including a few millet species (Akiyama et al., 2005; Awad et al., 2006; Matsuura et al., 2008; Xie et al., 2007, 2008, 2009a, b; Yoneyama et al., 2010; Kohlen et al., 2011; Ueno et al., 2011; Jamil et al., 2012).

More recently, several studies have provided evidence that SLs, which are carotenoid-derived signaling molecules, are the key cues toward the first location of the host and also provide relevant directional cues (Al-Babili and Bouwmeester, 2015). The structure of the major SLs (Fig. 6.3) has been confirmed by total synthesis by Zwanenburg et al. (2009) and reviewed by Al-Babili and Bouwmeester (2015).

In the meantime, around 20 known different SLs have been isolated from the root exudates of different plants, all acting with varying efficiency as *Striga* and other parasitic seed germination stimulants (Muller et al., 1992; Yokota et al., 1998; Mori et al., 1999; Xie et al., 2008). Apparently, during the evolution of roots parasitic plants, SLs were selected to facilitate the process of host root recognition. This deleterious (negative) effect of SLs has led to the question of why do plants produce SLs and why *Striga* and similar parasitic plants have selected such compounds as host-finding signals? This mystery had remained unrevealed until recently when it was discovered that

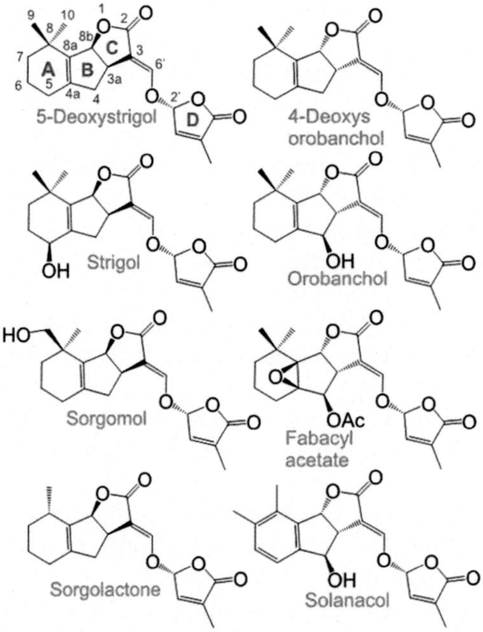

FIGURE 6.3

Structure and examples of major strigolactones (SLs). Canonical SLs consist of a tricyclic lactone (ABC-ring) connected to second lactone (D-ring) by an enol ether bridge. They are divided in strigol-like SLs with the C-ring in β orientation (left column) and orobanchol-like with the C-ring in α orientation (right column). C-atom numbering and the characteristic ABC-D-ring are shown in the structure of 5-deoxystrigol.

SLs are involved in establishing the beneficial arbuscular mycorrhizal symbiosis, by inducing hyphal branching in the fungal partner, a critical step in host recognition by arbuscular mycorrhizal fungi (AMF), hence demonstrating that SLs play also a role in mediating ecologically significant interactions with the fungi (Akiyama et al., 2005). In addition, SLs are also produced in nonmyco-trophic plants, suggesting that they exert other biological functions. Thus, SLs have been recognized as a new class of plant hormone that inhibits shoot branching (Umehara et al., 2008; Gomez-Roldan et al., 2008). Meanwhile, SLs have been demonstrated to be involved in many other aspects of plant development and to coordinate plant growth and architecture in response to availability of nutrients, such as phosphate (Al-Babili and Bouwmeester, 2015). Accordingly, these findings have unveiled that parasitic plants utilized the communication cues, delivered by plants to attract their arbuscular mycorrhizal symbionts, by sensing the SLs that are released by the roots as signals for their symbiosis with AMF (Bouwmeester et al., 2007).

Though the perception of SLs induces seed germination of *Striga*, it is however not sufficient to establish the parasitic interaction: strigol, for example, was first identified from cotton, a nonhost for *Striga* (Cook et al., 1972). That is evident by the fact that many nonhost plants (such as cotton) also release SLs. Thus, most stages of the parasite life showed a highly specific host-finding behavior and host specificity has been also determined at later stages following germination by other factors such as the haustorium-inducing factors, incompatibility reactions, or the presence of toxic metabolites (Estabrook and Yoder, 1998; Goldwasser et al., 1999; Serghini et al., 2001).

The mechanisms that enable the recognition of the host by *Striga* are still elusive. However, it can be assumed that such strategies have also evolved to achieve maximal chance of parasitism under various ecological conditions. More recently, the usage of modern omics technologies is expected to shed light on the recent developments in molecular analysis and the need to study the biology of the parasitic plants at the molecular level which will provide insight into the genetic changes associated with parasitism throughout evolution (Westwood et al., 2012).

6.3.2 SOURCES OF RESISTANCE

The course from identification of good sources of resistance to exploitation of these sources by developing improved varieties is never straightforward. The development of improved varieties with *Striga* resistance depends on many factors. The most obvious one is the availability of resistance traits, the type of approach used to assess such trait, and the power of detection. In the context of field experiments, the literature reported three types of response to *Striga*: resistance, tolerance, and susceptibility. Resistance to *Striga* has been defined as the ability of the host genotype to hinder *Striga* attachment, growth, and development when producing higher yield than the control susceptible genotype (Ramaiah, 1987; Ejeta and Butler, 1993). While, tolerance was reported as the ability of a host plant to maintain biomass and yield compared to the susceptible genotype under the same level of *Striga* parasitism, that is, when supporting comparable numbers of *Striga* plants (Haussmann et al., 2000; Rodenburg et al., 2005; Hearne, 2009). The search for resistance or tolerance often begins by the recognition of plants around which fewer or less vigorous *Striga* plants are observed or the crop yield is less affected relative to other plants growing around them in fields deliberately infested with parasitic weed seeds. This was followed by the development of several screening, including both field and laboratory, methods that allow the assessment of plant germplasm toward characterization for their reaction to *Striga* (Hess et al., 1992; Gurney et al., 2006; Mohamed et al., 2010; Amusan et al., 2011). The power and advantage of the laboratory-based approach is that it pushes the limit of what would traditionally (identified under field conditions) be considered good sources of resistance to *Striga*.

Over the past decades, research efforts have led to the discovery of new sources of resistance to *Striga* in several host crops including millets and other cereals (Lane et al., 1997; Gurney et al., 2003, 2006; Karaya et al., 2012; Wilson et al., 2000, 2004; Amusan et al., 2008; Cissoko et al., 2011). In sorghum, several cultivars and breeding lines have been reported to be resistant and/or tolerant to *Striga* (Haussmann et al., 2000; Rodenburg et al., 2006; Ejeta, 2007). Most of these donor sources have been summarized in Haussmann et al. (2000) and include Dobbs, Framida (SRN 4841), 555, N 13, SRN 6496, and SRN 39 (Ramaiah, 1987; Hess et al., 1992; Hess and Ejeta, 1992; Ezeaku and Gupta, 2004). Tesso et al. (2007) reported three sorghum cultivars (P-9401, P-9403, and PSL85061) to express resistance to *Striga*. More recently, new sorghum genotypes have been reported to be resistant to *Striga* (Robert, 2011). Resistance to *Striga* has been also documented to be present in wild accessions of *Sorghum versicolor*, *S. drummondii*, and *S. arundinaceum* (Lane et al., 1995; Gurney et al., 2002).

In pearl millet, however, only a few reports have documented the presence of *Striga* resistance (Ramaiah, 1984, 1987). Wilson et al. (2000) surveyed the presence of *Striga* resistance in wild accessions of *Pennisetum glaucum* subsp. *monodii*, a wild pearl millet relatives originating from Africa. Later, the same authors reported four accessions, PS 202 (also resistant to downy mildew), PS 637, PS 639, and PS 727, to be resistant to *Striga* (Wilson et al., 2004). Efforts to identify sources of resistance in cultivated pearl millet have followed and yielded in the identification of six resistant landraces (M141, M239, M029, M197, M017, and KBH), which show less sensitivity to the parasite and relatively higher yielding compared to susceptible controls (Kountche et al., 2013a).

Surprisingly, despite their importance as a source of nutrient-rich food grain for household farmers and the serious damage caused by *Striga* (Austin, 2006; Dida and Devos, 2006; Gigou et al., 2009; Bangoura et al., 2011; Rao et al., 2011; Mishra et al., 2014; Mannuramath et al., 2015; Muthamilarasan and Prasad, 2015), there are no similar examples of sources of resistance to *Striga* in small millets. However, because of their potential as smart crops meaning that they can provide alternative solutions in the context of climate change, the small millets are now gradually being rediscovered and considered by scientists for various needs. In general, genetic resources across all millets species exist which remain highly unexplored and underused and might contain valuable additional sources of resistance to *Striga*. There is still a need for further research to characterize and identify new sources of resistance to *Striga* in millets for future exploitation.

6.3.3 MECHANISMS OF RESISTANCE

At present, there is a strong consensus that resistance to a parasitic plant is a multicomponent event, resulting from a battery of host-plant defense acting at different levels of the host—parasite interaction (Haussmann et al., 2000; Rubiales, 2003; Ejeta, 2007). Where a potential host-plant defense is expressed, it can be narrowed to a specific stage in the parasite life cycle. This has been made possible through the development and extensive exploitation of advanced laboratory-based bioassays, which have provided weed researchers with fine phenotyping tools (Hess et al., 1992; Omanya et al., 2004; Ejeta, 2007). Based on the knowledge gained, it has been suggested that the overall expression of resistance to the parasites can be broken down into two major parts, preattachment and postattachment, in the establishment of the parasitic association. The preattachment resistance mechanisms involve host-plant defense acting before the *Striga* seeds germinate to the formation of the haustorium, while postattachment mechanisms start from the attachment of the haustorium to the host root to the subsequent parasite development (Scholes and Press, 2008; Irving and Cameron, 2009).

The first resistance mechanism observed in isolating genotypes with host-plant resistance is at the level of parasitic seed germination. Some crop genotypes have been identified to produce relatively low amounts of germination stimulants, and, therefore, to induce germination of less parasitic seeds and preventing the host plant from parasitism. This resistance phenotype, which is called low germination stimulant production (*lgs*), has been commonly found in *Striga*-resistant sorghum genotypes (Hess et al., 1992; Ejeta and Butler, 1993; Jamil et al., 2011; Robert, 2011). Resistance associated with low production of germination stimulant may not be related to low production of total SLs, but rather to the types of SLs released. The sorghum-resistant cultivar SRN 39 was found to produce lower amounts of 5-deoxystrigol but equal or even greater amounts of other SLs compared to a susceptible cultivar (Yoneyama et al., 2010). However, the low germination stimulant activity-based resistance phenotype has not been documented in pearl millet.

Furthermore, since this resistance phenotype can be threatened by individual parasites within *Striga* populations able to germinate in response to other SLs, additional resistance characters have been identified to be involved in host-plant defense systems. These include strong resistant reactions which were found to be expressed through reduced number of parasitic attachments, failure of attached parasite seedlings to establish vascular connections with the host, and by diminished growth or eventual death of the few parasites (Amusan et al., 2008, 2011). Some potential resistant host plants have been reported to lack the capacity to stimulate haustoria initiation of *Striga* germinated seeds or to produce compounds that inhibits the formation of the haustorium, an important organ toward parasite connection to host root (Gurney et al., 2003; Rich et al., 2004; Robert, 2011). Several authors demonstrated also the establishment of mechanical barriers to protect the host root from parasite penetration and access to the host-plant vascular system (Maiti et al., 1984; Gurney et al., 2006; Amusan et al., 2008; Yoshida and Shirasu, 2009). Another protecting mechanism is the recognition of the invasion as a threat and the rapid mobilization of defense response such as the hypersensitive response (HR). Two sorghum-resistant cultivars (Framida and Dobbs), three breeding lines (SAR 16, SAR 19, and SAR 33), and a wild sorghum accession (P47121) have been shown to exhibit a hypersensitive-like necrosis at the site of attachment of *S. asiatica* (Haussmann et al., 2000; Mohamed et al., 2003). Recently, hypersensitive response-based resistance phenotype has been observed in some breeding materials of cultivated pearl millet genotypes, SR-EC (for *Striga*-Resistant Epis Court), but this phenotype has not been confirmed yet. Alternatively, even if penetration of the host root is accomplished, some incompatibility response has been reported to prevent a parasite from thriving or surviving to growth and emerge above the soil. Amusan et al. (2011) demonstrated that SRN 39, a *Striga*-resistant cultivar of sorghum, expresses an incompatible response, in addition to the low SL production activity. Ejeta et al. (2000) reported other potential postgermination mechanisms of resistance that impede attachment and emergence of *Striga* in crops. It is possible that other host-plant defense mechanisms may be overlooked. However, to our knowledge none of the described resistance mechanisms has been surveyed yet in any of the small millets. There is therefore an urgent need to put efforts into the identification of resistance donor sources, the detailed characterization and understanding of specific resistance mechanisms in small millets in particular. Improved and accurate laboratory-based coculture approaches are now available, including the extended agar gel assay and the paper roll assay, that allow the determination of germination stimulant production in host roots (Hess and Ejeta, 1992; Ejeta et al., 2000), in addition to the rhizotron (root observation chamber) and microscopy, which are used to investigate postattachment resistance mechanisms (Gurney et al., 2006), and HPLC coupled to tandem mass spectrometry analysis (HPLC, LC−MS/MS) used for quantifying and characterizing the production and the type of SLs (Sato et al., 2003; Yoneyama et al., 2010; Jamil et al., 2011). Hence, when available, the deployment of all these sources of resistance in a single millet cultivar should provide more durable resistance to the parasitic weed *Striga*.

6.3.4 NATURE AND GENETIC BASIS OF RESISTANCE

From the previous section, it is evident that resistance to the parasitic weeds *Striga* appears to be complex, based on a chain of resistance mechanisms that are deployed by host-plant either alone or in combination throughout the parasite infection process. In general, the nature of resistance to parasitic weeds has been reported to be qualitative, usually a strong acting character controlled by one or two major genes (monogenic), or quantitative described as partial resistance inheritance governed by several alleles at multiple loci (polygenic). Efforts to understand the inheritance of traits associated with resistance to *Striga* have yielded very mixed results, with relatively good success in sorghum, only ephemeral progress in pearl millet, and with no efforts invested in small millets to date.

In sorghum, the low *Striga* germination stimulant activity (*lgs*), so far the best characterized resistance phenotype, has been reported to be under the control of a single nuclear recessive gene (Ramaiah et al., 1990; Hess and Ejeta, 1992; Vogler et al., 1996; Ezeaku and Gupta, 2004; Satish et al., 2012). Haussmann et al. (2001) indicated that one major gene and several minor genes appear to be involved in the stimulation of *Striga* seed germination using sorghum recombinant inbred populations. Although the lines showing low germination-inducing activity have been reported to express good resistance in bioassays, resistance mediated by low SL production has been, however, described to be less reliable when the *Striga* infestation pressure is high (Atera et al., 2011). In pearl millet, the heredity of host-plant resistance to *Striga* is not well described. Only a few pearl millet cultivars have been reported to be partially resistant or tolerant to *Striga*, and resistance (or at least less susceptibility) in some pearl millet materials was shown to be dominant (Ramaiah, 1987). However, the very existence of *Striga* resistance in cultivated pearl millet has been questioned by other authors (Chisi and Esele, 1997). Recently, partial quantitative resistance to *S. hermonthica* has been reported in wild pearl millet relatives in Africa (Wilson et al., 2000, 2004). Field measure of *Striga* resistance, most commonly emergence counts of *Striga* plants, represents a sum total of the entire parasitic association and is therefore a quantitative trait with polygenic inheritance. More recently, Kountche et al. (2013a) reported the presence of quantitative resistance to *S. hermonthica* in a diversified gene pool of cultivated pearl millet under field conditions.

In small millets, early studies toward germplasm characterization and breeding for crop improvement for agronomic traits have not been coupled with the investigation and assessment of the genetic variants for *Striga* resistance. Hence, characterization of the resistance mechanisms usually follows the identification of resistance sources (Rubiales et al., 2006).

Host-plant resistance to *Striga* involves physiological and genetic mechanisms, and requires a thorough understanding of the biophysical processes of the host–parasite association. In many cases, *S. hermonthica* seeds collected from one cereal host can infect other cereal species, for instance, there is evidence for some relaxed interspecies specificity, particularly with respect to the reciprocal infectivity of populations of *S. hermonthica* collected from sorghum and pearl millet (Vasudeva Rao and Musselman, 1987; Estep et al., 2011). A deeper understanding of the heredity of specific mechanism associated with resistance to *Striga* would facilitate the development of improved selection strategies to enhance *Striga* resistance in susceptible millet crops (Ejeta, 2007).

6.3.5 DEVELOPMENT AND USE OF *STRIGA*-RESISTANT MILLET CULTIVARS

Research toward combating *Striga* has been going on for more than 60 years, however, *Striga* control has remained a challenge. As a result of decades of intensive research and farmer experience, a wide range of *Striga* control strategies have been developed starting from those that relate to soil fertility improvement, as *Striga* are also good indicator of low soil fertility, to those that directly affect the parasite life (Oswald, 2005; Rector, 2009; Teka, 2014). This has provided the smallholder farmers with a diversity of options to control the parasites. Several researchers have extensively documented these potential options to combat *Striga* which have been classified into the most commonly used terms: cultural and mechanical control options including hand-pulling, crop rotation, trap-cropping, intercropping, appropriate improvement of soil fertility, and planting methods (Berner et al., 1995; Kuchinda et al., 2003; Hess and Williams, 1994; Hess and Dodo, 2004; Samake et al., 2006; Gworgwor, 2007; Khan et al., 2002, 2006, 2008; Eltayb et al., 2013; Hooper et al., 2015); chemical control method such as application of chemical herbicides (Kanampiu et al., 2003, 2007a, b); and biological control approaches using the pathogenic fungus *Fusarium oxysporum* as a mycoherbicide, or insects (Abbasher et al., 1995, 1998; Kroschel et al., 1996; Marley et al., 1999, 2005; Hess et al., 2002; Elzein and Kroschel, 2004; Yonli et al., 2006; Zahran et al., 2008; Venne et al., 2009; Zarafi et al., 2015; Watson, 2013). Recently, the potential of the plant growth promoting rhizobacteria of the genus *Bacillus* (*B. subtilis*, *B. amyloliquefaciens*) and *Burkholderia* (*B. phytofirmans*) as a *Striga* biocontrol agent has been surveyed by Mounde (2014). Although, the potential of these control options has been demonstrated in various research centers across *Striga*-prone regions, they have had only limited or partial impact since they do not adequately and consistently address the long-term management, which requires a significant, if not complete, depletion of the parasitic seed reserve. In some cases, their adoption has been limited as the most affected small farmers cannot afford them.

Striga appear to be difficult to manage, as the parasites inflict most of their damage under-ground, produce numerous tiny seeds, and are continuously increasing the seed load in soil (Oswald, 2005; Scholes and Press, 2008). Certainly, huge seed reserves and long-term seed viability (more than 20 years) pose acute problem in *Striga* management. Any control method that affects directly the germination of parasitic seeds and their attachment to the host root have been accepted to be more effective than those that affect later stages of the parasites development. In this sense, host-plant resistance has been widely recognized to be the most practical and sustainable method for the long-term control of the parasites that would be accessible to farmers (Ejeta, 2007; Hearne, 2009; Yoder and Scholes, 2010). The potential advantage of controlling *Striga* in subsistence agriculture via genetic resistance is high because, assuming that the resistant varieties are locally adapted and adopted, they fit within the varied agricultural practices where these weeds occur (Hearne, 2009).

6.3.5.1 Conventional breeding

Keeping in view the persistent problem of *Striga*, the key role of genetic resistance to combat *Striga* must be emphasized, even though progress toward the development and release of *Striga*-resistant varieties has been small. Development of genetically improved varieties with resistance to *Striga* is often straightforward, given that reliable sources of resistance are available and efficient and easily controlled and practical screening methods to provide sufficient selection pressure exist (Rubiales, 2003). Unfortunately, this has been seldom the case. From the previous sections, it can be admitted that resistance against *Striga* is difficult to develop due to the complex nature of the parasites, the scarcity of donor sources reported so far, and in some cases due to varying heritability (moderate to low) because of the numerous environmental effects and genetic variability of parasite populations (Rubiales, 2003; Aly, 2012). In spite of all these difficulties, significant success has been achieved. During the last few decades, research efforts have been devoted to identifying germplasm with resistance to *Striga*. Relatively few cultivars with a good level of *Striga* resistance have been identified in millet crops, mainly sorghum and pearl millet (Hess and Ejeta, 1992; Kountche et al., 2013a). Improved breeding approaches that promise good prospects for developing cultivars with resistance to *Striga* have been suggested to take advantage of the ever-growing knowledge of the complex nature of the host—parasite interactions (Ejeta and Butler, 1993; Haussmann et al., 2000; Omanya et al., 2004). Adoption of these strategies has resulted in the development of appropriate breeding populations and deployment of deliberate selection for resistance to *Striga*. Although the effort for breeding for *Striga* resistance in millets has been very limited, the conventional breeding work has generally been successful with the development and release of crop varieties with relatively high levels of resistance to *Striga*. A number of lines and improved varieties with *Striga* resistance has been selected and documented by ICRISAT and Purdue University (US) (Ejeta, 2007; Adugna, 2007; Teka, 2014). Tesso et al. (2007) and Mbwaga et al. (2007) demonstrated that resistant sorghum cultivars, which have been developed based on the low production of germination stimulants, have shown good level of *Striga* resistance. Three *Striga*-resistant varieties, P9401 (locally called Gubiye), P9403 (called Abshir), and PSL5061 (named Birhan), have been reported to be officially released for wide cultivation in *Striga*-endemic regions of Ethiopia (Elzein and Kroschel, 2004; Adugna, 2007). The *Striga*-resistant sorghum variety (ICSV1112BF, locally named Hormat) has been developed and released in India by ICRISAT (Adugna, 2007). More recently, breeding for *Striga* resistance under field conditions using phenotypic recurrent selection resulted in significant improvement in *Striga* resistance in cultivated pearl millet and the development of the first pearl millet *Striga*-resistant experimental varieties (Kountche et al., 2013a). These experimental varieties are currently being validated under farmer's field conditions. Unfortunately, in other millet species no sources of resistance have been surveyed and yet reported, and consequently no resistant cultivars have been selected so far.

The outlook for long-term stability of the released *Striga*-resistant varieties has been analyzed by many researchers and does not look bright. A high level of genetic variation has been reported to be maintained within populations from generation to generation, especially in *S. hermonthica* (Bharatha et al., 1990). Furthermore, the ability of this *Striga* species to quickly evolve new races and to overcome resistance has been documented (Rich and Ejeta, 2008). Thus, the use of genetically heterogeneous *Striga*-resistant open-pollinated cultivars, in which different plants carry different resistance alleles, has been reported to be a practical alternative for ensuring the stability of resistance to *Striga* over time (Kountche et al., 2013a). A key to long-term stable resistance is more likely when both quantitative and qualitative resistance genes are deliberately stacked in the improved cultivars such that multiple mutations would have to accumulate in the parasite populations to overcome the multiple resistance genes in the host (Rich and Ejeta, 2008).

6.3.5.2 Marker-assisted selection

Although conventional breeding has made significant contributions to millet crops improvement for *Striga* resistance, particularly in sorghum and pearl millet, this approach has been in general slowed when targeting the complex quantitative trait of resistance to *Striga*. The development of molecular markers associated with resistance to *Striga* has offered a promising way to rapidly accumulate several resistance genes since conventional breeding has shown to be time-consuming, largely if not completely dependent on climatic and environmental conditions, and therefore less effective

(Ejeta and Gressel, 2007; Rispail et al., 2007). It is interesting to note that in the most studied millet crops, sorghum and pearl millet, potential resistance phenotypes have been identified in wild relatives. Hence, molecular markers may facilitate the transfer of resistance genes into cultivars and facilitate pyramiding of multiple resistance genes into agronomically desirable elite and locally adapted *Striga*-susceptible varieties. Over the past decades, much progress has been made in the field of the development of molecular markers to assist and boost the development of *Striga*-resistant varieties. However, their effective application in a marker-aided breeding requires a thorough understanding of the inheritance of *Striga* resistance, which in turn requires the development of tailor-made populations.

Major advances in our understanding of the molecular genetic basis of mechanisms of *Striga* resistance have been achieved. Research effort invested to characterize the mechanisms behind and inheritance of resistance to *Striga* has been followed by the development of molecular markers. Recently, mapping of chromosomal regions affecting qualitative or quantitative traits has received growing attention in millets (Hess et al., 1992; Ejeta and Butler, 1993; Haussmann et al., 2001; Omanya et al., 2004; Rich et al., 2004; Mohamed et al., 2003; Kountche, 2013; Moumouni et al., 2015). In sorghum, several populations segregating for genes associated with low germination stimulant production (*lgs*), low haustorial initiation, mechanical barriers, HR, and incompatible response mechanisms of *Striga* resistance have been developed using the source materials (Ejeta et al., 2000; Haussmann et al., 2000; Grenier et al., 2001, 2007; Rodenburg et al., 2006; Omanya et al., 2004; Mutengwa et al., 2005). Quantitative trait loci (QTL) mapping for *Striga* resistance has resulted in identification of many genes/QTLs spread across the genome. Haussmann et al. (2004) mapped several QTLs for *Striga* resistance in sorghum and reported 9 and 11 QTLs explaining, respectively, 77% and 82% of the phenotypic variation in a recombinant inbred line population. The most significant QTL has been identified to correspond to the major gene locus *lgs*, which was recently fine-mapped on the chromosome 5 (Satish et al., 2012). More interestingly, validation of some of these QTLs has provided an opportunity to employ molecular marker-assisted breeding (MAB) for sorghum improvement for *Striga* resistance. MAB has been applied using the identified *Striga*-resistant QTLs for their introgression into elite farmer-preferred sorghum local varieties which lack *Striga* resistance (Grenier et al., 2007; Kapran et al., 2007). Four *Striga*-resistant sorghum lines developed through marker-assisted selection have been recently reported and released for cultivation in sub-Saharan Africa (Mohamed et al., 2014). Very recently, Yohannes et al. (2015) indicated that marker-assisted backcrossing (MABC) has been effective in introgressing five QTLs for *Striga* resistance from the donor source N13 to the genetic background of Hugurtay, a susceptible sorghum variety from Eritrea.

In pearl millet, although marker-aided selection (MAS) has not been practiced yet in breeding for *Striga* resistance, there have been extensive efforts toward development of DNA-based markers including simple sequence repeat (SSR) (Allouis et al., 2001; Budak et al., 2003; Mariac et al., 2006; Qi et al. 2001, 2004; Senthilvel et al., 2004, 2008; Jia et al., 2007; Yadav et al., 2007, 2008; Rajaram et al., 2013), diversity array technology (DArT) (Mace et al., 2008, 2009; Supriya et al., 2011), and single nucleotide polymorphism (SNP) (Bertin et al., 2005; Feltus et al., 2006; Kountche, 2013; Moumouni et al., 2015). In most of these studies, a number of genetic maps have been constructed using different crosses, mostly with the aim of mapping specific traits. More recently, Moumouni et al. (2015) reported the development of a relatively high density SNP-based linkage map using a cross between the wild resistant PS202 and SOSAT-C88, a susceptible open-pollinated improved variety (OPV), segregating for *Striga* resistance. While the original aim was to conduct a classical QTL analysis in this population, a lack of reliable phenotypic data due to environmental factors has hampered the marker-trait association for mapping *Striga* resistance in pearl millet. Even when it is well established, MABC has certain limits for introgressing quantitative trait controlled by many genes/QTLs with minor effects. The markers to be used in MABC are usually identified in a biparental segregating population, which limits the exploitation of the potentially wide range of allelic diversity present in the crop species. In addition, markers may be valid only for the genetic background in which they were identified (Varshney and Dubey, 2009). Because of these limitations, marker-assisted recurrent selection has emerged as an indispensable breeding strategy for developing germplasm with strong long-lasting resistance (Bernardo and Charcosset, 2006). However, the key component and the great challenge toward an efficient system for molecular breeding for *Striga* resistance in pearl millet is the identification of significant associations between genetic markers and the genes that determine resistance to the parasites. Interestingly, a genome-wide association study has been recently surveyed in pearl millet for *Striga* resistance using a diversified *Striga*-resistant gene pool of full-sib populations derived from five cycles of phenotypic recurrent selection (Kountche et al., 2013a; Kountche, 2013). These authors reported three major and several minor SNPs which have been identified to be significantly associated with the *Striga* resistance trait, each of which explain 15%, 18%, and 21% of the phenotypic variation, respectively, for the three major QTLs. From these preliminary results, they speculated that resistance to *Striga* in pearl millet is controlled by a small number of major loci which can easily be manipulated in a MABC program, as it was the case for sorghum. For this purpose, it is desirable to have molecular markers flanking the gene/QTL closely on both sites. Therefore, there is a need to map and validate these major QTLs across different genetic backgrounds and environments for their future deployment.

In the past decades, extensive efforts have been invested in small millets to develop genetic and genomic resources. Various DNA markers are routinely used to assess population structure and genetic diversity in small millets (Dwivedi et al., 2012; Barnaud et al., 2012). Interestingly, genetic maps of varying density level have been also documented for foxtail millet, finger millet, proso millet, and tef (Dwivedi et al., 2012). However, to the best of our knowledge, no specific study has reported on *Striga* resistance, its molecular genetic basis, and availability of molecular marker associated with resistance to *Striga* in small millets. The availability of these enormous genomic resources in small millets has now provided numerous scientific leads to proceed further toward crop improvement for *Striga* resistance in millets, particularly in small millets. National and international efforts need to be invested to assess and identify sources of *Striga* resistance, characterize and understand the specific resistance mechanisms underlying the resistance phenotype, and examine the inheritance of identified *Striga* resistance in small millet using phenomics and genomics resources. Notably, with the increased accessibility of low cost and high-throughput genotyping, genotyping-by-sequencing (GBS), and the availability of the whole genome sequence of some millet crops, such as sorghum (Paterson et al., 2009), foxtail millet (Bennetzen et al., 2012), and pearl millet (which will be completed very soon), it can be speculated that targeting specific *Striga* resistance genes and employment of genomic tools to tailor millets for *Striga* resistance will be less challenging (Kountche et al., 2013b). The combination of genetic mapping and gene expression studies should provide an integrated approach to pave the knowledge gap in millets, especially pearl millet and small millets. Moreover, comparative genomics can point to important resistance genes in the understudied millet crops as they are discovered in other millet crops like sorghum (Michelmore, 2000; Rispail et al., 2007).

6.3.5.3 Transgenics

To date, there is a consensus that genetic control of *Striga* through the use of resistant cultivars appears to be the central component of any concerted integrated *Striga* control approach (Ejeta, 2007). However, the deployment of host-plant resistance even in the most studied millet crops, particularly sorghum and pearl millet, has been largely hampered by the scarcity of reliable resistance donor sources (Aly, 2012; Kountche et al., 2013a). This has prompted the researchers to explore new horizons for developing new strategies that allow extending the sources of resistance further (also known as artificial resistance) to the very limited number of natural resistance sources that have been identified so far. Genetic engineering for enhancing host-plant resistance to *Striga* is a very promising approach and offers new opportunities for developing improved varieties. Thanks to the recent emergence of biotechnology techniques and genomic resources, the potential of artificial resistance in host plants makes the genetic engineering a viable strategy in this context. Indeed, studies aimed at the molecular characterization of the host plant−parasite interaction and host resistance through expression analysis of the genes, proteins, and metabolites involved in these processes are the subject of increasing interest and offer weed researchers the opportunity for deploying genetic transformation tools to enhance the control of the parasitic plants (Rispail et al., 2007; Aly, 2012). With the great initiative of the parasitic plant genome project (http://ppgp.huck.psu.edu/) (Westwood et al., 2012), genetic resistance based on silencing of a target gene in the host plant is now feasible. Yoder et al. (2009) and Runo et al. (2011) have investigated the use of RNA interference (RNAi) technology as a means for enhancing host resistance against parasitic weeds. However, this approach has not been successful in controlling maize−*Striga* parasite interaction (Yoder and Scholes, 2010). Nevertheless, understanding the biology of the early stages of parasitism by *Striga* will help identify potential barriers towards the success of this technology. Though significant progress has been achieved in genetic transformation, there have been very few reports on millet crops, which have trailed behind other cereals, such as maize, in the progress toward genetic transformation (Visarada and Kishore, 2015). To date, there are no transgenic *Striga*-resistant millets crops that have been reported and released for cultivation so far.

As an alternative approach to the control of parasitic weeds, Gressel (2009) suggested the development and use of herbicide-resistant crops. Green and Owen (2011) documented that herbicide-resistant sorghum varieties have been developed by the private sector in the USA using mutations. Interestingly, attempts are being made using the approach of the acetolactate synthase to develop genetically engineered herbicide-resistant sorghum crops. However, the success of generating resistant cultivars by genetic engineering requires an efficient gene transfer, stable integration, and predictable expression of the transgene. Thus, with the genetic transformation techniques based on recombinant DNA technology, it is now possible to insert genes that confer resistance to a number of biotic stresses, including parasitic plants such as *Striga*, into the susceptible host-plant genome (Rakshit and Patil, 2014). When identified and characterized, *Striga* resistance genes available across millet species can be deployed and incorporated into other species that lack resistance through gene transfer techniques along with reproducible tissue culture protocols to produce transgenic millet varieties with enhanced agronomic performance. Recently, Kohlen et al. (2011) emphasized that the discovery of SLs, the major *Striga* germination stimulants, provides a potential way to enhance *Striga* resistance by reducing the production of SLs in the host-plant roots using a transgenic approach—at least for the few millet crops, such as sorghum, foxtail millet, and pearl millet, for which a complete sequence of the genome is available.

Beside the potential advantages of these technologies, it appears that researchers have been reluctant toward the development of engineered resistance through genetic engineering since the transgenic crops may not be easily adopted by farmers due to the current controversial debate on food safety issues. The approach provides genetically modified organisms (GMO), therefore the use of transgenic crops in the center of origin of most millets crops may arise concern regarding the possible gene flow from transgenic crops to cultivated plants and their wild relatives. In this regards, innovative genetic modifications have been developed to keep away from the classification of GMOs and biosafety concerns. These modifications such as zinc finger nuclease technology, *cis*-genesis, RdDM, and other technologies are dependent on transgenic technology for the introduction of DNA segments (Visarada and Kishore, 2015). Science and the arts of tissue culture, genetic transformation, and more importantly, plant breeding has to be integrated to reach effective *Striga*-resistant millet cultivars.

6.3.6 INTEGRATED *STRIGA* MANAGEMENT

The various *Striga* control methods reviewed above have offered varying levels of *Striga* control and have not proved to be as effective, economical, and applicable as desired (Joel, 2000). In the past decades, several control measures have been developed and extensively deployed for few crops; however, when applied individually the control method has been affected by the diversity of farming systems and environmental conditions, and has often shown only little success. In addition, although these methods have helped in reducing damage caused by the parasites to host crops, it has been observed that they did not adequately address the long-term management of the root-parasitic weeds, as this requires the destruction of the *Striga* seed bank. It appears that there is no single control method that can effectively solve the *Striga* problem (Joel, 2000; Oswald, 2005). Thus, the most comprehensive and sustainable way to cope with the *Striga* is certainly through an integrated approach, incorporating a variety of measures in a concerted and smart manner (Oswald, 2005; Ejeta, 2007; Elzein et al., 2008). Haussmann et al. (2000) reported that an effective integrated control strategy must certainly include at least one (if not all) control method from each of the three major categories highlighted by the authors. Oswald (2005) suggested a combined action with containment and sanitation, using direct and indirect measures to prevent the damage caused by *Striga*, and with means to eradicate the *Striga* seed bank in infested soils. To ensure that the integrated *Striga* control technology is adopted by farmers, *Striga* control practices must improve crop yield, maintain soil fertility, and be practicable (Berner et al., 1995; Kroschel, 1998). To this end, adaptive and on-farm research has been invested, which must play an important role in identifying the most effective mix of component-practices. Berner et al. (1995) suggested crop rotation with a highly effective trap-crops as the key component of an integrated *Striga* control program. The combined use of the mycoherbicide, *Fusarium oxysporum* and *Striga*-resistant sorghum cultivars has been demonstrated to be effective in controlling *Striga* and led to increased sorghum yield in the field (Marley et al., 2004; Venne et al., 2009). Hess and Dodo (2004) reported the potential use of sesame as a trap-crop in integrated control of *Striga* in pearl millet. Tesso and Ejeta (2011) suggested a promising integrated control strategy based on the use of sorghum-resistant variety, tied-ridge tillage, and nitrogen fertilizer. More recently, ICRISAT and its collaborators embarked on participatory research on integrated *Striga* management through farmer field schools and came up with practical integrated *Striga* and soil fertility management practices for pearl millet and sorghum in West Africa.

Regardless of the global context of climate change, which will certainly have an impact on *Striga* infestation, the main obstacle in the long-term management of *Striga* in infested fields appears to be the persistent seed bank, with only a very limited annual depletion percentage induced by the current integrated approach. As long as the *Striga* seed bank is not controlled effectively, the need to apply means to control the parasite will persist. An integrated approach that combines genetic resistance and suicidal germination components has been suggested to be the best approach for seed bank demise (Kgosi et al., 2012; Zwanenburg and Pospisil, 2013; Kannan and Zwanenburg, 2014). While several research efforts have been invested in the field of development and deployment of host-plant resistance, the elimination of the *Striga* seed bank through the suicidal germination approach using synthetic analogs of SLs has gained very little attention. This approach may be appropriate and even more efficient if suitable candidates of synthetic SLs are identified. Hence, the development of highly efficient synthetic germination stimulant analogs, economically affordable by smallholder farmers, and environmentally friendly that can be used to induce suicidal germination of *Striga* seeds, when possible, will provide an important alternative component for the long-term control of *Striga* (Zwanenburg et al., 2009, 2013).

6.4 CONCLUSIONS

In general, research efforts invested during the last decades have dramatically improved our understanding of the host–parasite relationship in the *Orobanchaceae*, and more so, contributed to numerous methods for the control of the weed parasitic plants *Striga*. Albeit limited in efficacy in many cases, the control approaches developed so far represent today major progress in combating *Striga* in millet cropping systems. It is now evident that the cultivated millet crops can be protected by cultural methods, by biocontrol agents, and through genetic resistance. Moreover, it will be important to develop new strategies that can address the long-term control of the parasite. In this sense, fundamental research on key aspects of *Striga* parasitism should not only lead to a better understanding of this intricate pathosystem, but also provide new ideas for the development of novel methods for parasitic weed control, to increase millets productivity. From the above-cited knowledge on the millet–*Striga* pathosystem, it seems evident that multiple responses from specific host plant-dependent signals seem to be a part of a strategy to find and recognize host plants. SLs with their multifaceted biological roles can undeniably become a potent and valuable tool for developing new alternatives for breeding/managing for *Striga* resistance in millets. In this sense, it is important to understand some of the SLs' interactions at the biosynthesis level and elucidation of SLs' release/pattern seems essential for understanding developmental and defense-related processes mediated by specific families of these phytohormones. Finally, it is crucial when dealing with *Striga* that several resistances, genetic and/or engineered, should be pyramided in susceptible millet cultivars, to avoid the development of more virulent parasite populations and to overcome the resistance. This will be deliverable only when combined work at molecular, genetic, and physiological levels is continuously invested in millet crops.

6.5 FUTURE PERSPECTIVES AND PRIORITIES

This review emphasizes the enormous efforts to understand the host−*Striga* interactions and the development of genetically *Striga*-resistant cultivars in millets. We firmly believe that an increased research effort in millet species will lead to significant progress. Indeed, a deeper understanding of parasite population dynamics and of the genetic changes associated with parasitism should improve our understanding of the virulence of the parasite and of host resistance mechanisms, and the transfer of the knowledge to breeders. With regard to *Striga* control, elucidation of the SL biosynthesis pathway in millets will also offer possibilities for the development of chemical inhibitors of SL production and allow modulation of the amounts and types of released SLs. Thus, a comprehensive effort should be made using all available tools to improve *Striga* resistance/tolerance in millets so that they can continue to play the indispensable role in the welfare of the poor in the semiarid tropics of Africa and Asia.

Moreover, the many bioassay tools that have been developed for sorghum can be now used in other millets, to identify resistance sources and to investigate the mechanisms underlying the resistance phenotype in the understudied millet crops. Conventional breeding can be supported, in addition to MAS and innovative genetic modification strategies, by a mutagenesis approach to produce new resistant variants to tackle the *Striga* problem in millets. In this concern, ethyl methane sulfonate, a commonly used chemical mutagen, has been largely used because of its high effectiveness in generating new and desired traits.

It is now clear that the main impediment of the long-term control of *Striga* in infested fields lies in the huge seed bank. As long as the *Striga* seed bank is not controlled, the need to apply means to control the parasite will persist. With regard to depletion of *Striga* seed reserve, the development and testing of efficient synthetic germination stimulant analogs is a promising approach toward eliminating the *Striga* seed bank. Therefore, it will be of interest to further explore this opportunity in order to provide farmers with new means for combating the parasitic weeds *Striga*.

ACKNOWLEDGMENTS

Our grateful thanks are extended to the resource persons of the e-write shop organized by the Collaborative Crop Research Program of the McKnight Foundation. This work was supported by funding from King Abdullah University of Science and Technology (KAUST).

REFERENCES

Abbasher, A.A., Kroschel, J., Sauerborn, J., 1995. Microorganisms of *Striga hermonthica* in Northern Ghana with potential as biocontrol agents. Biol. Sci. Technol. 5, 157−161.

Abbasher, A.A., Hess, D.D., Sauerborn, J., 1998. Fungal pathogens for biocontrol of *Striga hermonthica* on sorghum and pearl millet in West Africa. Afr. Crop Sci. J. 6, 179−188.

Adugna, A., 2007. The role of introduced sorghum and millets in Ethiopian agriculture. J. SAT Agric. Res. 3, 1−4.

Akiyama, K., Matsuzaki, K., Hayashi, H., 2005. Plant sesquiterpenes induce hyphal branching in arbuscular mycorrhizal fungi. Nature 435, 824—827.

Al-Babili, S., Bouwmeester, H.J., 2015. Strigolactones, a novel carotenoid-derived plant hormone. Annu. Rev. Plant Biol. 66, 161—186.

Albrecht, H., Yoder, J.I., Phillips, D.A., 1999. Flavonoids promote haustoria formation in the root parasite *Triphysaria*. Plant Physiol. 119, 585—591.

Allouis, S., Qi, X., Lindup, S., Gale, M.D., Devos, K.M., 2001. Construction of BAC library of pearl millet, *Pennisetum glaucum*. Theor. Appl. Genet. 102, 1200—1205.

Aly, R., 2007. Conventional and biotechnological approaches for control of parasitic weeds. In Vitro Cell. Dev. Biol. Plant 43, 304—317.

Aly, R., 2012. Advanced technologies for parasitic weed control. Weed Sci. 60, 290—294.

Amusan, I.O., Rich, P.J., Menkir, A., Housley, T., Ejeta, G., 2008. Resistance to *Striga hermonthica* in a maize inbred line derived from *Zea diploperennis*. New Phytol. 178, 157—166.

Amusan, I., Rich, P., Housley, T., Ejeta, G., 2011. An in-vitro method for identifying post-attachment *Striga* resistance in maize and sorghum. Agron J. 103, 1472—1478.

Andrews, D.J., Kumar, K.A., 1992. Pearl millet for food, feed and forage. Adv. Agron 48, 89—139.

Atera, E., Itoh, K., Onyango, J.C., 2011. Evaluation of ecologies and severity of *Striga* weed on rice in sub-Saharan Africa. Agric. Biol. J. N. Am. 2, 752—760.

Atera, E.A., Itoh, K., Azuma, T., Ishii, T., 2012. Farmers' perspectives on the biotic constraint of *Striga hermonthica* and its control in western Kenya. Weed Biol. Manage. 12, 53—62.

Austin, D.F., 2006. Fox-tail millets (*Setaria*: Poaceae) — abandoned food in two hemispheres. Econ. Bot. 60 (2), 143—158.

Awad, A.A., Sato, D., Kusumoto, D., Kamioka, H., Takeuchi, Y., et al., 2006. Characterization of strigolactones, germination stimulants for the root parasitic plants *Striga* and orobanche, produced by maize, millet and sorghum. Plant Growth Regul. 48, 221—227.

Bangoura, M.L., Huiming, Z., Nsor-Atindana, J., Xue, Z.K., Tolno, M.B., Wei, P., 2011. Extraction and fractionation of insoluble fibres from foxtail millet (*Setaria italica* (L.) P. Beauv. Am. J. Food Technol. 6, 1034—1044.

Barnaud, A., Vignes, H., Risterucci, A.M., Noyer, J.L., Pham, J.L., Blay, C., et al., 2012. Development of nuclear microsatellite markers for the fonio, *Digitariaexilis* (Poaceae), an understudied West African cereal. Am. J. Bot. 99, e105—e107.

Bennetzen, J.L., et al., 2012. Reference genome sequence of the model plant *Setaria*. Nat. Biotech. 30, 555—561.

Bernardo, R., Charcosset, A., 2006. Usefulness of gene information in marker-assisted recurrent selection: a simulation appraisal. Crop Sci. 46, 614—621.

Berner, D.K., Kling, J.G., Singh, B.B., 1995. *Striga* research and control: a perspective from Africa. Plant Dis. 79, 652—660.

Berner, D.K., Winslow, M.D., Awad, A.E., Cardwell, K.F., Mohan Raj, D.R., Kim, S.K. (Eds.), 1997. *Striga* Research Methods—A Manual. second ed. International Institute of Tropical Agriculture (IITA), PMB 5320, Ibadan, Nigeria, <http://www.cgiar.org/iita/inform/striga.pdf>.

Berner, D.K., Schaad, N.W., Völksch, B., 1999. Use of ethylene-producing bacteria for stimulation of *Striga* spp. seed germination. Biol. Control 15 (3), 274—282.

Bertin, C., Yang, X.H., Weston, L.A., 2003. The role of root exudates and allelochemicals in the rhizosphere. Plant Soil 256, 67—83.

Bertin, I., Zhu, J.H., Gale, M.D., 2005. SSCP-SNP in pearl millet—a new marker system for comparative genetics. Theor. Appl. Genet. 110, 1467—1472.

Bharatha, L., Werth, C.R., Musselman, L.J., 1990. A study of genetic diversity among host-specific populations of the witchweed *Striga hermonthica* (*Scrophulariaceae*) in Africa. Plant Syst. Evol. 172, 1—12.

Botanga, C.J., Kling, J.G., Berner, D.K., Timko, M.P., 2002. Genetic variability of *Striga asiatica* (L.) Kuntz based on AFLP analysis and host-parasite interaction. Euphytica 128, 375−388.

Bouwmeester, H.J., Matusova, R., Sun, Z., Beale, M.H., 2003. Secondary metabolite signalling in host−parasitic plant interactions. Curr. Opin. Plant Biol. 6, 358−364.

Bouwmeester, H.J., Roux, C., Lopez-Raez, J.A., Becard, G., 2007. Rhizosphere communication of plants, parasitic plants and AM fungi. Trends Plant Sci. 12, 224−230.

Budak, H., Pedraza, F., Cregan, P.B., Baenziger, P.S., Dweikat, I., 2003. Development and utilization of SSRs to estimate the degree of genetic relationships in a collection of pearl millet germplasm. Crop Sci. 43, 2284−2290.

Butler, L.G., 1995. Chemical communication between the parasitic weed *Striga* and its crop host−a new dimension in allelochemistry. ACS Symp. Ser. Am. Chem. Soc. 582, 158−168.

Cardoso, C., Zhang, Y., Jamil, M., Hepworth, J., Charnikhova, T., et al., 2014. Natural variation of rice strigolactone biosynthesis is associated with the deletion of two MAX1 orthologs. Proc. Natl. Acad. Sci. U.S.A. 111, 2379−2384.

Chang, M., Netzly, D.H., Butler, L.G., Lynn, D.G., 1986. Chemical regulation of distance: characterization of the first natural host germination stimulant for *Striga asiatica*. J. Am. Chem. Soc. 108, 7858−7860.

Chemisquy, M.A., Giussani, L.M., Scataglini, M.A., Kellogg, E.A., Morrone, O., 2010. Phylogenetic studies favour the unification of *Pennisetum*, *Cenchrus* and *Odontelytrum* (Poaceae): a combined nuclear, plastid and morphological analysis, and nomenclatural combinations in *Cenchrus*. Ann. Bot. (Oxford) 106, 107−130.

Chisi, M., Esele, P., 1997. Discussion for Session VII-Breeding for resistance to other abiotic stresses and *Striga*. Proceeding International Conference: Genetic Improvement of Sorghum and Pearl Millet. INTSORMIL, Lincoln, NE, pp. 525−528.

Cissoko, M., Boisnard, A., Rodenburg, J., Press, M.C., Scholes, J.D., 2011. New Rice for Africa (NERICA) cultivars exhibit different levels of post-attachment resistance against the parasitic weeds *Striga hermonthica* and *Striga asiatica*. New Phytol. 192, 952−963.

Cochrane, V., Press, M.C., 1997. Geographical distribution and aspects of the ecology of the hemiparasitic angiosperm *Striga asiatica* (L.) Kuntze: a herbarium study. J. Trop. Ecol. 13 (3), 371−380.

Cook, C.E., Whichard, L.P., Turner, B., Wall, M.E., Egley, G.H., 1966. Germination of witchweed (*Striga lutea* Lour.): isolation and properties of a potent stimulant. Science 154, 1189−1190.

Cook, C.E., Whichard, L.P., Wall, M., Egley, G.H., Coggon, P., et al., 1972. Germination stimulants. II. Structure of strigol, a potent seed germination stimulant for witchweed (*Striga lutea*). J. Am. Chem. Soc. 94, 6198−6199.

De Groote, H., Wangare, L., Kanampiu, F., Odendo, M., Diallo, A., Karaya, H., et al., 2008. The potential of a herbicide resistant maize technology for *Striga* control in Africa. Agric. Syst. 97 (1−2), 83−94.

Dida, M.M., Devos, K.M., 2006. Finger millet. In: Kole, C. (Ed.), Genome Mapping and Molecular Breeding in Plants, Vol. 1, Cereals and Millets. Springer, Heidelberg, pp. 333−343.

Dwivedi, S.L., Upadhyaya, H., Senthilvel, S., Hash, C.T., 2012. Millets: genetic and genomic resources. In: Janick, J. (Ed.), Plant Breeding Reviews, (Chapter 5), vol. 36. John Wiley & Sons, Inc, Hoboken, NJ, USA, pp. 247−375.

Ejeta, G., 2007. Breeding for *Striga* resistance in sorghum: exploitation of an intricate host−parasite biology. Crop Sci. 47, S216−S227.

Ejeta, G., Butler, L.G., 1993. Host−parasite interactions throughout the *Striga* life cycle, and their contributions to *Striga* resistance. Afr. Crop Sci. J. 1, 75−80.

Ejeta, G., Gressel, J., 2007. Integrating New Technologies for *Striga* Control: Towards Ending the Witch-Hunt. World Scientific Publishing Co Pte Ltd, Singapore, p. 345.

Ejeta, G., Mohammed, A., Rich, P.J., Melake-Berhan, A., Housley, T.L., Hess, D.E., 2000. Selection for specific mechanisms of resistance to *Striga* in sorghum. In: Haussmann, B.I.G., et al. (Eds.), Breeding for *Striga* Resistance in Cereals. Proc. of a Workshop, IITA, Ibadan, Nigeria. 18−20. August 1999. Margraf, Weikersheim, Germany, pp. 29−37.

Eltayb, M.T.A., Magid, T.D.A., Mohamed, A.G., Dirar, A.M.A., 2013. Effect of *Acacia seyal* (Del) seeds powder on growth and yield of sorghum bicolor (L) in *Striga hermonthica* (Del.) infested area. J. Forest Prod. Ind. 2 (6), 18−25.

Elzein, A., Kroschel, J.K., 2004. *Fusarium oxysporum* Foxy 2 shows potential to control both *Striga hermonthica* and *S. asiatica*. Weed Res. 44, 433−438.

Elzein, A., Kroschel, J., Cadisch, G., 2008. Efficacy of Pesta granular formulation of *Striga*-mycoherbicide *Fusarium oxysporum* f. sp. strigae Foxy 2 after 5-year of storage. J. Plant Dis. Prot. 115, 259−262.

Estabrook, E.M., Yoder, J.I., 1998. Plant−plant communications: rhizosphere signalling between parasitic angiosperms and their hosts. Plant Physiol. 116, 1−7.

Estep, M., Van Mourik, T., Muth, P., Guindo, D., Parzies, H., Koita, O., et al., 2011. Genetic diversity of a parasitic weed, *Striga hermonthica*, on sorghum and pearl millet in Mali. Trop. Plant Biol. 4, 91−98.

Ezeaku, I.E., Gupta, S.C., 2004. Development of sorghum populations for resistance to *Striga hermonthica* in the Nigerian Sudan Savanna. Afr. J. Biotech. 3 (6), 324−329.

FAO, IFAD and WFP, 2013. The State of Food Insecurity in the World 2013. The Multiple Dimensions of Food Security. FAO, Rome.

Feltus, F.A., Singh, H.P., Lohithaswa, H.C., Schulze, S.R., Silva, T.D., Paterson, A.H., 2006. A comparative genomics strategy for targeted discovery of single-nucleotide polymorphisms and conserved-moncoding sequences in orphan crops. Plant Physiol. 140, 1183−1191.

Gethi, J.G., Smith, M.E., Mitchell, S.E., Kresovich, S., 2005. Genetic diversity of *Striga hermonthica* and *Striga asiatica* populations in Kenya. Weed Res. 45, 64−73.

Gigou, J., Stilmant, D., Diallo, T.A., Cisse, N., Sanogo, M.D., Vaksmann, M., et al., 2009. Fonio millet (*Digitariaexilis*) response to N, P and K fertilizers under varying climatic conditions in West. Afr. Expl. Agric. 45, 401−415.

Goldwasser, Y., Hershenhorn, J., Plakhine, D., Kleifeld, Y., Rubin, B., 1999. Biochemical factors involved in vetch resistance to *Orobanche aegyptiaca*. Physiol. Mol. Plant Pathol. 54, 87−96.

Gomez-Roldan, V., Fermas, S., Brewer, P.B., Puech-Pages, V., Dun, E.A., et al., 2008. Strigolactone inhibition of shoot branching. Nature 455, 189−194.

Green, J.M., Owen, M.D.K., 2011. Herbicide-resistant crops: utilities and limitations for herbicide-resistant weed management. J. Agric. Food Chem. 59, 5819−5829.

Grenier, C., Rich, P.J., Mohamed, A., Ellicott, A., Shaner, C., Ejeta, G., 2001. Independent inheritance of lgs and IR genes in sorghum. In: 7th international parasitic weed symposium. Nantes, France, pp. 220−223.

Grenier, C., Haussmann, B.I.G., Kiambi, D., Ejeta, G., 2007. Marker-assisted selection for *Striga* resistance in sorghum. In: Ejeta, G., Gressel, J. (Eds.), Integrating New Technologies for *Striga* Control: Towards Ending the Witch-Hunt. World Scientific Publishing Company Pvt. Ltd., Singapore, pp. 159−172.

Gressel, J., 2009. Crops with target-site herbicide resistance for *Orobanche* and *Striga* control. Pest Manage. Sci. 65, 560−565.

Gressel, J., Hanafi, A., Head, G., Marasas, W., Obilana, B., Ochanda, J., et al., 2004. Major heretofore intractable biotic constraints to African food security that may be amenable to novel biotechnological solutions. Crop Prot. 23, 661−689.

Gurney, A.L., Press, M.C., Scholes, J.D., 2002. Can wild relatives of sorghum provide new sources of resistance or tolerance against *Striga* species? Weed Res. 42, 317−324.

Gurney, A.L., Grimanelli, D., Kanampiu, F., Hoisington, D., Scholes, J.D., Press, M.C., 2003. Novel sources of resistance to *Striga hermonthica* in *Tripsacumdactyloides*, a wild relative of maize. New Phytol. 160, 557−568.

Gurney, A.L., Slate, J., Press, M.C., Scholes, J.D., 2006. A novel form of resistance in rice to the angiosperm parasite *Striga hermonthica*. New Phytol. 169, 199−208.

Gworgwor, N., 2007. Trees to control weeds in pearl millet. Agron. Sustain. Dev. 27, 89−94.

Hamrick, J., 1982. Plant population genetics and evolution. Am. J. Bot. 69 (10), 1685−1693.

Hash, C.T., Yadav, R.S., Cavan, G.P., Howarth, C.J., Liu, H., Qi, X., et al., 2000. Marker-assisted backcrossing to improve terminal drought tolerance in pearl millet. In: Ribaut, J.M., Poland, D. (Eds.), Molecular Approaches for the Genetic Improvement of Cereals for Stable Production in Water-Limited Environments. A Strategic Planning Workshop held at CIMMYT, El Batan, Mexico, 21−25 June 1999. International Maize and Wheat Improvement Center (CIMMYT), Mexico, D.F., Mexico, pp. 114−119.

Hauck, C., Muller, S., Schildknecht, H., 1992. A germination stimulant for parasitic flowering plants from sorghum bicolor, a genuine host plant. J. Plant Physiol. 139, 474−478.

Haussmann, B.I.G., Hess, D.E., Welz, H.G., Geiger, H.H., 2000. Improved methodologies for breeding *Striga*-resistant sorghums. Field Crops Res. 66, 195−211.

Haussmann, B.I.G., Hess, D.E., Omanya, G.O., Reddy, B.V.S., Welz, H.G., Geiger, H.H., 2001. Major and minor genes for stimulation of *Striga hermonthica* seed germination in sorghum, interaction with different *Striga* populations. Crop Sci. 41, 1507−1512.

Haussmann, B.I.G., Hess, D.E., Omanya, G.O., Folkertsma, R.T., Reddy, B.V.S., Kayentao, M., et al., 2004. Genomic regions influencing resistance to the parasitic weed *Striga hermonthica* in two recombinant inbred populations of sorghum. Theor. App. Genet. 109, 1005−1016.

Hearne, S.J., 2009. Control − the *Striga* conundrum. Pest Manage. Sci. 65, 603−614.

Hess, D.E., Dodo, H., 2004. Potential for sesame to contribute to integrated control of *Striga hermonthica* in the West African Sahel. Crop Prot. 23, 515−522.

Hess, D.E., Ejeta, G., 1992. Inheritance of resistance to *Striga* in sorghum genotype SRN39. Plant Breed 109, 233−241.

Hess, D.E., Williams, J.H., 1994. Influence of planting date on *Striga* infestation and yield of pearl millet. Phytopathology 84, 1104.

Hess, D.E., Ejeta, G., Butler, L.G., 1992. Selecting sorghum genotypes expressing a quantitative biosynthetic trait that confers resistance to *Striga*. Phytochemistry 31, 493−497.

Hess, D.E., Thakur, R.P., Hash, C.T., Sérémé, P., Magill, C.W., 2002. Pearl millet downy mildew: problems and control strategies for a new millennium. In: Leslie, J.F. (Ed.), Sorghum and Millets Pathology 2000. Iowa State Press, Ames, Iowa, USA, pp. 37−42.

Hirsch, A.M., Bauer, W.D., Bird, D.M., Cullimore, J., Tyler, B., Yoder, J.I., 2003. Molecular signals and receptors: controlling rhizosphere interactions between plants and other organisms. Ecology 84, 858−868.

Hooper, A.M., Caulfield, J.C., Hao, B., Pickett, J.A., Midega, C.A.O., Khan, Z.R., 2015. Isolation and identification of *Desmodium* root exudates from drought tolerant species used as intercrops against *Striga hermonthica*. Phytochemistry 117, 380−387.

Huang, K., Whitlock, R., Press, M.C., Scholes, J.D., 2011. Variation for host range within and among populations of the parasitic plant *Striga hermonthica*. Heredity 1−9.

Irving, L.J., Cameron, D.D., 2009. You are what you eat: interactions between root parasitic plants and their hosts. Adv. Bot. Res. 50, 87−138.

Jamil, M., Rodenburg, J., Charnikhova, T., Bouwmeester, H.J., 2011. Pre-attachment *Striga hermonthica* resistance of New Rice for Africa (NERICA) cultivars based on low strigolactone production. New Phytol. 192, 964−975.

Jamil, M., Kanampiu, F.K., Karaya, H., Charnikhova, T., Bouwmeester, H.J., 2012. *Striga hermonthica* parasitism in maize in response to N and P fertilisers. Field Crops Res. 134, 1−10.

Jia, X.P., Shi, Y.S., Song, Y.C., Wang, G.Y., Wang, T.Y., Li, Y., 2007. Development of EST-SSR in foxtail millet (*Setaria italica*). Genet. Resour. Crop Evol. 54, 233−236.

Joel, D.M., 2000. The long-term approach to parasitic weeds control: manipulation of specific developmental mechanisms of the parasite. Crop Prot. 19, 753−758.

Joel, D.M., Hershenhorn, J., Eizenberg, H., Aly, R., Ejeta, G., Rich, P.J., et al., 2007. Biology and management of weedy root parasites. Hort. Rev. 33, 267−349.

Kanampiu, F.K., Kabambe, V., Massawe, C., Jasi, L., Friesen, D., Ransom, J.K., et al., 2003. Multi-site, multi-season field tests demonstrate that herbicide seed-coating herbicide-resistance maize controls *Striga* spp. and increases yields in several African countries. Crop Prot. 22, 679−706.

Kanampiu, F., Diallo, A., Karaya, H., 2007a. Herbicide-seed coating technology: a unique approach for *Striga* control in maize. Afr. Crop Sci. Conf. Proc. 8, 1095−1098.

Kanampiu, F., Diallo, A., Burnet, M., Karaya, H., Gressel, J., 2007b. Success with the low biotech of seed-coated imidazolinone-resistant maize. In: Ejeta, G., Gressel, J. (Eds.), Integrating New Technologies for *Striga* Control: Towards Ending the Witch-Hunt. World Scientific Publishing Company Pvt. Ltd., Singapore, pp. 145−158.

Kannan, C., Zwanenburg, B., 2014. A novel concept for the control of parasitic weeds by decomposing germination stimulants prior to action. Crop Prot. 61, 11−15.

Kapran, I., Grenier, C., Ejeta, G., 2007. Introgression of *Striga* resistance into African sorghum landraces. In: Ejeta, G., Gressel, J. (Eds.), Integrating New Technologies for *Striga* Control: Towards Ending the Witch-Hunt. World Scientific Publishing Company Pvt. Ltd., Singapore, pp. 129−142.

Karaya, H., Kiarie, N., Mugo, S., Kanampiu, F., Ariga, E., Nderitu, J., 2012. Identification of new maize inbred lines with resistance to *Striga hermonthica* (Del.) Benth. J. Crop Prot. 1 (2), 131−142.

Kgosi, R.L., Zwanenburg, B., Mwakaboko, A.S., Murdoch, A.J., 2012. Strigolactone analogues induce suicidal germination of *Striga* spp. in soil. Weed Res. 52, 197−203.

Khan, Z.R., Hassanali, A., Overholt, W., Khamis, T.M., Hooper, A.M., Pickett, J.A., et al., 2002. Control of witchweed *Striga hermonthica* by intercropping with *Desmodium* spp., and the mechanism defined as allelopathic. J. Chem. Ecol. 28, 1871−1885.

Khan, Z.R., Pickett, J.A., Wadhams, L.J., Hassanali, A., Midega, C.A.O., 2006. Combined control of *Striga hermonthica* and stemborers by maize-*Desmodium* spp. intercrops. Crop Prot. 25, 989−995.

Khan, Z.R., Pickett, J.A., Hassanali, A., Hooper, A.M., Midega, C.A.O., 2008. Desmodium species and associated biochemical traits for controlling *Striga* species: present and future prospects. Weed Res. 48, 302−306.

Kohlen, W., Charnikhova, T., Liu, Q., Bours, R., Domagalska, M.A., Beguerie, S., et al., 2011. Strigolactones are transported through the xylem and play a key role in shoot architectural response to phosphate deficiency in nonarbuscular mycorrhizal host Arabidopsis. Plant Physiol. 155, 974−987.

Kountche, B.A., 2013. Resistance to *Striga hermonthica* (Del.) Benth.: From Development of Resistant Populations Towards Understanding the Molecular Genetic Basis in Pearl Millet [*Cenchrus americanus* (L.) Morrone]. (Ph.D. thesis). Montpellier SupAgro, Montpellier, France, p. 143.

Kountche, B.A., Hash, C.T., Dodo, H., Oumarou, L., Sanogo, M.D., Amadou, T., et al., 2013a. Development of a pearl millet *Striga*-resistant genepool: response to five cycles of recurrent selection under *Striga*-infested field conditions in West Africa. Field Crops Res. 154, 82−90.

Kountche, B.A., Kane, N., Ousseini, I.S., Vigouroux, Y., 2013b. Crop adaptation to biotic and abiotic conditions: going wild with next generation sequencing technologies. Agrotechnology 2, Available from: http://dx.doi.org/10.4172/2168-9881.1000e103.

Koyama, M.L., 2000. Molecular markers for studying pathogen variability: implications for breeding for resistance to *Striga hermonthica*. In: Haussmann, B.I.G., Hess, D.E., Koyama, M.L., Grivet, L., Rattunde, H.F.W., Geiger, H.H. (Eds.), Breeding for *Striga* Resistance in Cereals. MargrafVerlag, Weikersheim, Germany, pp. 227−245.

Kroschel, J., 1998. Striga − How will it affect African agriculture in the future? − An ecological perspective. In: Martin, K., Müther, J., Auffarth, A. (Eds.), Agroecological, Plant Protection and the Human Environment: Views and Concepts. Plits 16 (2), Verlag W. and S Koch, Am Eichenhain, Germany, pp. 137−158.

Kroschel, J., Hundt, A., Abbasher, A.A., Sauerborm, J., 1996. Pathogenicity of fungi collected in northern Ghana to *Striga hermonthica*. Weed Res. 36, 515—520.

Kuchinda, N.C., Kureh, I., Tarfa, B.D., Shinggu, C., Omolehin, R., 2003. On-farm evaluation of improved maize varieties intercropped with some legumes in the control of *Striga* in the northern Guinea savanna of Nigeria. Crop Prot. 22, 533—538.

Lane, J.A., Child, D.V., Moore, T.H.M., Bailey, J.A., 1995. Wild relatives of cereals as new sources of resistance to *Striga*. Parasitic Plants: Biology and Resistance. International Workshop. IACR-Long Ashton Research Station, Bristol, UK, 30 May—2 June 1995, p. 13.

Lane, J.A., Child, D.V., Reiss, G.C., Entcheva, V., Bailey, J.A., 1997. Crop resistance to parasitic plants. In: Crute, I.R., Holub, E.B., Burdon, J.J. (Eds.), The Gene-for-Gene Relationship in Plant Parasite Interactions. CAB International, Oxfordshire, UK, pp. 81—97.

Logan, D.C., Stewart, G.R., 1991. Role of ethylene in the germination of the hemiparasite, *Striga hermonthica*. Plant Physiol. 97, 1435—1438.

Mace, E.S., Xia, L., Jordan, D.R., Halloran, K., Parh, D.K., Huttner, E., et al., 2008. DArT markers: diversity analyses and mapping in *Sorghum bicolor*. BMC Genomics 9, 26. Available from: http://dx.doi.org/10.1186/1471-2164-9-26.

Mace, E.S., Rami, J.F., Bouchet, S., Klein, P.E., Klein, R.R., Kilian, A., et al., 2009. A consensus genetic map of sorghum that integrates multiple component maps and high-throughput diversity array technology (DArT) markers. BMC Plant Biol. 9, 13. Available from: http://dx.doi.org/10.1186/1471-2229-9-13.

Maiti, R.K., Ramaiah, K.V., Bisen, S.S., Chidley, V.L., 1984. A comparative study of the haustorial development of *Striga asiatica* (L.) Kuntze on sorghum cultivars. Ann. Bot. (Lond.) 54, 447—457.

Mannuramath, M., Yenagi, N., Orsat, V., 2015. Quality evaluation of little millet (*Panicum miliare*) incorporated functional bread. J. Food Sci. Technol. 52 (12), 8357—8363.

Mariac, C., Luong, V., Kapran, I., Mamadou, A., Sagnard, F., Deu, M., et al., 2006. Diversity of wild and cultivated pearl millet accessions [*Pennisetum glaucum* (L.) R. Br.] in Niger assessed by microsatellite markers. Theor. Appl. Genet. 114, 49—58.

Marley, P.S., Ahmed, S.M., Shebayan, J.A.Y., Lagoke, S.T.O., 1999. Isolation of Fusarium oxysporum with potential for biocontrol of the witchweed (*Striga hermonthica*) in the Nigerian Savanna. Bio. Sci. Technol. 9, 159—163.

Marley, P.S., Aba, D.A., Shebayan, J.A.Y., Musa, R., Sanni, A., 2004. Integrated management of *Striga hermonthica* in sorghum using a mycoherbicide and host plant resistance in the Nigerian Sudano-Sahelian savanna. Weed Res. 44, 157—162.

Marley, P.S., Kroschel, J., Elzein, A., 2005. Host specificity of *Fusarium oxysporum* Schlecht (isolate PSM 197), a potential mycoherbicide for controlling *Striga* spp. in West Africa. Weed Res. 45, 407—412.

Matsuura, H., Ohashi, K., Sasako, H., Tagawa, N., Takano, Y., et al., 2008. Germination stimulant from root exudates of *Vignaunguiculata*. Plant Growth Regul. 54, 31—36.

Matusova, R.K., Rani, F.W., Verstaen, A., Franssen, M.C.R., Beale, M.H., Bouwmeester, H.J., 2005. The strigolactone germination stimulants of the plant-parasitic *Striga* and *Orobanches* are derived from the carotenoid pathway. Plant Physiol. 139, 920—934.

Mayer, A.M., 2006. Pathogenesis by fungi and by parasitic plants: similarities and differences. Phytoparasitica 34 (1), 3—16.

Mbwaga, A.M., Riches, C., Ejeta, G., 2007. Integrated *Striga* management to meet sorghum demand in Tanzania. In: Ejeta, G., Gressel, J. (Eds.), Integrating New Technologies for *Striga* Control: Towards Ending the Witch-Hunt. World Scientific Publishing Co. Pvt. Ltd., Singapore, pp. 253—264.

Michelmore, R., 2000. Genomic approaches to plant diseases resistance. Curr. Opin. Plant Biol. 3, 125—131.

Mishra, V., Yadav, N., Pandey, S., Puranik, V., 2014. Bioactive components and nutritional evaluation of underutilized cereals. Ann. Phytomed. 3 (2), 46—49.

Mohamed, K.I., Musselman, L.J., Riches, C.R., 2001. The genus *Striga* (*Scrophulariaceae*) in Africa. Ann. Missouri Bot. Garden 88, 60–103.

Mohamed, A., Ellicott, A., Housley, T.L., Ejeta, G., 2003. Hypersensitive response to *Striga* infection in *Sorghum*. Crop Sci. 43, 1320–1324.

Mohamed, K., Bolin, J., Musselman, L., Peterson, A., 2007. Genetic diversity of *Striga* and implications for control and modeling future distributions. In: Ejeta, G., Gressel, J. (Eds.), Integrating New Technologies for *Striga* Control: Towards Ending the Witch-Hunt. World Scientific Publishing Co. Pvt. Ltd, Singapore, pp. 71–84.

Mohamed, A.H., Housley, T.L., Ejeta, G., 2010. An *in vitro* technique for studying specific *Striga* resistance mechanisms in sorghum. Afr. J. Agric. Res. 5, 1868–1875.

Mohamed, A., Ali, R., Elhassan, O., Suliman, E., Mugoya, C., Masiga, C.W., et al., 2014. First products of DNA marker-assisted selection in sorghum released for cultivation by farmers in sub-saharan Africa. J. Plant Sci. Mol. Breed 3, 3. Available from: http://dx.doi.org/10.7243/2050-2389-3-3.

Mori, K., Matsui, J., Yokota, T., Sakai, H., Bando, M., et al., 1999. Structure and synthesis of orobanchol, the germination stimulant for *Orobanche minor*. Tetrahedron Lett. 40, 943–946.

Moumouni, K.H., Kountche, B.A., Jean, M., Hash, C.T., Vigouroux, Y., Haussmann, B.I.G., et al., 2015. Construction of a genetic map for pearl millet, *Pennisetum glaucum* (L.) R. Br., using a genotyping-by-sequencing (GBS) approach. Mol. Breed. 35, 5. Available from: http://dx.doi.org/10.1007/s11032-015-0212-x.

Mounde, L.G., 2014. Understanding the Role of Plant Growth Promoting Bacteria on Sorghum Growth and Biotic Suppression of *Striga* Infestation. (Ph.D. thesis). University of Hohenheim, Stuttgart, Germany, p. 113.

Muller, S., Hauck, C., Schildknecht, H., 1992. Germination stimulants produced by *Vignaunguiculata* Walpcv saunders upright. J. Plant Growth Regul. 11, 77–84.

Mutengwa, C.S., Tongoona, P.B., Sithole-Niang, I., 2005. Genetic studies and a search for molecular markers that are linked to *Striga asiatica* resistance in sorghum. Afr. J. Biotechnol. 4, 1355–1361.

Muthamilarasan, M., Prasad, M., 2015. Advances in *Setaria* genomics for genetic improvement of cereals and bioenergy grasses. Theor. Appl. Genet. 128, 1–14.

Obilana, A.T., Ramaiah, K.V., 1992. *Striga* (witchweeds) in sorghum and millet: knowledge and future research needs. In: de Milliano, W.A.J., Frederiksen, R.A., Bengston, G.D. (Eds.), Sorghum and Millets Diseases: A Second World Review. International Crops Research Institute for the Semi-Arid Tropics Patancheru, India, pp. 187–201.

Olivier, A., 1995. Le *Striga*, mauvaise herbe parasite des céréales africaines: biologie et méthodes de lutte. Agronomie 15, 517–525.

Olmstead, R.G., DePamphilis, C.W., Wolfe, A.D., Young, N.D., Elisons, W., Reeves, P.A., 2001. Disintegration of the *Scrophulariaceae*. Am. J. Bot. 88, 348–361.

Omanya, G.O., Haussmann, B.I.G., Hess, D.E., Reddy, B.V.S., Kayentao, M., Welz, H.G., et al., 2004. Utility of indirect and direct selection traits for improving *Striga* resistance in two sorghum recombinant inbred populations. Field Crops Res. 89, 237–252.

Oswald, A., 2005. *Striga* control – technologies and their dissemination. Crop Prot. 24, 333–342.

Parker, C., 2009. Observations on the current status of *Orobache* and *Striga* problems worldwide. Pest Manage. Sci. 65, 453–459.

Parker, C., 2012. Parasitic weeds: a world challenge. Weed Sci. 60, 269–276.

Parker, C., Riches, C.R., 1993. Parasitic Weeds of the World: Biology and Control. CAB International, Wallingford, UK, p. 332.

Paterson, A.H., et al., 2009. The *Sorghum bicolor* genome and the diversification of grasses. Nature 457, 551–556.

Pennisi, E., 2010. Armed and dangerous. Science 327, 804–805.

Pennisi, E., 2015. How crop-killing witchweed senses its victims. Science 350, 146–147.

Qi, X., Lindup, S., Pittaway, T.S., Allouis, S., Gale, M.D., Devos, K.M., 2001. Development of simple sequence repeat markers from bacterial artificial chromosomes without subcloning. Biotechniques 31, 355−362.

Qi, X., Pittaway, T.S., Lindup, S., Liu, H., Waterman, E., Padi, F.K., et al., 2004. An integrated genetic map and a new set of simple sequence repeat markers for pearl millet, *Pennisetum glaucum*. Theor. Appl. Genet. 109, 1485−1493.

Rai, K.N., Govindaraj, M., Rao, A.S., 2012. Genetic enhancement of grain iron and zinc content in pearl millet. Qual. Assur. Saf. Crops Foods 4 (3), 119−125. Available from: http://dx.doi.org/10.1111/j.1757-837X.2012.00135.x.

Rajaram, V., Nepolean, T., Senthilvel, S., Varshney, R.K., Vadez, V., Srivastava, R.K., et al., 2013. Pearl millet [*Pennisetum glaucum* (L.) R. Br.] consensus linkage map constructed using four RIL mapping populations and newly developed EST-SSRs. BMC Genomics 14, 159. Available from: http://www.biomedcentral.com/1471-2164/14/159.

Rakshit, S., Patil, J.V., 2014. Sorghum. In: Chopra, V.L. (Ed.), Breeding Field Crops II: Advances. Studium Press (LLC), Houston, pp. 79−96.

Ramaiah, K.V., 1984. Patterns of *Striga* resistance in sorghum and millets with special emphasis on Africa. Striga: Biology and Control, Proceedings Workshop on the Biology and Control of Striga, 14−17 Nov 1983. ICSU Press/IDRC, Dakar, Senegal, pp. 71−92.

Ramaiah, K.V., 1987. Breeding cereal grains for resistance to witchweed. In: Musselman, L.J. (Ed.), Parasitic Weeds in Agriculture, Vol I, *Striga*. CRC Press, Boca Raton, FL, pp. 227−242.

Ramaiah, K.V., Chidley, V.L., House, L.R., 1990. Inheritance of *Striga* seed-germination stimulant in sorghum. Euphytica 45, 33−38.

Rao, P.S., Reddy, P.S., Rathore, A., Reddy, B.V.S., Panwar, S., 2011. Application GGE biplot and AMMI model to evaluate sweet sorghum (*Sorghum bicolor*) hybrids for genotype × environment interaction and seasonal adaptation. Indian J. Agric. Sci. 81, 438−444.

Rector, B.G., 2009. Molecular biology approaches to control of intractable weeds: new strategies and complements to existing biological practices. Plant Sci. 175, 437−448.

Rich, P.J., Ejeta, G., 2008. Towards effective resistance to *Striga* in African maize. Plant Signal. Behav. 3, 618−621.

Rich, P.J., Grenier, C., Ejeta, G., 2004. *Striga* resistance in the wild relatives of sorghum. Crop Sci. 44, 2221−2229.

Rispail, N., Dita, M.A., González-Verdejo, C., Pérez-de-Luque, A., Castillejo, M.A., Prats, E., et al., 2007. Plant resistance to parasitic plants: molecular approaches to an old foe. New Phytol. 173, 703−712.

Robert, O.J., 2011. Genetic Analysis of *Striga hermonthica* Resistance in Sorghum (*Sorghum bicolor*) Genotypes in Eastern Uganda. (Ph.D. thesis). University of KwaZulu-Natal, Pietermaritzburg, South Africa.

Rodenburg, J., Bastiaans, L., Weltzien, E., Hess, D.E., 2005. How can field selection for *Striga* resistance and tolerance in sorghum be improved? Field Crops Res. 93, 34−50.

Rodenburg, J., Bastiaans, L., Kropff, M.J., Van Ast, A., 2006. Effects of host plant genotype and seedbank density on *Striga* reproduction. Weed Res. 46, 251−263.

Rubiales, D., 2003. Parasitic plants, wild relatives and the nature of resistance. New Phytol. 160, 459−461.

Rubiales, D., Pérez-de-Luque, A., Fernandez-Aparico, M., Sillero, J.C., Roman, B., Kharrat, M., et al., 2006. Screening techniques and sources of resistance against parasitic weeds in grain legumes. Euphytica 147, 187−199.

Runo, S., Alakonya, A., Machuka, J., Sinha, N., 2011. RNA interference as a resistance mechanism against crop parasites in Africa: a 'Trojan orse' approach. Pest Manage. Sci. 67, 129−136.

Samake, O., Stomph, T.J., Kropff, M.J., Smaling, E.M.A., 2006. Integrated pearl millet management in the Sahel: effects of legume rotation and fallow management on productivity and *Striga hermonthica* infestation. Plant Soil 286, 245−257.

Satish, K., Gutema, Z., Grenier, C., Rich, P.J., Ejeta, G., 2012. Tagging and validation of microsatellite markers linked to the low germination stimulant gene (*lgs*) for *Striga* resistance in sorghum [*Sorghum bicolor* (L.) Moench]. Theor. App. Genet. 124, 989–1003.

Sato, D., Awad, A.A., Chae, S.H., Yokota, T., Sugimoto, Y., Takeuchi, Y., et al., 2003. Analysis of strigolactones, germination stimulants for *Striga* and *Orobanche*, by high-performance liquid chromatography/tandem mass spectrometry. J. Agric. Food Chem. 51, 1162–1168.

Scholes, J.D., Press, M.C., 2008. *Striga* infestation of cereal crops—an unsolved problem in resource-limited agriculture. Curr. Opin. Plant Biol. 11, 180–186.

Senthilvel, S., Mahalakshmi, V., Sathish, K.P., Reddy, A.R., Markandeya, G., Reddy, M.K., et al., 2004. New SSRs markers for pearl millet from data mining of expressed sequence tags. 4th International Crop Science Congress, Brisbane-Australia, 26 Sep–1 Oct 2004. <www.cropscience.org.au>.

Senthilvel, S., Jayashree, B., Mahalakshmi, V., Sathish, K.P., Nakka, S., Nepolean, T., et al., 2008. Development and mapping of simple sequence repeat markers for pearl millet from data mining of expressed sequence tags. BMC Plant Biol. 8, 119. Available from: http://dx.doi.org/10.1186/1471-2229-8-119.

Serghini, K., Perez-de-Luque, A., Castejon-Munoz, M., Garcıa-Torres, L., Jorrın, J.V., 2001. Sunflower (*Helianthus annuus* L.) response to broomrape (*Orobanche cernua* Loefl.) parasitism: induced synthesis and excretion of 7-hydroxylated simple coumarins. J. Exp. Bot. 52, 2227–2234.

Shen, H., Ye, W., Hong, L., Huang, H., Wang, Z., Deng, X., et al., 2006. Progress in parasitic plant biology: host selection and nutrient transfer. Plant Biol. 8, 175–185.

Siame, B.A., Weerasuriya, Y., Wood, K., Ejeta, G., Butler, L.G., 1993. Isolation of strigol, a germination stimulant for *Striga* asiatica, from host plants. J. Agric. Food Chem. 41, 1486–1491.

Spallek, T., Mutuku, J.M., Shirasu, K., 2013. The genus *Striga*: a witch profile. Mol. Plant Pathol. 14, 861–869.

Steinkellner, S., Lendzemo, V., Langer, I., Schweiger, P., Khaosaad, T., Toussaint, J.P., et al., 2007. Flavonoids and strigolactones in root exudates as signals in symbiotic and pathogenic plant-fungus interactions. Molecules 12, 1290–1306.

Supriya, S., Senthilvel, S., Nepolean, T., Eshwar, K., Rajaram, V., Shaw, R., et al., 2011. Development of a molecular linkage map of pearl millet integrating DArT and SSR markers. Theor. Appl. Genet. 123, 239–250.

Tank, D.C., Beardsley, P.M., Kelchner, S.A., Olmstead, R.G., 2006. Review of the systematics of *Scrophulariaceae* and their current disposition. Aust. Syst. Bot. 19, 289–307.

Teka, H.B., 2014. Advance research on *Striga* control: a review. Afr. J. Plant Sci. 8 (11), 492–506.

Tesso, T.T., Ejeta, G., 2011. Integrating Multiple Control Options Enhances *Striga* Management and Sorghum Yield on Heavily Infested Soils. Agron. J. 103, 1464–1471.

Tesso, T., Gutema, Z., Deressa, A., Ejeta, G., 2007. An integrated *Striga* management option offers effective control of *Striga* in Ethiopia. In: Ejeta, G., Gressel, J. (Eds.), Integrating New Technologies for *Striga* Control: Towards Ending the Witch-Hunt. World Scientific Publishing Co. Pvt. Ltd., Singapore, pp. 199–212.

Ueno, K., Nomura, S., Muranaka, S., Mizutani, M., Takikawa, H., et al., 2011. Ent-2'-epi-orobanchol and its acetate, as germination stimulants for *Striga gesnerioides* seeds isolated from cowpea and red clover. J. Agric. Food Chem. 59, 10485–10490.

Umehara, M., Hanada, A., Yoshida, S., Akiyama, K., Arite, T., et al., 2008. Inhibition of shoot branching by new terpenoid plant hormones. Nature 455, 195–200.

Van Mourik, T.A., 2007. *Striga Hermonthica* Seed Bank Dynamics: Process Quantification and Modelling. (Ph.D. thesis). Wageningen University, Wageningen, The Netherlands, p. 123.

Varshney, R.K., Dubey, A., 2009. Novel genomic tools and modern genetic and breeding approaches for crop improvement. J. Plant Biochem. Biotechnol. 18, 127–138.

Vasey, R.A., Scholes, J.D., Press, M.C., 2005. Wheat (*Triticumaestivum*) is susceptible to the parasitic angiosperm *Striga hermonthica*, a major cereal pathogen in Africa. Phytopathology 95, 1294–1300.

Vasudeva Rao, M.J., Musselman, L.J., 1987. Host specificity in *Striga* spp. and physiological strains. In: Musselman, L.J. (Ed.), Parasitic Weeds in Agriculture, Vol. I, Striga. CRC Press, Boca Raton, FL, pp. 13–25.

Venne, J., Beed, F., Watson, A., 2009. Integrating *Fusarium oxysporum* f. sp. *strigae* into cereal cropping systems in Africa. Pest Manage. Sci. 65, 572–580.

Visarada, K.B.R.S., Kishore, N.S., 2015. Advances in Genetic Transformation. In: Madhusudhana, R., et al., (Eds.), Sorghum Molecular Breeding. Springer (India) Pvt. Ltd., pp. 199–215.

Vogler, R.K., Ejeta, G., Butler, L.G., 1996. Inheritance of low production of *Striga* germination stimulant in sorghum. Crop Sci. 36, 1185–1191.

Watson, A.K., 2013. Biocontrol. In: Joel, D.M., Gressel, J., Musselman, L.J. (Eds.), Parasitic Orobancheceae. Springer Verlag, Berlin Heidelberg, pp. 469–497.

Westwood, J.H., Yoder, J.I., Timko, M.P., Pamphilis, C.W., 2010. The evolution of parasitism in plants. Trend. Plant Sci. 15 (4), 227–235.

Westwood, J.H., dePamphilis, C.W., Das, M., Fernandez-Aparicio, M., Honaas, L.A., Timko, M.P., et al., 2012. The parasitic plant genome project: new tools for understanding the biology of *Orobanche* and *Striga*. Weed Sci. 60, 295–306.

Wilson, J.P., Hess, D.E., Hanna, W.W., 2000. Resistance to *Striga hermonthica* in wild accessions of the primary gene pool of *Pennisetum glaucum*. Phytopathology 90, 1169–1172.

Wilson, J.P., Hess, D.E., Hanna, W.W., Kumar, K.A., Gupta, S.C., 2004. *Pennisetum glaucum* subsp. *monodii* accessions with *Striga* resistance in West Africa. Crop Prot. 23, 865–870.

Xie, X., Kusumoto, D., Takeuchi, Y., Yoneyama, K., Yamada, Y., 2007. 2′-epi-Orobanchol and solanacol, two unique strigolactones, germination stimulants for root parasitic weeds, produced by tobacco. J. Agric. Food Chem. 55, 8067–8072.

Xie, X., Yoneyama, K., Kusumoto, D., Yamada, Y., Yokota, T., et al., 2008. Isolation and identification of alectrol as (+)-orobanchyl acetate, a germination stimulant for root parasitic plants. Phytochemistry 69, 427–431.

Xie, X.N., Yoneyama, K., Harada, Y., Fusegi, N., Yamada, Y., et al., 2009a. Fabacyl acetate, a germination stimulant for root parasitic plants from Pisusativum. Phytochemistry 70, 211–215.

Xie, X.N., Yoneyama, K., Kurita, J.Y., Harada, Y., Yamada, Y., et al., 2009b. 7-Oxoorobanchyl acetate and 7-oxoorobanchol as germination stimulants for root parasitic plants from Flax (*Linumusitatissimum*). Biosci. Biotechnol. Biochem. 73, 1367–1370.

Xie, X., Yoneyama, K., Yoneyama, K., 2010. The strigolactone story. Annu. Rev. Phytopathol. 48, 93–117.

Yadav, O.P., Mitchell, S.E., Zamora, A., Fulton, T.M., Kresovich, S., 2007. Development of new simple sequence repeat markers for pearl millet. J. SAT Agric. Res. 3 (1), 34–37.

Yadav, O.P., Mitchell, S.E., Fulton, T.M., Kresovich, S., 2008. Transferring molecular markers from sorghum, rice and other cereals to pearl millet. J. SAT Agric. Res. 6.

Yoder, J.I., Scholes, J.D., 2010. Host plant resistance to parasitic weeds; recent progress and bottlenecks. Curr. Opin. Plant Biol. 13, 478–484.

Yoder, J.I., Gunathilake, P., Wu, B., Tomilova, N., Tomilov, A.A., 2009. Engineering host resistance against parasitic weeds with RNA interference. Pest Manage. Sci. 65, 460–466.

Yohannes, T., Abraha, T., Kiambi, D., Folkertsma, R., Hash, C.T., Ngugi, F., et al., 2015. Marker-assisted introgression improves *Striga* resistance in an Eritrean farmer-preferred sorghum variety. Field Crops Res. 173, 22–29.

Yokota, T., Sakai, H., Okuno, K., Yoneyama, K., Takeuchi, Y., 1998. Alectrol and orobanchol, germination stimulants for *Orobanche minor*, from its host red clover. Phytochemistry 49, 1967–1973.

Yoneyama, K., Awad, A.A., Xie, X., Yoneyama, K., Takeuchi, Y., 2010. Strigolactones as germination stimulants for root parasitic plants. Plant Cell Physiol. 51, 1095–1103.

Yonli, D., Traoré, H., Hess, D.E., Sankara, P., Sérémé, P., 2006. Effect of growth medium, *Striga* seed burial distance and depth on efficacy of *Fusarium* isolates to control *Striga hermonthica* in Burkina Faso. Weed Res. 46, 73−81.

Yoshida, S., Shirasu, K., 2009. Multiple layers of incompatibility to the parasitic witchweed, *Striga hermonthica*. New Phytol. 183, 180−189.

Zahran, E., Kohlschmid, E., Sauerborn, J., Abbasher, A.A., Muller-Stover, D., 2008. Does an application as seed coating stabilize the efficacy of biocontrol agents against *Striga hermonthica* under field conditions? J. Plant Dis. Prot. S21, 467−471.

Zarafi, A.B., Elzein, A., Abdulkadir, D.I., Beed, F., Akinola, O.M., 2015. Host range studies of *Fusarium oxysporum* f.sp. strigae meant for the biological control of *Striga hermonthica* on maize and sorghum. Arch. Phytopathol. Plant Prot. 48 (1), 1−9. Available from: http://dx.doi.org/10.1080/03235408.2014.880580.

Zwanenburg, B., Pospisil, T., 2013. Structure and activity of strigolactones: new plant hormones with a rich future. Mol. Plant 6, 38−62.

Zwanenburg, B., Mwakaboko, A.S., Reizelman, A., Anilkumar, G., Sethumadhavan, D., 2009. Structure and function of natural and synthetic signaling molecules in parasitic weed germination. Pest Manage. Sci. 65, 478−491.

Zwanenburg, B., Nayak, S.K., Charnikhova, T.V., Bouwmeester, H.J., 2013. New strigolactone mimics: structure−activity relationship and mode of action as germinating stimulants for parasitic weeds. Bioorg. Med. Chem. Lett. 23, 5182−5186.

WEED PROBLEM IN MILLETS AND ITS MANAGEMENT

J.S. Mishra

ICAR Research Complex for Eastern Region, Patna, India

7.1 INTRODUCTION

The millets are major crops of the semiarid regions of the world, and have the potential to contribute substantially for food, fodder, and the nutritional security of these regions. Because of their drought tolerance, millets can be cultivated in the areas that are often too hot and dry for other crops. Weeds are the major obstacles in increasing the productivity of millets especially during the rainy season. Burnside and Wicks (1969) reported that weed competition had a greater effect on sorghum yield than row-spacing or plant population. The millets are relatively poor competitors against weeds especially during the early growth stages (first few weeks) of the crop. During this phase, millets grow slowly compared to the weed species and this creates better conditions for weed growth. It takes up to the mid-season for the millets to develop a good canopy that can cover and shade the space between the rows and discourage weed growth. Planting in wider rows to facilitate interrow cultivation and/or ditch furrow irrigation worsens the problems. The weed competes with the millets for the light, soil moisture, and nutrients and reduces crop yield and quality. Therefore, appropriate weed management would help improve productivity and input use-efficiency of these crops. When an improved agricultural technology is adopted, an efficient weed management becomes more important, otherwise the weeds rather than the crops will benefit from the costly inputs.

7.2 WEEDS OF MILLETS AND THEIR IMPORTANCE

7.2.1 WEEDS OF MILLETS

In general terms, weeds are undesired plants. About 200 plant species in the world act as weeds, of which around 80 species are troublesome enough to humans. Of the 80, around 45 are broadleaf, 28 are grasses, 5 are sedges, and 2 are ferns (Holm, 1978). A mixed population of broad-leaved, grasses, and sedges grows with the millet crops under different agro-climatic conditions. The composition of weed species, their intensity, and competition offered by them to the crop vary with the geographic regions, soil and weather conditions, and the field and crop management practices

(Stahlman and Wicks, 2000; Klaij and Hoogmoed, 1996; Mashingaidze et al., 2012). A well tilled field faces relatively less weed problems than a field with minimum tillage. The farmers practicing minimum tillage face the biggest and often the most difficult challenge for the management of weed (Gowing and Palmer, 2008). The weeds can be classified based on their plant type (monocotyledonous or grassy species, dicotyledonous or broadleaf species), duration of life span (annual, perennial), and their mode of nutrition (autotrophic, parasitic). Most of the weeds of millets are autotrophic by nature, although *Striga* spp. is a parasitic one. *Striga* is a major biotic constraint in the subsistence agriculture and causes considerable crop damage in millets in the semiarid tropics. It depends upon the host plant not only for metabolic inputs such as water, minerals, and energy, but also for developmental signals. *Striga* requires chemical signals exuded by the host roots for seed germination and haustoria formation. The major weeds associated with millets are presented in Table 7.1. *Echinochloa colona* (L.) Link. (Jungle rice), *Echinochloa crus-galli* (L.) Beauv. (Barnyard grass), *Eleusine indica* (L.) Gaertn. (Goose grass), *Digitaria sanguinalis* (L.) Scop. (Crab grass), and *Sorghum halepense* L. Pers. (Johnson grass) among grasses; *Amaranthus palmeri* S. Wats (Palmer amaranth), *Amaranthus retroflexus* L. (Redroot pigweed), *Celosia argentea* L. (White cock's comb), *Trianthema portulacastrum* L. (Horse weed), *Tribulus terrestris* L. (Puncture vine), *Boerhaavia diffusa* L. (Hog weed), and *Acanthospermum hispidum* DC. (Bristly starbur) among broadleaf; *Cyperus rotundus* L. among sedges, and *Striga asiatica* (L.) Kuntze. and *Striga hermonthica* (Del.) Benth. (Witch weed) are the most common weeds of millets worldwide. Reddy et al. (2014) reported Canada thistle (*Cirsium arvense* [L.] Scop.), Kochia (*Kochia scoparia* [L.] Schrad.), Redroot pigweed (*Amaranthus retroflexus* L.), Green foxtail (*Setaria viridis* [L.] Beauv.), and Palmer amaranth (*Amaranthus palmeri* S. Watson) as the most common troublesome weeds interfering with millet crops in the US central Great Plains.

7.2.2 LOSSES DUE TO WEEDS

Weeds compete with crops for nutrients, soil moisture, sunlight, and space when they are limiting, resulting in reduced yields, lower grain quality, and increased production costs. The magnitude of losses depends on crop cultivars, nature and intensity of weeds, spacing, duration of weeds infestation, environmental conditions, and management practices. Grain yield loss due to weeds in sorghum, pearl millet, and finger millet varies from 5−94% depending on growing conditions (Table 7.2).

Weeds are one of the major biological deterrents in increasing the grain sorghum productivity and quality (Geier et al., 2009). Among the millet farmers in the African countries like Sudan, Ethiopia, Mali, Uganda, Zimbabwe, and Ghana weeds are one of the most important production constraints. They take up nutrients from the soil and make nutrients less efficient to the millets by lowering use-efficiency, which adversely affects the production and productivity of the millets. The problem is greater during the rainy season. An unhindered growth of the weed during the initial growth stage of the millets helps it to draw nutrients at a faster rate than the crop. As an example, an uncontrolled weed growth in grain sorghum removed $29.94-51.05$ kg ha^{-1} nitrogen, $5.03-11.58$ kg ha^{-1} phosphorus, and $48.74-74.34$ kg ha^{-1} potassium from the soil (Satao and Nalamwar, 1993). As most of

Table 7.1 Major Weeds of Millet

Scientific Name	English Name	Family
Grasses		
Cynodon dactylon Pers.	Bermuda grass	Poaceae
Brachiaria ramosa L.	Brown top millet	Poaceae
Digitaria sanguinalis (L.) Scop.	Crab grass	Poaceae
Dactyloctenium aegyptium L.	Crowfoot grass	Poaceae
Dinebra retroflexa Vahl.	Viper grass	Poaceae
Chloris barbata Sw.	Peacock plume grass	Poaceae
Eleusine indica (L.) Gaertn.	Goose grass	Poaceae
Echinochloa colona Link.	Jungle rice	Poaceae
Sorghum halepense (L.) Pers.	Johnson grass	Poaceae
Setaria glauca Beauv.	Yellow foxtail	Poaceae
Setaria viridis L.	Green foxtail	Poaceae
Panicum repens L.	Tarpedo grass	Poaceae
Paspalum paspaloides (Mischx.) Scribner	Hilo grass, sour grass	Poaceae
Broadleaf species		
Convolvulus arvensis L.	Field bind weed	Convolvulaceae
Acanthospermum hispidum DC.	Bristly starbur	Asteraceae
Achyranthes aspera L.	Prickly chaff flower	Amaranthaceae
Commelina benghalensis L.	Tropical spider wort	Commelinaceae
Ageratum conyzoides L.	Bill goat weed	Compositae
Amaranthus viridis L.	Pigweed	Amaranthaceae
Amaranthus palmeri S. Wats.	Palmer amaranth	Amaranthaceae
Amaranthus retroflexus L.	Redroot pigweed	Amaranthaceae
Boerhaavia diffusa L.	Hog weed	Nyctaginaceae
Celosia argentea L.	White cock's comb	Amaranthaceae
Cleome viscosa L.	Cleome	Capparaceae
Digera arvensis Forsk.	False amaranth	Amaranthaceae
Kochia scoparia (L.) Schrad.	Kochia	
Portulaca oleracea L.	Common purslane	Portulacaceae
Euphorbia hirta L.	Pill pod spurge	Euphorbiaceae
Eclipta alba Hassk.	False daisy	Compositae
Corchorus acutangulus Lamk.	Jew's mallow	Tiliaceae
Ipomoea haderacea Jack.	Morning glory	Convolvulaceae
Portulaca oleracea L.	Purselane	Portulaceae
Salsola iberica Sennan & Pau	Russian thistle	
Trianthema portulacastrum L.	Horse purslane	Aizoaceae
Tridax procumbens L.	Coat buttons, tridax daisy	Compositae

(Continued)

Table 7.1 Major Weeds of Millet *Continued*

Scientific Name	English Name	Family
Tribulus terrestris L.	Puncture vine	Zygophyllaceae
Xanthium strumarium L.	Common cocklebur	Asclepiadaceae
Sedges		
Cyperus rotundus L.	Purple nut sedge	Cyperaceae
Parasitic species		
Striga spp.	Witch weed	Scrophulariaceae

Table 7.2 Losses Due to Weed of Millet

Crops	Reduction in Grain Yield (%)	References
Sorghum	15−83	Mishra (1997), Stahlman and Wicks (2000)
Pearl millet	55	Banga et al. (2000)
	35−90	Umrani et al. (1980)
	31−46	Kaushik and Gautam (1984)
	16−94	Balyan et al. (1993)
	40	Sharma and Jain (2003)
	46	Choudhary and Lagoke (1981)
Finger millet	55−61	Ramachandra Prasad et al. (1991)
	5−70	Rao et al. (2015)

the millet-lands suffer from poor soil fertility due to their marginal nature, the removal of nutrients by the weed further deteriorates the situation. There is a difference between broadleaf weed species and grasses in the amount of nutrient uptake. Broadleaf weeds consume more nutrients than sorghum for production of per unit dry matter (Stahlman and Wicks, 2000).

Weeds use soil-water for their growth and deplete it from the soil through transpiration, resulting in soil moisture deficit in the field. As millets are mostly grown in the dryland situations, availability of water is meager and uncertain. Millets have an inherent capacity of drought tolerance. This implies that they can grow and survive in soil moisture conditions where many other plants cannot grow. There are a few weeds of millets that can thrive in lower amounts of water than the millets. For example, the water requirement for production of unit dry matter for barnyard grass is less than that of many types of millets (Shiplet and Wiese, 1969). Depletion of soil-water by weeds, however, may create severe moisture deficit conditions for the millets to grow.

The parasitic weed *Striga* spp. causes severe loss in yield of millets. An estimated 44 million ha were considered to be "at risk" of *Striga* attack in Africa, and the total loss of revenue from maize, pearl millet, and sorghum "could total" US$ 2.9 billion. More recent figures suggest that 50 million ha and 300 million farmers are affected by *Striga* species in Africa, with losses of US$ 7 billion

(Ejeta, 2007). In India, incidence of *Striga* alone caused 75% reduction in grain yield of sorghum (Nagur et al., 1962; Rao, 1978). In sub-Saharan Africa, *S. hermonthica* caused 70−100% crop loss in sorghum and pearl millet (Emechebe et al., 2004).

Many weeds act as alternate or collateral host for the plant pathogens and pests and thus help in the perpetuation and spread of the pests and diseases. The rust, smut, ergot, and downy mildew pathogens of various millets infect weed species like *Cynodon dactylon, Sorghum halepense, Oxalis corniculata, Digitaria marginata, Pennisetum* sp., and *Eragrostis tenuifolia* and help them overwinter (Frederiksen, 1984; Marley, 1995; Reed et al., 2000). Weeds are an important plant resource for insects, although feeding by insects on weeds can have both a positive and a negative effect on the crop productivity (Capinera, 2005). Sorghum shoot fly (*Atherigona soccata*) and gall midge (*Stenodiplosis sorghicola*) infest weeds like *Brachiaria distachya, Panicum repens, Setaria intermedia, Cyperus rotundus,* and *Sorghum halepense* and survive therein until new crops come (Nwilene et al., 1998; Bilbro, 2008).

7.2.3 CRITICAL PERIOD OF CROP−WEED COMPETITION

The weed emerges continuously throughout the crop growth period especially during the rainy season. Therefore, the timing of weed removal is as important as removal *per se* to keep crop−weed competition at bare minimum. The "critical period" of crop−weed competition defines the maximum period the weeds can be tolerated without affecting final crop yields (Zimdahl, 1980). This provides information on the active duration when the presence of weeds make their deleterious effect on the crops. There is a little variation in the critical period among the millets (Table 7.3). Millets are very susceptible to competition from weeds early in the life of the crop. Therefore, efficient weed control at the pre- and early postemergence stages is essential. Once the crop reaches approximately 50 cm in height, weed control no longer affects the yields in the millets. The competition between millet and weed is largely influenced by moisture availability. Wiese et al. (1964) obtained a higher yield for irrigated sorghum in narrow rows without cultivation than in wide rows with cultivation, where as in dryland, plants in wide rows were more able to compete for limited soil moisture.

7.2.4 CLIMATE CHANGE AND WEED COMPETITION

Like every living being weeds also get impacted by changes in climatic conditions. Changes in temperature and carbon dioxide are likely to have significant influence on weed biology *vis-à-vis* crop−weed interactions. Ziska (2003) studied the effect of elevated CO_2 concentration on interactions of the dwarf sorghum (C_4) with and without the presence of a C_3 weed (velvetleaf; *Abutilon theophrasti*)

Table 7.3 Critical Period of Crop−Weed Competition in Millets

Crops	Critical Periods (Days After Sowing)	References
Sorghum	28−42	Sundari and Kumar (2002)
Pearl millet	15−30	Labrada et al. (1994)
Finger millet	25−42	Sundraseh et al. (1975)

and a C_4 weed (redroot pigweed; *Amaranthus retroflexus*). He reported that in a weed-free environment, increased CO_2 concentration significantly increased the leaf weight and leaf area of sorghum but there was no significant effect on seed yield or total above-ground biomass relative to the ambient CO_2 concentration. An increase in velvet leaf biomass in response to an increasing CO_2 concentration reduced the yield and biomass of sorghum. Similarly, significant losses in both seed yield and total biomass were observed for sorghum—redroot pigweed competition as CO_2 concentration increased. Increased CO_2 was not associated with a significant increase in redroot pigweed biomass. These results indicated a potentially greater loss in a widely grown C_4 crop from weed competition as atmospheric CO_2 increases. In another experiment, Ziska (2001) observed that the vegetative growth, competition, and potential yield of sorghum (C_4) could be reduced by the co-occurrence of common cocklebur (*Xanthium strumarium*: C_3) as the atmospheric CO_2 increases. Watling and Press (1997) investigated the effects of CO_2 concentrations (350 and 700 μmol mol^{-1}) in sorghum with and without *Striga* infestation. They observed that a high CO_2 concentration resulted in taller sorghum plants, and greater biomass, photosynthetic rates, water-use efficiencies, and leaf areas; and lower *Striga* biomass/host plant. The stomata conductance was not responsive to CO_2 concentration. The witch weed emerged above the ground and flowered earlier under the lower CO_2 concentration.

7.3 MANAGEMENT STRATEGIES

As weeds are undesirable in an agricultural production system, they need to be managed so as to alleviate their negative impacts on the crop. Different strategies are used for the management of weeds in millets. The main aims of these strategies are to reduce the competition between the crop and the weeds for common agricultural inputs. The lower the competition at the early stage of the crop, the greater is the benefit. The weed management practices in millets include various mechanical, cultural, and chemical methods. In practice, integration of these methods, rather than a particular single one, yield the best result.

7.3.1 MECHANICAL METHODS

Manual and mechanical weeding is by far the most widely followed method of weed control in millets. The weed produces numerous seeds, which fall on the ground and remain there until further germination. Some weeds produce underground plant parts like bulbs, rhizomes, and swollen roots, which serve as resting structures for survival and spread. The mechanical methods of weed control aim to reduce the weed propagule so that their population and competitive ability are lessened. Different mechanical methods like fallow-season tillage, preplant tillage, and limited postplant tillage are used for this purpose. These practices not only kill the weeds and the weed-seeds but also save soil nutrients by minimizing their growth. Repeated tilling of the soil helps to reduce the total number of weed seeds and their storage parts in the soil and thus lessens the chances of weed growth in the crop. As millets are mostly grown in semiarid environments, these tillage practices also help to conserve soil moisture. Klaij and Hoogmoed (1996) reported that presowing tillage had reduced seasonal weed growth, increased crop yield, and reduced the amount of crop weed in pearl millet in a sandy soil of the West African Sahelian zone. Hand-weeding or interrow cultivation provides reasonable weed control. But during the rainy season, there are not many clear days and as a result, interculture operations have to be delayed and this help the weeds to overtake the crops and cause severe reduction in yield. Also with rising labor wages and nonavailability of adequate labor right at the time of the requirement, it is becoming a serious problem to control weeds manually on a larger area in time.

7.3.2 CULTURAL METHODS

The cultural methods of weed control are practiced on a standing crop. In addition to tillage practices, they include crop rotation, manipulation or row spacing, plant stand, and timing of fertilizer application and placement techniques. Growing of mungbean, groundnut, cowpea, and soybean as intercrops in sorghum or pearl millet could exert a suppressing effect on the weeds. Similarly, narrow row spacing, use of higher seed rate, early application of nitrogen, and its placement near to plants can help in increasing the vigor of the crop and exert a smothering effect on the weeds. Narrow row spacing ($<$30 cm) was found beneficial in reducing weed competition and increasing yield of foxtail and proso millets (Nelson, 1977; Agdag, 1995). Sorghum roots exude a potent bioherbicide known as "sorgoleone," which is produced in living root hairs and is phytotoxic to broadleaf and grassy weeds at concentrations as low as 10 μmol (Yang et al., 2004). Crop management practices that can induce enhanced production of this bioherbicide should be worked on for ecofriendly management of sorghum and weeds.

7.3.3 CHEMICAL METHODS

The use of herbicide saves labor that otherwise have to be used for hand-weeding and thus helps in diverting the labor to more important and productive activities. Depending upon the nature of the herbicides they may be applied before the planting of the crop (preplanting, eg, fluchloralin), after the planting but before the emergence of the crop (preemergence, eg, atrazine, metolachlor, pendimethalin) or after the emergence of the crop (postemergence, eg, 2,4-D). In no-till conditions, herbicides are becoming a major component of weed management in grain sorghum as they improve weed control and production efficiency (Brown et al., 2004).

Several herbicides have been evaluated for weed control efficacy and crop safety in sorghum. However, in other millets, specifically small millets, the herbicide recommendations have been limited. The recommended herbicides for millets, their time of application, doses, and weeds controlled by them are listed in Table 7.4. At present atrazine is the only herbicide most commonly used as a preemergence application for weed control in millets. One supplementary weeding at 30 days after sowing following preemergence herbicides is required for broad-spectrum weed control and higher yields. Lower doses of saflufenacil (50 g ha^{-1}) may be safely applied as near as seven days before planting of pearl millet and proso millet. If the situation demands, saflufenacil (36 g ha^{-1}) can also be applied as a preemergence application to these crops with risk of some crop injury (Reddy et al., 2014). Foxtail millet, however, lacks tolerance to saflufenacil.

Most of the presently available herbicides provide control only on a narrow spectrum of weeds. A herbicide mixture may allow control of a wider spectrum of weeds with a lower amount of total chemical. Ramakrishna et al. (1991) reported that preemergence application of metolachlor in grain sorghum (1.0−1.25 kg ha^{-1}) or combination of atrazine + metolachlor or sequential application of

Table 7.4 Recommended Herbicides for Millets

Millets	Herbicide	Dose (kg a.i. ha^{-1})	Time of Application	Weeds Controlled	Remarks
Sorghum	Atrazine	0.75–1.0	Preemergence, early postemergence	Broad-spectrum weed control, some grasses are tolerant	For sole crop only. Does not control *Acrachne racemosa*, *Brachiaria reptans*, and *Commelina benghalensis* (Walia et al., 2007)
	Pendimethalin	0.75–1.0	Preemergence	Effective control of grasses	Suitable for intercropping, higher doses may cause phytotoxicity
	Alachlor	1.5–2.0	Preemergence	Effective control of grasses	Suitable for intercropping
	Metolachlor	1.0–1.5	Preemergence	Effective control of grasses	Suitable for intercropping
	2,4-D	0.50–0.75	Postemergence	Effective against broad-leaved weeds	For sole crop only. To be applied between 4 and 6 weeks after planting. Good as sequential application to preemergence herbicides
	Atrazine + pendimethalin	0.75 + 0.75	Preemergence	Broad-spectrum weed control	For sole crop only
	Atrazine + alachlor	0.75 + 0.75	Preemergence	Broad-spectrum weed control	For sole crop only
	Atrazine + metolachlor	0.75 + 0.50	Preemergence	Broad-spectrum weed control	For sole crop only
Pearl millet	Atrazine + Hand weeding	0.50 + 1 hand weeding	Preemergence, early postemergence (10 DAS), hand weeding at 30 DAS	*Trianthema portulacastrum*, *Echinochloa colona*	For sole crop only (Banga et al., 2000; Ramakrishna, 1994)

Crop	Herbicide	Rate	Time of application	Remarks/effectiveness	Reference
	2,4-D	0.50–0.75	Postemergence	Effective against broad-leaved weeds	For sole crop only. To be applied between 4 and 6 weeks after planting. Good as sequential application to preemergence herbicides
	Pendimethalin	1.0	Preemergence	Broad-spectrum weed control	Each supplemented with one hand weeding at 45 DAS (Ram et al., 2005)
	Oxadiazone	1.0	Preemergence	Broad-spectrum weed control	Each supplemented with one hand weeding at 45 DAS (Ram et al., 2005)
Finger millet	Oxadiazone	0.75–1.0	Preemergence	Broad-spectrum weed control	Singh and Yadav (1990)
	Isoproturon	0.50–0.75	Preemergence		Ashok et al. (2003)
	Butachlor	0.75	Preemergence		Prasad et al. (2010)
Kodo millet	Isoproturon + Intercultivation + Hand weeding	0.50 + 1 + 1	Preemergence, 20 DAS, 40 DAS	Broad-spectrum weed control	Prajapati et al. (2007)
Proso millet	Atrazine	0.28–0.56	Preemergence	Broad-spectrum weed control	Anderson and Greb (1987)
	Propazine	0.28–0.56	Preemergence	Broad-spectrum weed control	Anderson and Greb (1987)
	2,4-D	0.56	Postemergence (4–6 leaf stage)	Effective against broad-leaved weeds	Grabouski (1971)

metolachlor and bentazon, atrazine (0.75 kg ha^{-1}) yielded as good as repeated weeding. Jadhav et al. (1988) found oxyfluorfen (0.15 kg ha^{-1}) and atrazine (0.75 kg ha^{-1}) as safe herbicides for postrainy sorghum as a preemergence application. Kalyansundaram and Kuppuswamy (1999) reported that a mix application of butachlor (0.75 kg ha^{-1}) + atrazine (0.75 kg ha^{-1}) followed by one hand weeding at 45 DAS controlled the weeds effectively and produced the highest grain yield. Wu et al. (2004) reported that soil incorporation of atrazine mixed with metolachlor at sorghum planting provided effective seasonal control of barnyard grass (*E. colona*). Late postemergence application (weeds and sorghum about 10−15 cm tall) of atrazine + pendimethalin or trifluralin resulted in 99% control of tumble pigweed (*Amaranthus albus*) with less than 3% sorghum stunting (Grichar et al., 2005). Ishaya et al. (2007) observed that pretilachlor + dimethametryne (2.5 kg ha^{-1}) or cinosulfuron (0.05 kg ha^{-1}) or piperophos + cinosulfuron (1.5 kg ha^{-1}) effectively controlled weeds, increased crop vigor, plant height, reduced plant injury, and produced higher grain yield of sorghum. In proso millet, Lyon et al. (2008) observed slight leaf injury with carfentrazone applied at 18 g ai ha^{-1}, but leaves emerging after treatment were healthy and grain yields were not affected. Preemergence application of the premix formulation of bensulfuron methyl (0.6 % G) + pretilachlor (6.0 % G) at 7.5 kg ha^{-1} resulted in lower weed population and higher grain yield of finger millets (Kumar et al., 2015). Herbicides such as isoproturon, nitrofen, and neburon are also very effective in controlling weeds in finger millet (Naik et al., 2001). As the pearl millet and finger millets are more sensitive to most herbicides, a mixture of safeners could increase the margin of selectivity (Rao, 2000). Reddy et al. (2014) indicated that foxtail millet lacked tolerance to saflufenacil, but up to a dose of 50 g ha^{-1} of saflufenacil may be safely applied as near as seven days before planting proso or pearl millets. If the situation demands, saflufenacil (36 g ha^{-1}) can also be applied preemergence to either crop with risk of some crop injury.

7.3.4 INTERCROPPING

Weed management strategies in an intercropping system differ from a sole crop system. Growing of intercrops in widely spaced rows of the main crop not only reduces intensity of weeds but also gives additional yield. Although intercropping may reduce weed infestation and growth, there is still a need for some degree of weed management in most cases. Unlike the sole crop the intercrops do not require the second weeding. After the first weeding the canopy coverage of the row-spaces in an intercropping is almost complete and this checks weed growth. As most of the herbicides are crop-specific, often they cannot be applied in an intercropping system. Manual or mechanical weed control is, therefore, the main method in this system. The use of a few chemicals like pendimethalin (0.75−1.0 kg ha^{-1}), metolachlor (1.0 kg ha^{-1}), and butachlor (0.75−1.0 kg ha^{-1}) have been found safe and effective in intercropping systems. Metolachlor was, however, not effective against *Celosia argentea* and pendimethalin at 1.0 kg ha^{-1} and was toxic for seed germination of sorghum (Ponnuswami et al., 2003).

7.3.5 SEQUENCE CROPPING/DOUBLE CROPPING

Weed management in sequential cropping is a little different from those in intercropping systems. The continuous presence of the crop-cover, the residual toxicity of the herbicides applied to the previous crops on the succeeding crops, and the changing weed flora with the season—all these need different approaches for weed management practices. Selective herbicides are available for sole crops but the residual effects of these herbicides have to be carefully assessed before using them in a crop sequence. Very little attempt has been made in this direction. In a three-year study with a fixed three crop-rotation (cotton–sorghum–finger millet) raised under zero tillage conditions with chemical weed control, *Cynodon dactylon* became a major problem after the second year and was difficult to control (Palaniappan, 1988). In the sorghum–cotton cropping sequence, preemergence application of atrazine (0.25 kg ha^{-1}) in sorghum and pendimethalin (1.0 kg ha^{-1}) in cotton was effective for control of broad-leaved weeds. The following cotton crop was not affected (Palaniappan and Ramaswamy, 1976). Atrazine applied as preemergence (0.50 kg ha^{-1}) gave effective control of the weed in sorghum but the establishment of the subsequent legumes such as green gram and groundnut was poor. In a sorghum–safflower sequence cropping, Giri and Bhosle (1997) observed that preemergence application of atrazine (0.75 kg ha^{-1}) alone or atrazine (0.50 kg ha^{-1}) combined with one weeding and hoeing six weeks after sowing were as effective as two weedings and hoeings at three and six weeks after sowing in controlling weeds without any phytotoxic effect on succeeding safflower.

7.3.6 MANAGEMENT OF *STRIGA* SPP.

Prevention of seed production by *Striga* by any means has a major effect on the control of this obnoxious weed. Hand pulling is the most common control measure used by the small-scale farmers, but it is only effective when the *Striga* population is low. Cattle should not be fed with this weed as its seeds pass through the cattle and are distributed in the manure. Cultural practices such as stubble cleaning in millets fields after harvest, crop rotation with nonhosts and with catch crops, mixed cropping without host crops, fertilizer management with high doses of nitrogen as top dressing, and use of resistant or tolerant varieties help in reducing *Striga* infestation. Trap crops stimulate seed germination of this weed but are not themselves attacked. It is, therefore, possible to rotate the trap crops with millets to induce suicidal germination. Crop rotation with the trap or catch crops like soybean and cotton; intercropping with groundnut, soybean, and cowpea; and green manuring with sunhemp help in reducing the problem of *Striga*. Chemical control of *Striga* is difficult due to the lack of a good selective herbicide. Since *Striga* is a broad-leaved plant, use of preplant or preemergence herbicides such as atrazine and oxyfluorfen show some effect, though are not efficient. Postemergence application of 2,4-D is effective when sprayed on their leaves. However, sorghum is vulnerable to stalk twisting and lodging if 2,4-D is sprayed into the leaf whorl. Hence, proper precautions should be taken while spraying. Further details on *Striga* have been discussed in Chapter 6, *Striga*: A Persistent Problem on Millets.

7.4 HERBICIDE RESISTANCE

Herbicide has become an indispensable weapon for the farmers practicing intensive agriculture especially in the developed nations. Unavailability of human labor and its high cost have directed them increasingly toward this chemical option. Many herbicides act by inhibiting a specific enzyme present in the weed. Inhibitors of acetyl-CoA carboxylase (ACCase) and acetolactate synthase

(ALS) are frequently used as herbicides worldwide. However, repeated and indiscriminate use of herbicides has led to the development of herbicide tolerant weed flora in many countries. Presently around 200 weed species have become resistant to ACCase and ALS inhibitors. The ALS-herbicides, that is, nicosulfuron and nimsulfuron, are widely used to control broadleaf and grassy weeds in corn (*Zea mays*). They were not used in sorghum as it was susceptible to these herbicides. However, recently by transferring a major resistance gene from a wild sorghum relative, researchers at Kansas State University, USA, developed a grain sorghum line that is resistant to several ALS-inhibiting herbicides like Steadfast (nicosulfuron), Accent (nicosulfuron), Resolve (rimsulfuron), and Ally (metsulfuron) (Tuinstra and Al-Khatib, 2007; Tuinstra et al., 2009). A similar approach may be adopted for other millets for the development of herbicide resistant millet-lines. However, all precautions must be taken to ensure that the resistant gene should not get transferred to the weed species.

Herbicide tolerance through transgenic technology has not been addressed worldwide because of the opinion of the development of "Super weeds." It is understood that crops and related wild or weedy plants can and will exchange genes through pollen transfer, if provided with the opportunity, and have been doing so ever since there have been crops and weeds (Harlan, 1982). The transfer of a herbicide-tolerant gene to Johnson grass from cultivated sorghum is considered a threat if the hybrid develops due to their cross-compatibility. Iwakami et al. (2015) reported the existence of a multiple-herbicide resistant biotype of *Echinochloa crus-galli* var. *formosensis* at Okayama, Japan, that showed resistance to cyhalofop-butyl and several ALS inhibitors.

7.5 CONCLUSIONS

Weed management is a challenging task especially during the rainy season due to the emergence of weeds in flushes, the unpredictability of rains, the nonworkable soil conditions, and the nonavailability of timely labor. Considering the diversity of the weed problem, no single method of weed control, whether manual, mechanical, or chemical would be sufficient to provide season-long weed control under all situations. An integrated weed management system as a part of an integrated crop management practice would be an effective, economical, and ecofriendly approach for weed management in millets. A combination of preemergence herbicides with manual or mechanical weeding or intercropping of smothering crops like cowpea would be required for effective weed management in millets. Sequential application of pre and postemergence herbicides may provide broad-spectrum weed control. Considering the several advantages of using the genetically modified herbicide-tolerant corn, it is worthwhile exploring the possibility of herbicide-tolerant grain sorghum.

7.6 FUTURE THRUSTS

Millets have now been emphasized as nutricereals and will play a major role in crop diversification and food and nutritional security under the changing climate scenario. As these crops are grown as subsistence crops mainly during the rainy season by resource-poor farmers on marginal lands with low inputs, efficient weed management is a major challenge. Most of the minor millets are the improved species of the most troublesome grassy weeds. Hence, it is very difficult to identify and control them in the early stages of growth. In general, weeds in millets are removed manually using hand tools and implements at the stage when they attain a good amount of biomass and are used as source of animal fodder. But the crop yield reduces drastically due to severe competition for nutrients and moisture. Therefore, the critical period of crop−weed competition, especially for the minor millets, needs to be identified and weeds should be managed during that period. There is a need to develop energy-efficient small weeding tools for different agro-ecological regions. Herbicides, although very effective, are rarely used in millets except in sorghum and pearl millets. As these crops are also used as a major fodder source for animals, farmers fear that the use of herbicides may deteriorate the fodder quality and animal health. Hence, they should be educated and trained about the use of herbicides in millets. As the millets are grown in moisture stress conditions, the efficacy of preemergence herbicides like atrazine is reduced. Hence, there is a need for exploring potential postemergence herbicides for safe and effective weed control. Millets are mainly grown as intercrop with pulses and oilseeds. Under such conditions, safe and effective broad-spectrum herbicides need to be developed and evaluated. Herbicide residues in soil and plant (grain and stover) need to be studied in different situations. More investigations are needed on integrated weed management, especially in minor millets.

REFERENCES

Agdag, M.I., 1995. Row Spacing in Proso Millet. M.Sc. thesis. University of Nebraska, Lincoln, USA.

Anderson, R.L., Greb, B.W., 1987. Residual herbicides for weed control in proso millet (*Panicum miliaceum* L.). Crop Prot. 6, 61−63.

Ashok, E.G., Chandrappa, M., Kadalli, G.G., Kirankumar, Mathed, V., Krishnegowda, K.T., 2003. Integrated weed control in drill-sown rainfed finger millet. Indian J. Agron. 48, 290−293.

Balyan, R.S., Kumar, S., Malik, R.K., Panwar, R.S., 1993. Post-emergence efficacy of atrazine in controlling weeds in pearl millet. Indian J. Weed Sci. 25, 7−11.

Banga, R.S., Yadav, A., Malik, R.K., Pahwa, S.K., Malik, R.S., 2000. Evaluation of tank mixture of acetochlor and atrazine or 2,4-D Na against weeds in pearl millet. Indian J. Weed Sci. 32, 194−198.

Bilbro, J.D., 2008. Grain Sorghum Producers Contend With Many Insect Pests. Southwest Farm Press, September 4, 2008, <www.southwestfarmpress.com>.

Brown, D.W., Al-Khatib, K., Regehr, D.L., Stahlman, P.W., Loughin, T.M., 2004. Safening grain sorghum injury from metsulfuron with growth regulator herbicides. Weed Sci. 52, 319−325.

Burnside, O.C., Wicks, G.A., 1969. Influence of weed competition on sorghum growth. Weed Sci. 17, 332−334.

Capinera, J.L., 2005. Relationship between insect pests and weeds: an evolutionary perspective. Weed Sci. 53, 892−901.

Choudhary, A.H., Lagoke, S.T.O., 1981. Weed control in pearl millet in the Savanna zone of Nigeria. Trop. Pest Manage. 27, 465−471.

Ejeta, G., 2007. The *Striga* scourge in Africa: a growing problem. In: Ejeta, G., Gressel, J. (Eds.), Integrating New Technologies for *Striga* Control: Toward Ending the Witchhunt. World Scientific Publishing Co., Hackensack, NJ, pp. 3−16.

Emechebe, A.M., Ellis Jones, J., Schulz, S., Chikoye, D., Douthwaite, B., Kureh, I., et al., 2004. Farmers perception of the *Striga* problem and its control in Northern Nigeria. Exp. Agr. 40, 215−232.

Frederiksen, R.A., 1984. Anthracnose stalk rot. In: Mughogho, L.K. (Ed.), Sorghum Root and Stalk Rots: A Critical Review. International Crop Research Institute for the Semi-Arid Tropics, Patancheru, AP, pp. 37—42.

Geier, P.W., Stahlman, P.W., Regehr, D.L., Olson, B.L., 2009. Pre-emergence herbicide efficacy and phytotoxicity in grain sorghum. Weed Technol. 23, 197—201.

Giri, A.N., Bhosle, R.H., 1997. Weed management in sorghum (*Sorghum bicolor*)-safflower (*Carthamus tinctorius*) sequence. Indian J. Agron. 42, 214—219.

Gowing, J.W., Palmer, M., 2008. Sustainable agriculture development in sub-Saharan Africa: the case for a paradigm shift in land husbandry. Soil Use Manage. 24, 92—99.

Grabouski, P.H., 1971. Selective control of weeds in proso millet with herbicides. Weed Sci. 19, 207—209.

Grichar, W.J., Besler, B.A., Brewer, K.D., 2005. Weed control and grain sorghum response to post-emergence application of atrazine/pendimethalin and trifluralin. Weed Technol. 19, 999—1003.

Harlan, J.R., 1982. Relationship Between Crops and Weeds. American Society of Agronomy, Madison, WI, 295 p.

Holm, L.G., 1978. Some characteristics of weed problems in two worlds. Proc. West. Soc. Weed Sci. 31, 3—12.

Ishaya, D.B., Dadari, S.A., Shebayan, J.A.Y., 2007. Evaluation of herbicides for weed control in sorghum (*Sorghum bicolor*) in Nigeria. Crop Prot. 26, 1697—1701.

Iwakami, S., Hashimoto, M., Matsushima, K., Watanabe, H., Hamamura, K., Uchino, A., 2015. Multiple-herbicide resistance in *Echinochloa crus-galli var formosensis*, an allohexaploid weed species, in dry-seeded rice. Pestic. Biochem. Physiol. 119, 1—8, http://dx.doi.org/10.1016/j.pestbp.2015.02.007. Epub 2015 Feb 25.

Jadhav, N.S., Shelke, D.K., Bhosle, R.H., 1988. Screening of herbicides for rabi sorghum. Indian J. Weed Sci. 20, 88—89.

Kalyansundaram, D., Kuppuswamy, G., 1999. Effect of different weed control methods on the performance of sorghum and soil health. In: Abstracts, 8th Biennial Conference, Indian Society of Weed Science, February, 5—7, 1999. Varanasi, India, 37 p.

Kaushik, S.K., Gautam, R.C., 1984. Weed control studies in pearl millet under rainfed condition. Indian J. Agron. 29, 31—36.

Klaij, M.C., Hoogmoed, W.B., 1996. Weeding method and pre-sowing tillage effects on weed growth and pearl millet yield in a sandy soil of the West African Sahelian zone. Soil Tillage Res. 39, 31—43.

Kumar, M.K.P., Shekara, B.G., Sunil, C.M., Yamuna, B.G., 2015. Response of drill sown finger millet (*Eleusine coracana*) to pre and post-emergent herbicides. Bioscan 10, 299—302.

Labrada, R., Caseley, J.C., Parker, C., 1994. Weed management for developing countries. FAO plant production and protection paper, 120. FAO, Rome. ISSN No. 0259—2517.

Lyon, D.J., Burgener, P.A., DeBoer, K.L., 2008. Producing and marketing proso millet in the Great Plains. University of Nebraska-Lincoln Extension Circular 137. Available from: <http://www.ianrpubs.unl.edu/epublic/live/ec137/build/ec137.pdf> (accessed 31.12.09.).

Marley, P.S., 1995. *Cynodon dactylon*: an alternate host for *Sporisorium sorghi*, the causal organism of sorghum covered smut. Crop Prot. 14, 491—493.

Mashingaidze, N., Madakadze, I.C., Twomlow, S., 2012. Response of weed flora to conservation agriculture systems and weeding intensity in semi-arid Zimbabwe. J. Agric. Res. 36, 5069—5082.

Mishra, J.S., 1997. Critical period of weed competition and losses due to weeds in major field crops. Farmers Parliam. 33, 19—20.

Nagur, T., Sriramulu, C., Sivaramakrishnai, M.A., 1962. *Striga* resistant culture No. 109. Andhra Agric. J. 9, 145—148.

Naik, C.D., Muniyappa, T.V., Dinesh Kumar, M., 2001. Integrated weed management studies in drill sown finger millet. Kar. J. Agric. Sci. 14, 900–904.

Nelson, L.A., 1977. Influence of various row widths on yields and agronomic characteristics of proso millet. Agron. J. 69, 351–353.

Nwilene, F.E., Nwanze, K.F., Reddy, Y.V.R., 1998. Effect of sorghum ecosystem diversification and sowing date on shoot fly, stem borer and associated parasitoids. Crop Res. 16, 239–245.

Palaniappan, S.P., 1988. Cropping Systems in the Tropics: Principles and Management. Wiley Eastern Limited, New Delhi.

Palaniappan, S.P., Ramaswamy, R., 1976. Residual effect of atrazine applied to sorghum on the succeeding crops. Madras Agric. J. 65, 230–232.

Ponnuswami, K., Santhi, P., Sankaran, N., 2003. Effect of intercrops and herbicides on weeds and productivity of rainfed sorghum. In: Abstracts, Biennial Conference, Indian Society of Weed Science, March, 12–14 2003, GBPUAT, Pantnagar, India. 43 p.

Prajapati, B.L., Upadhyay, V.B., Singh, R.P., 2007. Integrated weed management in rainfed kodo millet (*Paspalum scrobiculatum*). Indian J. Agron. 52, 67–69.

Prasad, T.V., Ramachandra, Kiran Kumar, V.K., Denesh, G.R., Sanjay, M.T., 2010. Long-term herbicide usage on weed shift and productivity in transplanted finger millet-groundnut cropping system in southern Karnataka. J. Crop Weed 6, 44–48.

Ram, B., Chaudhary, G.R., Jat, A.S., Jat, M.L., 2005. Effect of integrated weed management and intercropping systems on growth and yield of pearl millet. Indian J. Agron. 50, 254–258.

Ramachandra Prasad, R.V., Narasimha, N., Dwarkanath, N., Munigowda, M.K., Krishnamurthy, K., 1991. Integrated weed management in drilled finger millet. Mysore J. Agric. Sci. 25, 13–17.

Ramakrishna, A., 1994. Efficacy of herbicides for weed control in pearl millet. Indian J. Plant Prot. 22, 202–206.

Ramakrishna, A., Ong, C.K., Reddy, S.L.N., 1991. Studies on integrated weed management in sorghum. Trop. Pest Manage. 37, 159–161.

Rao, A.N., 1978. Ecophysiological Responses of Crops and Weeds Against Herbicides and Their Residues. (Ph. D. thesis). Vikram University, Ujjain, MP.

Rao, V.S., 2000. Principles of Weed Science. Science Publishers, Inc., Enfield, NH.

Rao, A.N., Ladha, J.K., and Wani, S.P., 2015. Weeds and weed control in fingermillet in India—A review. pp. 114. In: Shetty, S.V.R., Prasad, T.V.R., Reddy, M.D., Rao, A.N., Mishra, J.S., Gita Kulshreshta and Abraham C.T. (Eds.), Proceedings, Volume II (Oral Papers). 25th Asian-Pacific Weed Science Society Conference, Hyderabad, India. Indian Society of Weed Science, Jabalpur.

Reed, J.D., Ramundo, B.A., Claflin, L.E., Tuinstra, M.R., 2000. Analysis of resistance to ergot in sorghum and potential alternate hosts. Crop Sci. 42, 1135–1138.

Reddy, S.S., Stahlman, P.W., Geier, P.W., Charvat, L.D., Wilson, R.G., Moechnig, M.J., 2014. Tolerance of foxtail, proso and pearl millets to saflufenacil. Crop Prot. 57, 57–62.

Satao, R.N., Nalamwar, R.V., 1993. Studies on uptake of nitrogen, phosphorus and potassium by weeds and sorghum as influenced by integrated weed control. Integrated weed management for sustainable agriculture. In: Proceedings of an Indian Society of Weed Science International Symposium, Hisar, India, November 18–20, 1993 Vol. III, 103–107.

Sharma, O.L., Jain, N.K., 2003. Integrated weed management in pearl millet (*Pennisetum glaucum*). Indian J. Weed Sci. 35, 34–35.

Shiplet, J.L., Wiese, A.F., 1969. Economics of weed control in sorghum and wheat. Bull. MP-909. Texas Agricultural Experiment Station, College Station, TX.

Singh, R., Yadav, S.K., 1990. Time and method of weed control in pearl millet. Exp. Agr. 26, 319–326.

Stahlman, P.W., Wicks, G.A., 2000. Weeds and their control in grain sorghum. In: Smith, C.W., Frederiksen, R.A. (Eds.), Sorghum Origin, History, Technology and Production. John Wiley & Sons, Inc, New York, NY, pp. 535−590.

Sundari, A., Kumar, S.M., 2002. Crop-weed competition in sorghum. Indian J. Weed Sci. 34, 311−312.

Sundraseh, H.N., Rajappa, M.G., Lingegowda, B.K., Krishnashastry, K.S., 1975. Critical stages of crop-weed competition in ragi (*Eleusine coracana*) under rainfed conditions. Mysore J. Agric. Sci. 9, 582−585.

Tuinstra, M.R., Al-Khatib, K., 2007. New herbicide tolerance traits in grain sorghum. In: Proceedings of the 2007 Corn, Sorghum and Soybean Seed Research Conf. and Seed Expo. Chicago, IL: Am. Seed Trade Assoc.

Tuinstra, M.R., Soumana, S., Al-Khatib, K., Kapran, I., Toure, A., Ast, A., et al., 2009. Efficacy of herbicide seed treatments for controlling *Striga* infestation of sorghum. Crop Sci. 49, 923−929.

Umrani, M.K., Bhoi, P.G., Patil, N.B., 1980. Effect of weed competition on growth and yield of pearl millet. J. Maharashtra Agric. Univ. 5, 56−57.

Walia, U.S., Singh, S., Singh, B., 2007. Integrated control of hardy weeds in maize. Indian J. Weed Sci. 39, 17−20.

Watling, J.R., Press, M.C., 1997. How is the relationship between the C_4 cereal *Sorghum bicolor* and the C_3 root hemi-parasites *Striga hermonthica* and *Striga asiatica* affected by elevated CO_2? Plant Cell Environ. 20, 1292−1300.

Wiese, A.F., Collier, J.W., Clark, L.E., Havelka, U.D., 1964. Effect of weeds and cultural practices on sorghum yields. Weeds 209−211.

Wu, H.W., Walker, S., Osten, V., Taylor, I., Sindel, B., 2004. Emergence and persistence of barnyard grass (Echinochloa colona (L.) Link) and its management options in sorghum. Weed management: balancing people, planet, profit. In: Proceedings of 14th Australian Weeds Conference, Wagga, New South Wales, Australia, September 6−9, 2004, 538−541.

Yang, X.H., Scheffler, B.E., Weston, L.A., 2004. SOR1 a gene associated with herbicide production in sorghum root hairs. J. Exp. Bot. 55, 2251−2259.

Zimdahl, R.L., 1980. Crop-weed competition: a review. International Plant Protection Centre, Oregon State University, Corovallis, OR.

Ziska, L.H., 2001. Changes in competitive ability between a C_4 and a C_3 weed with elevated carbon dioxide. Weed Sci. 4, 622−627.

Ziska, L.H., 2003. Evaluation of yield loss in field sorghum from a C_3 and C_4 weed with increasing CO_2. Weed Sci. 51, 914−918.

Index

Note: Page numbers followed by "*f*" and "*t*" refer to figures and tables, respectively.

A

Abiotic constraints, 15
Abutilon theophrasti, 209–210
Acanthospermum hispidum DC. (Bristly starbur), 205–206
Acetolactate synthase (ALS), 215–216
Acetyl-CoA carboxylase (ACCase), 215–216
Acidovorax avenae ssp. *avenae*
   and bacterial leaf blight, 88
Acigona ignefusalis, 150
Aethus taticotlis, 154
Afidentula minima, 148
AFLP markers, 50, 52–53
Agonoscells spp., 154
Agriculture
   climate resilient, 12–13
   dryland, 8–10
Agrobacterium, 54
   role in host-plant resistance to insects, 132
Agromyza spp., 148
Alternaria alternate
   and grain mold, 25–26
Alternaria spp
   and head blight, 26
   and leaf blight, 84
   and seedling blight, 84
Amaranthus palmeri S. Wats (Palmer amaranth), 205–206
Amaranthus retroflexus, 205–206, 209–210
Amplified fragment length polymorphism (AFLP), 34–35
Amsacta albistriga, 149
Anomala, 153–154
Anthracnose
   sorghum, 28–29, 42, 46–47, 49–51
Antibiosis
   effect on mechanisms of resistance, 119–121
   pearl millet, 160
   small millet, 160
Aphids, 109
   genetics of resistance, 118*t*
   mechanisms of resistance, 118*t*
   screening of resistance, 114
Arbuscular mycorrhizal fungi (AMF), 179–180
Area production and productivity, 5–8, 6*f*, 7*t*
Army worms
   pearl millet, 149
   small millet, 149

Ascochyta sorghina
   and leaf spots, 33*t*
Aspavia armigera, 149–150
Aspergillus flavus
   and kodo millet poisoning, 87
Aspergillus giganteus, 55
Aspergillus spp
   and grain mold, 25–26
Aspergillus tamari
   and kodo millet poisoning, 87
Atherigona, 147–148
Atherigona approximata, 147–148
Atherigona atripalpis, 147–148
Atherigona falcate, 147–148
Atherigona miliaceae, 148
Atherigona pulla, 125, 147–148
Atherigona simplex, 147–148
Atherigona soccata, 105–106, 112, 148, 209
Autoba silicula, 153

B

Bacillus, 184–185
Bacillus amyloliquefaciens, 184–185
Bacillus subtilis, 184–185
Bacillus thuringiensis (Bt)
   role in host-plant resistance to insects, 132
Bacterial diseases
   pearl millet, 75
   small millet, 87–88
   sorghum, 36–37, 36*t*, 43
Bagrada cruciferarum L., 154
Balclutha spp., 149–150
Banded sheath blight
   small millet, 86–87
Barnyard millet
   common diseases of, 81*t*
   and leaf blight, 84
   and seedling blight, 84
   sources of resistance, 158*t*
Biochemical factors of resistance, 125
Bioenergy production, 12
Biotic constraints, 14–15
Bipolaris panici-miliacei
   and seedling blight, 84
   and leaf blight, 84

Bipolaris setariae
  and grain mold, 25—26
  and leaf spot, 70*t*
Bipolaris turcica
  and leaf blight, 32
Blasts
  pearl millet, 72—73, 76—77, 79—80
  small millet, 82—83
Blissus leucopterus, 148
Boerhaavia diffusa L. (Hog weed), 205—206
Borers, 107—108
  Sesamia inferens, 108
  spotted stemborer, 107
Brachiaria distachya, 209
Bracon hebetor, 152
Brown spot
  small millet, 84
Burkholderia phytofirmans, 184—185
Busseola fusca, 112, 150

C

Cage-screening technique, 114
Callidea spp., 149—150
Calocoris angustatus, 105, 110, 115—116, 154
Camnula pellucida, 106
Caterpillars, 110
  hairy, 107
  head, 153
  red hairy, 107
Celama spp., 153
Celosia argentea L. (White cock's comb), 205—206
  intercropping, 214
Cercospora eleusinis
  and leaf spot, 83—84
Cercospora leaf spot
  small millet, 83—84
Cercospora penniseti
  and leaf spot, 70*t*
Cercospora sorghi
  and leaf spots, 33*t*
Cerebella spp
  and ergot, 27
Cetonia, 153
Chaetocnema indica, 148
Chaetocnematibialis, 148
Chaetosorghum, 115—116
Charcoal rot
  sorghum, 34—35, 43, 47, 49, 54
Chemical cues, influence on orientation behavior, 112

Chilo, 150
Chilo infuscatella, 150
Chilo partellus, 105, 112, 114, 150—151
Chiloloba acuta, 153
Chirothrips mexicanus, 149—150
Chrotogonus spp., 149
Cicadulina bipunctella
  and ragi mottle streak, 89
Cicadulina chinai
  and ragi streak, 89—90
Cirsium arvense [L.] Scop., 205—206
Cladosporium spp
  and grain mold, 25—26
Claviceps africana
  and ergot, 27, 70*t*
Claviceps fusiformis
  and ergot, 70*t*, 74
Claviceps sorghi
  and ergot, 27
Claviceps sorghicola
  and ergot, 27
Climate change
  and host-plant resistance to insects, 122
  resilient agriculture, 12—13
  and weed competition, 209—210
Cobb's scale, 42
Cochliobolus nodulosus
Cochliobolus setariae
  and leaf blight, 84
  and seedling blight, 84
Colemania spheneroides, 149
Colletotrichum circinans
  mechanisms of disease resistance, 94
Colletotrichum graminicola
  and anthracnose, 29
  sources of resistance, 46—47
Colletotrichum spp
  and grain mold, 25—26
Colletotrichum sublineolum
  and anthracnose, 29
Coniesta ignefusalis, 151
Contarinia sorghicola, 109—110, 113
Contarinia sorghii, 152
Conventional breeding
  insect pests, 130
  sorghum, 51—52
  Striga-resistant millet cultivars, 185—186
Crop—weed competition, critical period of, 209, 209*t*
Cryptoblabes gnidiella, 110, 153
Cryptoblabes spp., 153

Curvularia lunata
  and grain mold, 25—26
  and leaf blight, 84
  mechanisms of resistance, 48—49
  molecular breeding, 53
  and seedling blight, 84
Curvularia penniseti
  and leaf spot, 70*t*
Curvularia spp
  and grain mold, 25—26
  and head blight, 26
  sources of resistance, 45*t*
Cyamopsis tragonoloba, 154—155
Cyaneolytta, 153
Cyaneolytta aceton, 148
Cyanogenic glycosides
  role in host-plant resistance to insects, 125
Cylindrothorax, 153
Cymbopogon spp., 105—106
Cynodon dactylon, 209
  sequence cropping/double cropping, 214—215
  small millet, 86
Cyperus rotundus, 205—206, 209

D

Dactuliophora elongate
  and leaf spot, 70*t*
Dasyproctus agilis, 150
5-Deoxysorbanchol, structure of, 180*f*
5-Deoxystrigol, 182—183
  structure of, 180*f*
Diatraea grandiosella, 150
Dicraeus pennisetivora, 153
Digitaria marginata, 209
  and panicle diseases, 85
Digitaria sanguinalis (L.) Scop. (Crab grass), 205—206
Diploxys spp., 154
Distribution of millets, 4—5
Diversity array technology (DArT), 187—188
Dolycoris indicus, 149—150
Double cropping, 214—215
Downy mildew, 30—31
  mechanisms of resistance, 49
  pearl millet, 71—72, 76—77, 78*t*, 79—80
  qualitative resistance, 50
  screening for resistance, 42—43
  small millet, 85
  sources of resistance, 47

Drechslera, 23
Drechslera dematioidea
  and leaf spot, 70*t*
Drechslera nodulosa
  and leaf blight, 84
  and seedling blight, 84
Drechslera nodulosum
  and leaf blight, 84
  and seedling blight, 84
Drechslera poae
  and fungal diseases, 82
Drechslera turcica
  and leaf blight, 32
Dryland agriculture, 8—10
Dysdercus spp., 154

E

E. procera, 105—106
E. silicula, 153
Earhead bugs, 110
Earhead pests
  pearl millet, 152—154
  small millet, 152—154
Earwigs
  pearl millet, 154
  small millet, 154
Echinochloa colona, 108—109, 205—206
Echinochloa colonum, 105—106
Echinochloa crusgalli L., 108—109, 205—206
Echinochloa crus-galli var. *formosensis*, 215—216
Echinochloa frumentacea, 154
Edaphic factors
  effect on host-plant resistance to insects, 122
Eleusine indica, 108—109, 205—206
  and panicle diseases, 85
ELISA, for downy mildew, 76
Ephelis oryzae
  small millet, 86
Epicuticular lipids
  role in host-plant resistance to insects, 126
Eragrostis pectinacea
  and panicle diseases, 85
Eragrostis tenuifolia, 209
Ergostis tenufolia
  small millet, 86
Ergot
  pearl millet, 74—75, 77, 79
  sorghum, 26—27

Erwinia chrysanthemi, 14
   and stalk rot disease, 36*t*
Eryxia holoserica, 148
Estigmene lactinea, 149
Eublemma, 105, 153
Eublemma gayneri, 153
Euchlaena spp
   and downy mildew, 30
Euvipio spp., 151
Exserohilum monoceras
   and leaf blight, 84
   and seedling blight, 84
Exserohilum rostratum
   and leaf blight, 70*t*
Exserohilum spp
   and grain mold, 25−26
Exserohilum turcicum
   and leaf blight, 32

F

Fabacyl acetate, structure of, 180*f*
Finger millet (Ragi), 3
   and bacterial leaf blight, 88
   bacterial leaf streak of, 88
   and blast disease, 82
   common diseases of, 81*t*
   foot rot of, 87
   and grain smut, 86
   herbicides for, 212*t*
   nutrition, 11−12
   origin and distribution of, 4
   and ragi severe mosaic, 88−89
   resistant sources and utilization, 93
   and rust, 84
   sources of resistance, 158*t*
Flavonoids
   role in host-plant resistance to insects, 127−128
   in sorghum grains, 128*t*
Foliage pests
   pearl millet, 149
   small millet, 149
Foliar diseases
   pearl millet, 71−74
   small millet, 82−85
   sorghum, 28−33, 42
Fonio (*Digitaria* spp.), 3
Food security, 10−12
Foot rot, of finger millet, 87
Forficuta senegalensis, 154

Foxtail millet (*Setaria italica* (L.) Beauv), 3
   and blast disease, 82
   common diseases of, 81*t*
   and leaf blight, 84
   origin and distribution of, 5
   and rust, 84
   and seedling blight, 84
   sources of resistance, 158*t*
   and udbatta, 86
Fungal diseases
   pearl millet, 70*t*, 71−75
   small millet, 82−87
   sorghum, 24−35
Fusarium andiyazi
   and grain mold, 25−26
Fusarium moniliforme
   and grain mold, 25−26
   and head blight, 26
   molecular breeding, 53
   and stalk rot disease, 35
   and top rot, 70*t*
Fusarium nygamai
   and grain mold, 25−26
Fusarium oxysporum, 184−185, 190−191
Fusarium proliferatum
   and grain mold, 25−26
Fusarium spp
   and grain mold, 25−26
   sources of resistance, 45*t*
   stalk rot disease, 34−35
Fusarium thapsinum
   and grain mold, 25−26
   mechanisms of resistance, 48−49
Fusarium verticillioides
   and grain mold, 25−26

G

Gene expression
   role in host-plant resistance to insects, 129−130
Genetic basis of resistance
   Striga spp., 183−184
Genetically modified organisms (GMO), 190
Genetics of resistance
   pearl millet, 79, 155−159
   small millet, 93, 155−159
   sorghum, 49−51
Geromyia penniseti, 152
Gleocercospora sorghi
   and leaf spots, 33*t*, 70*t*

Glomerella graminicola Politis
  and anthracnose, 29
Glossiness
  role in host-plant resistance to insects, 123
Goniozus procerae, 151
Gossypium hirsutum L., 178–179
Grain midges
  pearl millet, 152–153
  small millet, 152–153
Grain mold
  mechanisms of resistance, 48–49
  molecular breeding, 53
  quantitative resistance, 51
  screening for resistance, 40–41
  sorghum, 25–26
  sources of resistance, 44–46, 45t
Grain smut
  finger millet, 86
Grasshoppers, 106
  pearl millet, 149
  small millet, 149
Greenbug
  inheritance of resistance, 117t
  quantitative trait loci, 131t

H

Hairy caterpillars, 107
  red, 107
Haplothrips ganglebauri, 154
Head beetles
  pearl millet, 153
  small millet, 153
Head blight
  sorghum, 26
Head bugs
  genetics of resistance, 118t
  inheritance of resistance, 117t
  mechanisms of resistance, 118t
  pearl millet, 154
  screening of resistance, 115
  small millet, 154
  sources of resistance, 116t
Head caterpillars
  pearl millet, 153
  small millet, 153
Head smut
  kodo millet, 86
Helicoverpa armigera, 105, 110
Heliocheilus, 160

Heliocheilus albipunctella, 152, 160
Heliothis armigera, 153
Helminthosporium poae
  and fungal diseases, 82
Helminthosporium turcicum
  and leaf blight, 32
Herbicide(s)
  for millets, 212t
  resistance, 215–216
Heterosorghum, 115–116
Hieroglyphus banian, 149
Hieroglyphus nigrorepletus, 149
Holotrichia, 154
Holotrichia consanguinea, 154
Holotrichia longipennis, 154
Host finding and orientation
  insect pest, 111–113
Host-plant resistance
  insect pest, 111–132
  pearl millet, 76–80, 155–161
  small millet, 90–94, 155–161
  sorghum, 39–55, 45t, 53t
  Striga spp., 178–191
Hysteroneura setariae, 149–150, 161

I

Indian barnyard millet (*Echinochloa frumentacea* Roxb.,
  Link), 3
  origin and distribution of, 5
Infester row technique, 115
Insect pest's resistance, in sorghum, 105
  biology, 105–111
    borers, 107–108
    leaf feeders, 106–107
    panicle pests, 109–110
    root pests, 110–111
    seedling pest, 105–106
    sucking pests, 108–109
  future priorities, 133
  host-plant resistance, 111–132
    genetics of resistance, 117
    host finding and orientation, 111–113
    inheritance of resistance, 117
    mechanisms of resistance, 117–130, 118t, 127t, 128t
    pest-resistant cultivars, development and use of,
      130–132, 131t
    screening techniques, 113–115, 113t
    sources of pest resistance, 115–116, 116t, 117t
Intercropping, 214
Interlard-fish-meal technique, 114

J

Japanese barnyard millet (*Echinochloa utilis* Ohwi et Yabuno), 3
Johnson grass mosaic virus (JGMV), 37

K

Kochia scoparia [L.] Schrad., 205–206
Kodo millet (*Paspalum scrobiculatum* L.), 3
  common diseases of, 81*t*
  head smut, 86
  herbicides for, 212*t*
  and leaf blight, 84
  nutrition, 11–12
  origin and distribution of, 5
  poisoning, 87
  and rust, 84
  and seedling blight, 84
  sheath rot of, 87
  sources of resistance, 158*t*
  and udbatta, 86

L

L. armata Fab., 148
Lasioptera spp., 152
Leaf blights
  bacterial, 88
  pearl millet, 74
  small millet, 84
  sorghum, 31–32
Leaf caterpillars
  pearl millet, 149
  small millet, 149
Leaf feeders, 106–107
  grasshoppers, 106
  hairy caterpillars, 107
  weevils, 106–107
Leaf spots
  bacterial, 87
  Cercospora, 83–84
  pearl millet, 74
  sorghum, 32–33, 33*t*, 51
Leaf streak
  bacterial, 88
Leaf surface wetness (LSW)
  role in host-plant resistance to insects, 123–124
Lema planifrons, 148
Leptocorisa acuta, 154
Lestodiplosis spp., 152

Little millet (*Panicum sumentranse* Roth.ex Roem. and schultz), 3
  common diseases of, 81*t*
  and leaf blight, 84
  origin and distribution of, 5
  and rust, 84
  and seedling blight, 84
  sources of resistance, 158*t*
  and udbatta, 86
Longitasus spp., 148
Longiunguis sacchari
  and ragi severe mosaic, 89
Losses due to weed, 206–209, 208*t*
Lygaeus spp., 154
Lytta tenuicollis, 153

M

M. albistigma, 149
M. frugalis, 149
Macrophomina phaseolina
  and charcoal rot, 34–35
Magnaporthe grisea
  and blast disease, 72–73, 82–83
Maize dwarf mosaic virus (MDMV), 37, 55
Maize mosaic virus (MMV), 37–39
Maize stripe virus (MStV, MStpV), 37–39, 43–44, 55
Management strategies, for weed problems, 210–215
  chemical methods, 211–214, 212*t*
  cultural methods, 211
  intercropping, 214
  mechanical methods, 210–211
  sequence cropping/double cropping, 214–215
  Striga spp., management of, 215
Marker-assisted selection (MAS), 130–131, 131*t*
  Striga-resistant millet cultivars, 186–189, 192
Mechanisms of resistance
  insect pest, 117–130, 118*t*, 127*t*, 128*t*
  pearl millet, 80, 159–161
  small millet, 93–94, 159–161
  sorghum, 48–49
  Striga spp., 182–183
Melanaphis sacchari, 105, 109, 125
Melanoplus bivitattus, 106
Melanoplus differentialis, 106
Melanoplus femurrubrum, 106
Melanoplus sanguinipes, 106
Melanopsichium eleusinis
  and smut, 86

Menolepta spp., 148
Midges, 109–110
  genetics of resistance, 118*t*
  inheritance of resistance, 117*t*
  mechanisms of resistance, 118*t*
  quantitative trait loci, 131*t*
  screening of resistance, 115
  sources of resistance, 116*t*
Millet stemborer
  pearl millet, 151
  small millet, 151
Minor millets, 10–11
  and smuts, 70*t*, 75
Molecular breeding
  sorghum, 52–54
Monolepta senegalensis, 148
Morphological structures
  role in host-plant resistance to insects, 126
Morpho-physiological traits
  role in host-plant resistance to insects, 123–124
Mylabris, 153
Mylabris pustulata, 153
Myllocerus cardoni, 148
Myllocerus discolor, 106–107, 148
Myllocerus maculosus, 106–107
Myllocerus subfasciatus, 106–107
Myllocerus undecimpunctulus maculosus, 148
Myllocerus viridanus, 106–107, 148
Myrothecium roridum
  and leaf spot, 70*t*
Mythimna separata, 105, 149

N

Namatrare, 122
Nature of resistance
  small millet, 90
  Striga spp., 183–184
Neomaskellia bergii, 149–150
Nezara viridula, 149–150, 154
No-choice headcage technique, 115
Nonpreference
  pearl millet, 160
  small millet, 160
  sorghum, 118–119
Nutritional elements
  role in host-plant resistance to insects, 125
Nutritional security, 10–12
Nysius raphanus, 148
Nystus ericae, 154

O

Oedaleus senegalensis, 149
Origin of millets, 4–5
Orius spp., 152–153
Orobanchaceae, 173, 178–179
Orobanchol, structure of, 180*f*
Oryza sativa L., 173
Oulema downsei, 148
Ovipositional antixenosis mechanisms of resistance, 118–119
Oxalis corniculata, 209
  and rust, 31
Oxidative enzymes
  role in host-plant resistance to insects, 129

P

P. vestita, 153
Panicle diseases
  pearl millet, 74–75
  small millet, 85–86
  sorghum, 24–28
Panicle pests, 109–110
  caterpillars, 110
  earhead bugs, 110
  midges, 109–110
Panicum miliaceae, 160
Panicum repens, 209
Panicum spp
  and downy mildew, 30
Papaver sp., 114
Parasorghum, 115–116
Paspalum scrobiculatum, 105–106
Pearl millet, 3
  area production and productivity, 5–6, 6*f*, 7*t*, 8, 9*f*
  climate resilient agriculture, 13
  disease resistance. *See* Pearl millet, disease resistance in
  dryland agriculture, 10
  herbicides for, 212*t*
  insect pest's resistance. *See* Pearl millet, insect pest's resistance in
  nutrition, 11–12
  origin and distribution of, 4
Pearl millet, disease resistance in, 70–80
  bacterial diseases, 75
  fungal diseases, 70*t*, 71–75
    foliar diseases, 70*t*, 71–75
    panicle diseases, 74–75
  future priorities, 95
  host-plant resistance, 76–80
    genetics of resistance, 79

Pearl millet, disease resistance in (*Continued*)
    mechanisms of resistance, 80
    resistant sources and utilization, 77–79
    screening for resistance, 76–77
    viral diseases, 75–76
Pearl millet, insect pest's resistance in, 147
    earhead pests, 152–154
    foliage pests, 149
    host plant resistance, 154–155
        genetics of resistance, 155–159
        mechanisms of resistance, 159–161
        sources of resistance, 155, 156*t*, 158*t*
        utilization of resistance, 161
    seedling pests, 147–148
    soil dwelling insects, 154–155
    stemborers, 150–151
    sucking pests, 149–150
Pellicularia rolfsii
    and foot rot of finger millet, 87
Pennisetum glaucum, 105–106
Pennisetum glaucum subsp. *monodii*, 182
Pennisetum sp., 209
    small millet, 86
    and downy mildew, 30
Peregrinus maidis, 38–39, 105, 108–109, 125, 149–150
Peronosclerospora sorghi
    and downy mildew, 30
    and leaf spots, 33*t*
Pest-resistant cultivars, development and use of, 130–132, 131*t*
    conventional breeding, 130
    marker-assisted selection, 130–131, 131*t*
    transgenics, 132
Phenolic acids, in sorghum grains, 127*t*
Phoma sorghina
    and grain mold, 25–26
Phomopsis paspali
    and kodo millet poisoning, 87
Phyllachora penniseti
    and leaf spot, 70*t*
Phyllachora sacchari
    and leaf spots, 33*t*
Phyllosticta penicillariae
    and leaf blight, 70*t*
Phyllotreata chotanica, 148
Pink stemborer, 108
Pink stemborer
    pearl millet, 151
    small millet, 151
Plant defense traits
    role in host-plant resistance to insects, 125–130

Poaceae, 174–176
Podagris spp., 148
Polyphenols
    role in host-plant resistance to insects, 127
Poophilus costalis, 148
Potyvirus, characteristics of, 37*t*
Premature seed rotting, 25
Proanthocyanidins, in sorghum grains, 128*t*
Production constraints, 13–16
    abiotic constraints, 15
    biotic constraints, 14–15
    socioeconomic factors, 16
Proso millet (*Panicum miliaceum* L.), 3
    common diseases of, 81*t*
    herbicides for, 212*t*
    and leaf blight, 84
    origin and distribution of, 5
    and seedling blight, 84
    sources of resistance, 158*t*
Prosophis juliflor
    mechanisms of disease resistance, 94
Proteins
    role in host-plant resistance to
        insects, 129
Psalydolytta fusca, 160
Psalydolytta rouxi, 153
Pseudocolapsis setulosa, 148
Pseudomonas
    disease resistance, 95
Pseudomonas andropogoni
    and leaf stripe, 36*t*
Pseudomonas avenae
    and bacterial leaf stripe, 75
Pseudomonas chlororaphis
    and charcoal rot, 35
Pseudomonas eleusinae
    and bacterial leaf streak, 88
Pseudomonas syringae
    and bacterial spot, 75
Pseudomonas syringae pv. *syringae*
    and leaf spot, 36*t*
Pseudo-resistance
    pearl millet, 161
    small millet, 161
Puccinia penniseti
    and rust, 73–74
Puccinia purpurea
    and rust, 31
Puccinia substriata
    and rust, 70*t*, 73–74, 84–85

Pyricularia
  and blast, 72–73
  nature of resistance, 90
Pyricularia grisea
  and blast disease, 70*t*, 72–73, 82
Pyricularia setariae
  and blast disease, 82–83
Pyrilla perpusilla, 149–150
Pyroderces simplex, 153, 160

Q

Qualitative disease resistance
  sorghum, 50–51
Quantitative trait loci (QTL), 50, 52–54, 53*t*, 79, 187–188
  insect resistant, 113, 131*t*

R

R. laeviceps, 153
R. meridionalis var. *puncticollis*, 153
Ragi mottle streak, 89
Ragi severe mosaic, 88–89
Ragi streak, 89–90
Ramulispora andropogonis
  and leaf spots, 33*t*
Ramulispora sorghi
  and leaf spots, 33*t*
Ramulispora sorghicola
  and leaf spots, 33*t*
Random amplified polymorphic DNA (RAPD), 34–35
RAPD markers, 50
Red hairy caterpillars, 107
Resistant sources and utilization
  pearl millet, 77–79
  small millet, 91–93, 92*t*
Restriction fragment length polymorphism (RFLP), 34–35
Rhabdovirus, characteristics of, 37*t*
Rhinyptia infuscata, 153
Rhinyptia spp., 153
Rhizoctonia solani
  and banded sheath blight, 86
  and leaf blight, 70*t*
Rhizoctonia zeae
  and leaf blight, 70*t*
Rhopalosiphum maidis, 109, 149–150
  and ragi severe mosaic, 89
RNA interference (RNAi) technology, 189
Root and stalk diseases
  small millet, 86–87
  sorghum, 34–35

Root pests, 110–111
Rottboellia cochinchinensis, 108–109
Rust
  pearl millet, 73–74, 77–80
  small millet, 84–85
  sorghum, 31, 50, 54

S

Saccharum officinarum L., 173
Sarocladium oryzae
  and sheath rot of kodo millet, 87
Scizaphis graminum, 119, 129–130
Sclerophthora macrospora
  and panicle diseases, 85
Sclerospora graminicola
  and downy mildews, 70*t*, 71–72, 76
  and panicle diseases, 85
Sclerotium bataticola
  and charcoal rot, 34–35
Sclerotium rolfsii
  and foot rot of finger millet, 87
  and leaf blight, 70*t*
Screening for resistance
  insect pest, 113–115, 113*t*
  pearl millet, 76–77
  small millet, 90–91
  sorghum, 39–44, 53*t*
Secondary metabolites
  role in host-plant resistance to insects, 127
Seedling blight
  small millet, 84
Seedling pest, 105–106
Seedling vigor
  role in host-plant resistance to insects, 123
Sequence cropping, 214–215
Serena, 122
Sesamia calamistis, 150
Sesamia inferens, 150–151
  role in host-plant resistance to insects, 132
Sesamia spp., 151
Setaria intermedia, 209
Setaria parviflora, 108–109
Setaria viridis [L.] Beauv., 205–206
Sheath rot of kodo millet, 87
Shoot bugs, 108–109
  genetics of resistance, 118*t*
  mechanisms of resistance, 118*t*
  screening of resistance, 114
  sources of resistance, 116*t*

Shoot flies, 105–106
  genetics of resistance, 118*t*
  inheritance of resistance, 117*t*
  mechanisms of resistance, 118*t*
  pearl millet, 147–148
  quantitative trait loci, 131*t*
  screening of resistance, 114
  small millet, 147–148
  sources of resistance, 116*t*
Silicula, 153
Simple sequence repeats (SSR), 34–35
Single nucleotide polymorphism (SNP), 187–188
Small millet, insect pest's resistance in, 147
  earhead pests, 152–154
  foliage pests, 149
  host plant resistance, 154–155
    genetics of resistance, 155–159
    mechanisms of resistance, 159–161
    sources of resistance, 155, 156*t*, 158*t*
    utilization of resistance, 161
  seedling pests, 147–148
  soil dwelling insects, 154–155
  stemborers, 150–151
  sucking pests, 149–150
Small millets, disease resistance in, 81–94
  bacterial diseases, 87–88
    bacterial blight, 88
    bacterial leaf spot, 87
    bacterial leaf streak, 88
  fungal diseases, 82–87
    foliar diseases, 82–85
    panicle diseases, 85–86
    root & stalk diseases, 86–87
  future priorities, 95
  host-plant resistance, 90–94
    genetics of resistance, 93
    mechanisms of resistance, 93–94
    nature of resistance, 90
    resistant sources and utilization, 91–93, 92*t*
    screening for resistance, 90–91
  viral diseases, 88–90
    ragi mottle streak, 89
    ragi severe mosaic, 88–89
    ragi streak, 89–90
Small millets
  area production and productivity, 5–6, 6*f*, 7*t*, 8, 9*f*
  climate resilient agriculture, 13
  common diseases of, 81*t*
  disease resistance. *See* Small millets, disease resistance in
  dryland agriculture, 10

  insect pest's resistance. *See* Small millet, insect pest's resistance in
  nutrition, 11–12
Smuts
  pearl millet, 75, 79
  small millet, 85–86
  sorghum, 27–28, 42, 51
Soil dwelling insects
  pearl millet, 154–155
  small millet, 154–155
Soil fertility, 15
Soil salinity, 15
Solanacol, structure of, 180*f*
Solanum melongena
  and rust, 73–74
Sorghum (*Sorghum bicolor* (L.) Moench), 3
  area production and productivity, 5, 6*f*, 7*t*
  bioenergy production from, 12
  climate resilient agriculture, 13
  disease resistance. *See* Sorghum, disease resistances in
  dryland agriculture, 10
  insect pest's resistance. *See* Sorghum, insect pest's resistance in
  nutrition, 11
Sorghum, disease resistances in, 21, 24*t*
  bacterial diseases, 36–37, 36*t*
  fungal diseases, 24–35
    foliar diseases, 28–33
    panicle diseases, 24–28
    root and stalk diseases, 34–35
  future research need, 56–57
  herbicides for, 212*t*
  host-plant resistance
    genetics of, 49–51
    mechanisms of, 48–49
    screening for, 39–44, 53*t*
    sources of, 44–48, 45*t*
    utilization of, 51–55
  viral diseases, 37–39
Sorghum, insect pest's resistance in, 105
  biology, 105–111
    borers, 107–108
    leaf feeders, 106–107
    panicle pests, 109–110
    root pests, 110–111
    seedling pest, 105–106
    sucking pests, 108–109
  future priorities, 133
  host-plant resistance, 111–132
    genetics of resistance, 117

host finding and orientation, 111–113

inheritance of resistance, 117

mechanisms of resistance, 117–130, 118*t*, 127*t*, 128*t*

pest-resistant cultivars, development and use of, 130–132, 131*t*

screening techniques, 113–115, 113*t*

sources of pest resistance, 115–116, 116*t*, 117*t*

Sorghum amplum, 119

Sorghum angustum, 115–116, 119

Sorghum arundinaceum, 181

Sorghum australience, 115–116

Sorghum bicolor, 4, 105, 108–109, 119, 126, 174–176

and leaf spots, 32–33

Sorghum brevicallosum, 115–116

Sorghum bulbosum, 119

Sorghum downy mildew (SDM), 30–31, 42–43, 56–57

mechanisms of resistance, 49

qualitative resistance, 50

screening for resistance, 42–43

sources of resistance, 47

Sorghum drummondii, 181

Sorghum ecarinatum, 115–116

Sorghum extans, 115–116

Sorghum halepense, 107, 115–116, 119, 129, 205–206, 209

and leaf spots, 32–33

Sorghum interjectum, 115–116

Sorghum intrans, 115–116

Sorghum laxiflorum, 115–116, 119

Sorghum macrospermum, 115–116, 119

Sorghum matarankense, 115–116

Sorghum mosaic virus (SrMV), 37

Sorghum nitidium, 115–116, 119

Sorghum purpureosericeum, 115–116

and leaf spots, 32–33

Sorghum stipoideum, 115–116

Sorghum timorense, 115–116

Sorghum versicolor, 115–116, 181

Sorgolactone, structure of, 180*f*

Sorgomol, structure of, 180*f*

Sorosporium paspali-thunbergii

and smut, 86

Sources of resistance

insect pest, 115–116, 116*t*, 117*t*

pearl millet, 155, 156*t*

small millet, 155, 158*t*

sorghum, 44–48, 45*t*

Striga spp., 181–182

Spike worms

pearl millet, 152

small millet, 152

Spilosoma obliqua, 149

Spilostefhus pandurus, 154

Spilostethus spp., 154

Spodoptera frugiperda, 149

Sporisorium cruenta

and smuts, 28

Sporisorium reilianum

and smuts, 28

Sporisorium sorghi

and smuts, 28

Spotted stemborer, 107

genetics of resistance, 118*t*

inheritance of resistance, 117*t*

mechanisms of resistance, 118*t*

pearl millet, 150

quantitative trait loci, 131*t*

screening of resistance, 114

small millet, 150

sources of resistance, 116*t*

SRSV, 37–39

Stemborers

pearl millet, 150–151

small millet, 150–151

Stenodiplosis sorghicola, 105, 209

Stenodiplosis spp., 152

Stiposorghum, 115–116

Striga, 14, 70

Striga asiatica, 173–174, 205–206

world distribution of, 175*f*

Striga hermonthica, 173–174, 205–206, 208–209

genetic basis of resistance, 184

life cycle of, 178

nature of resistance, 184

world distribution of, 175*f*

Striga spp., 171, 205–206, 208–209

biology of, 176–178

future perspectives and priorities, 192

host plant resistance and heredity, 178–191

genetic basis of resistance, 183–184

mechanisms of resistance, 182–183

nature of resistance, 183–184

sources of resistance, 181–182

Striga-resistant millet cultivars, development and use of, 184–190

strigolactones, role of, 178–181

importance of, 174–176

life cycle of, 177*f*

management of, 215

Striga-resistant millet cultivars, development and use of, 184–190

conventional breeding, 185–186

integrated management, 190–191

Striga-resistant millet cultivars, development and use of
(*Continued*)
    marker-assisted selection, 186—189
    transgenics, 189—190
Strigol, structure of, 180*f*
Strigolactones (SLs), 176—177, 180*f*, 182—183, 189—190
    role in host finding and orientation, 178—181
Sucking pests, 108—109
    aphids, 109
    pearl millet, 149—150
    shoot bugs, 108—109
    small millet, 149—150
Sugarcane aphid
    genetics of resistance, 118*t*
    mechanisms of resistance, 118*t*
    screening of resistance, 114
Sugarcane mosaic virus (SCMV), 37—39
    and ragi severe mosaic, 88—89
Sugars
    role in host-plant resistance to insects, 125
Surgery disease. *See* Ergot
Syzeuctus spp., 151

T

Tannins
    role in host-plant resistance to insects, 128—129
Tanymecus indicus, 148
Tef millet, 3
    common diseases of, 81*t*
    and rust, 84
Tenuivirus, characteristics of, 37*t*
Tetrastichus spp., 152—153
Thanatephorus cucumeris
    and banded sheath blight, 86
Thrips
    pearl millet, 154
    small millet, 154
Thrips hawaiiensis, 154
Tillering capacity
    pearl millet, 161
    small millet, 161
Tolerance/recovery resistance, 121—122
Tolyposporium ehrenbergii
    and smuts, 28
Tolyposporium penicillariae
    and smut, 75
Transgenic crops
    role in host-plant resistance to insects, 132
    Striga-resistant millet cultivars, 189—190
    utilization of host resistance, 54—55

Trianthema portulacastrum L. (Horse weed), 205—206
Tribulus terrestris L. (Puncture vine), 205—206
Trichoderma
    disease resistance, 95
Trichomes
    role in host-plant resistance to insects, 123, 126—127
Trichometasphaeria turcica
    and leaf blight, 32
Tripsacum dactyloides, 108—109
Triticum aestivum L., 173

U

Udbatta
    small millet, 86
Uromyces
    and rust, 84
Uromyces eragrostidis
    and rust, 84
    and rust, 84
Uromyces linearis
    and rust, 84—85
Uromyces setariae-italicae
    and rust, 84—85
Ustilago
    and smut, 86
Ustilago crameri
    and smut, 86
Ustilago panici-frumentacei
    and smut, 86
Utilization of host resistance
    pearl millet, 161
    small millet, 161
    sorghum, 51—55

V

Vigna radiata, 154—155
Vigna ungiculata, 154—155
Viral diseases
    pearl millet, 75—76
    small millet, 88—90
    sorghum, 37—39, 43—44
Visual stimuli, influence on orientation behavior, 113
Volatiles
    role in host-plant resistance to insects, 129

W

Weed competition, 209—210
    climate change and, 209—210

Weed problems, in millets, 205
  climate change and weed competition, 209−210
  crop−weed competition, critical period of, 209, 209*t*
  future thrusts, 216
  herbicide resistance, 215−216
  losses due to weed, 206−209, 208*t*
  management strategies, 210−215
    chemical methods, 211−214, 212*t*
    cultural methods, 211
    intercropping, 214
    mechanical methods, 210−211
    sequence cropping/double cropping, 214−215
    Striga spp., management of, 215
Weeds of millets, 205−206, 207*t*
Weevils, 106−107
White grubs, 110−111
  pearl millet, 154−155
  small millet, 154−155

X
Xanthium strumarium, 209−210
Xanthomonas axonopodis
  and bacterial leaf blight, 88
  and bacterial leaf streak, 75
Xanthomonas campestris pv. *holcicola*
  and leaf streak, 36*t*
Xanthomonas eleusinae
  and bacterial leaf spot, 87

Z
Zea mays, 173, 215−216
  and downy mildew, 30

Printed in the United States
By Bookmasters